Biological Attachment Systems and Biomimetics

Guest Editors
Stanislav N. Gorb
Department Functional
Morphology and Biomechanics
Zoological Institute of
the University of Kiel
Kiel
Germany

Thies Büscher
Department Functional
Morphology and Biomechanics
Zoological Institute of
the University of Kiel
Kiel
Germany

Editorial Office
MDPI AG
Grosspeteranlage 5
4052 Basel, Switzerland

This is a reprint of the Special Issue, published open access by the journal *Biomimetics* (ISSN 2313-7673), freely accessible at: https://www.mdpi.com/journal/biomimetics/special_issues/biological_attachment_systems.

For citation purposes, cite each article independently as indicated on the article page online and as indicated below:

Lastname, A.A.; Lastname, B.B. Article Title. *Journal Name* **Year**, *Volume Number*, Page Range.

ISBN 978-3-7258-3834-9 (Hbk)
ISBN 978-3-7258-3833-2 (PDF)
https://doi.org/10.3390/books978-3-7258-3833-2

Cover image courtesy of Stanislav N. Gorb

Contents

About the Editors . . . . **vii**

Thies H. Büscher and Stanislav N. Gorb
Biological Attachment Systems and Biomimetics—In Memory of William Jon P. Barnes
Reprinted from: *Biomimetics* **2025**, *10*, 220, https://doi.org/10.3390/biomimetics10040220 . . . . **1**

Hanns Hagen Goetzke, Malcolm Burrows and Walter Federle
Mantises Jump from Smooth Surfaces by Pushing with "Heel" Pads of Their Hind Legs
Reprinted from: *Biomimetics* **2025**, *10*, 69, https://doi.org/10.3390/biomimetics10020069 . . . . . **5**

Sepideh Fakhari, Clémence Belleannée, Steve J. Charrette and Jesse Greener
A Microfluidic Design for Quantitative Measurements of Shear Stress-Dependent Adhesion and Motion of *Dictyostelium discoideum* Cells
Reprinted from: *Biomimetics* **2024**, *9*, 657, https://doi.org/10.3390/biomimetics9110657 . . . . . **20**

Peng Xi, Yanqi Qiao, Qian Cong and Qingliang Cui
Experimental Study on the Adhesion of Abalone to Surfaces with Different Morphologies
Reprinted from: *Biomimetics* **2024**, *9*, 206, https://doi.org/10.3390/biomimetics9040206 . . . . . **35**

Yunsong Kong, Shuanhong Ma and Feng Zhou
Bioinspired Interfacial Friction Control: From Chemistry to Structures to Mechanics
Reprinted from: *Biomimetics* **2024**, *9*, 200, https://doi.org/10.3390/biomimetics9040200 . . . . . **47**

Iakov A. Lyashenko, Valentin L. Popov and Vadym Borysiuk
Indentation and Detachment in Adhesive Contacts between Soft Elastomer and Rigid Indenter at Simultaneous Motion in Normal and Tangential Direction: Experiments and Simulations
Reprinted from: *Biomimetics* **2023**, *8*, 477, https://doi.org/10.3390/biomimetics8060477 . . . . . **67**

Julian Thomas, Stanislav N. Gorb and Thies H. Büscher
Characterization of Morphologically Distinct Components in the Tarsal Secretion of *Medauroidea extradentata* (Phasmatodea) Using Cryo-Scanning Electron Microscopy
Reprinted from: *Biomimetics* **2023**, *8*, 439, https://doi.org/10.3390/biomimetics8050439 . . . . . **92**

May Gonen and Haytam Kasem
Effect of the Mechanical Properties of Soft Counter-Faces on the Adhesive Capacity of Mushroom-Shaped Biomimetic Microstructures
Reprinted from: *Biomimetics* **2023**, *8*, 327, https://doi.org/10.3390/biomimetics8030327 . . . . . **115**

Zhen Lin, Kangjian Xiao, Lijun Li, Yurong Zhang, Xiaolong Zhang, Daobing Chen and Longjian Xue
The Influence of Temperature on Anisotropic Wettability Revealed by Friction Force Measurement
Reprinted from: *Biomimetics* **2023**, *8*, 180, https://doi.org/10.3390/biomimetics8020180 . . . . . **128**

Thies H. Büscher and Stanislav N. Gorb
Convergent Evolution of Adhesive Properties in Leaf Insect Eggs and Plant Seeds: Cross-Kingdom Bioinspiration
Reprinted from: *Biomimetics* **2022**, *7*, 173, https://doi.org/10.3390/biomimetics7040173 . . . . . **141**

Sofia Seabra, Theresa Zenleser, Alexandra L. Grosbusch, Bert Hobmayer and Birgit Lengerer
The Involvement of Cell-Type-Specific Glycans in *Hydra* Temporary Adhesion Revealed by a Lectin Screen
Reprinted from: *Biomimetics* **2022**, *7*, 166, https://doi.org/10.3390/biomimetics7040166 . . . . . **161**

Luc M. van den Boogaart, Julian K. A. Langowski and Guillermo J. Amador
Studying Stickiness: Methods, Trade-Offs, and Perspectives in Measuring Reversible Biological Adhesion and Friction
Reprinted from: *Biomimetics* **2022**, *7*, 134, https://doi.org/10.3390/biomimetics7030134 . . . . . **178**

Constanze Grohmann, Anna-Lisa Cohrs and Stanislav N. Gorb
Underwater Attachment of the Water-Lily Leaf Beetle *Galerucella nymphaeae* (Coleoptera, Chrysomelidae)
Reprinted from: *Biomimetics* **2022**, *7*, 26, https://doi.org/10.3390/biomimetics7010026 . . . . . . **197**

About the Editors

Stanislav N. Gorb

Prof. Dr. Stanislav N. Gorb is a full professor at Kiel University. His scientific interests focus mainly on biological attachment, functional morphology, biomechanics, biotribology, and biomimetics. He now serves as the Editor-in-Chief of *Biomimetics*.

Thies Büscher

Dr. Thies Büscher is a postdoctor at Department of Functional Morphology and Biomechanics, Zoological Institute, Kiel University. His scientific interests include the functional morphology of attachment systems in insects, especially stick and leaf insects.

biomimetics

MDPI

Editorial

Biological Attachment Systems and Biomimetics—In Memory of William Jon P. Barnes

Thies H. Büscher * and Stanislav N. Gorb

Functional Morphology and Biomechanics, Zoological Institute, Kiel University, D-24118 Kiel, Germany; sgorb@zoologie.uni-kiel.de
* Correspondence: tbuescher@zoologie.uni-kiel.de

Received: 20 March 2025
Accepted: 31 March 2025
Published: 2 April 2025

Citation: Büscher, T.H.; Gorb, S.N. Biological Attachment Systems and Biomimetics—In Memory of William Jon P. Barnes. *Biomimetics* **2025**, *10*, 220. https://doi.org/10.3390/biomimetics10040220

1. Introduction to the Special Issue

Any system preventing the separation of two surfaces may be defined as an attachment system. Such systems are common in nature, aiding in locomotion [1], settlement [2], mating [3], and many more functions. These biological attachment systems (BASs) are used to either temporarily or permanently attach an organism to substrates, to other organisms, or for the temporary interconnection of body parts within an organism [4]. For this, BASs can either employ entirely mechanical principles, or additionally, rely on surface chemistry and incorporate fluids in the contact region. The structure and physical mechanisms of BASs vary enormously and are subject to different functional loads, due to their specific areas of application. Because of this, many functional solutions have evolved independently in different lineages of organisms [5]. Many species of animals and plants exhibit diverse BASs that differ in their morphology depending on the biology of the species and the particular function in which the corresponding BAS is involved [6]. However, all BASs rely on similar physical and chemical principles. This connection between specific problem solving and the usage of general physical principles renders BASs a promising field of research for biomimetics.

This Special Issue provides recent insights into state-of-the-art basic research on BASs and derived biomimetic studies. It showcases the width of research in the field of attachment systems across biological taxa and disciplines. We appreciate the diversity of contributions to this Special Issue and would like to thank the colleagues that kindly accepted our invitation. Their dedication enabled this collection of articles on biological attachment phenomena from a wide range of perspectives. The published articles cover topics from a range of biological taxa to experimental studies of their adhesive mechanisms, including *Dictyostelium* cells [7], cnidarians [8], molluscs [9], insects [10–12] and plant seeds [13]. Because of the diversity of functions in BASs, biology could provide interesting inspirations for the design and fabrication of biomimetic attachment devices. Furthermore, biological studies are complemented with experiments and simulations investigating the properties of artificial materials involved in adhesive contact formation [14] and interfacial mechanisms [15]. The combination of such original biological, physical, and engineering studies is the foundation for biomimetic innovation, as demonstrated in contributions on mushroom-shaped biomimetic microstructures in this Special Issue [16].

The experimental insights in this collection are accompanied by two review articles from both perspectives, elaborating on the methodological considerations for experiments on biological systems [17] and the structure–function relationships of friction control in bioinspired systems [18]. In summary, the reader can expect a wide overview of attachment-related phenomena, spanning from insights into the mechanisms of diverse taxa to bioinspired engineering.

2. Dedication

This Special Issue is dedicated to Prof. Dr. Jon Barnes. With sadness, we took note that William Jon Peter Barnes suddenly passed on 16 April 2024 in his home in Glasgow at the age of 83 years. Jon Barnes (Figure 1) was a well-known and enthusiastic specialist in the field of biological adhesion. He published various articles that were important milestones for our understanding of adhesion and contact phenomena in biological systems.

Figure 1. Jon Barnes in August 2011, giving a lecture at Kiel University, Germany.

In his research, Jon Barnes combined various morphological and experimental approaches, field studies, and modern microscopical and force measurement techniques that significantly promoted knowledge in the field of biological adhesion. His work was dedicated to the understanding of tree frog adhesion, and contributed to the in-depth understanding of this particular adhesive system. His contributions gave rise to various applied aspects that were integrated in large-scale industrial developments, such as the development of bioinspired profiles for winter tires [19,20]. In addition to the major impact of his research on the community, Jon Barnes left significant marks in the fields of neuroethology, animal physiology, and behavior.

Jon Barnes was well respected in the adhesion community, and his elegant combination of high-quality basic research and applied aspects were highly influential for many scientists. As a leading scientist with an established reputation worldwide, he passed his knowledge onto colleagues in various conferences and by organizing several symposia on his related research topics. In this manner, his achievements were acknowledged in the symposium '*Biomechanics of arboreal locomotion—a tribute to Jon Barnes*', organized by Walter Federle at the 2007 annual meeting of the Society for Experimental Biology in Glasgow, where Jon Barnes resided. At the University of Glasgow, he was a valued colleague and teacher who received excellent assessments from his students over the years and demonstrated a strong ability to influence, motivate, and inspire his students, and from October 2006, he carried on his commitment as Honorary Research Fellow. As well as his position in Glasgow, Jon Barnes spent several Alexander von Humboldt Fellowships at various universities over the years, such as the universities of Konstanz (1974/5), Frankfurt (1986), and Würzburg (2005). Alongside his academic passion, he dedicated himself to local nature conservation with the Scottish Wildlife Trust over the decades. He was actively involved with this organization from the late 1960s, and served as member of the Trust's Council and Convenor of the Conservation Committee for decades, including periods as Vice-Chairman (1985 to 2003) and Chairman (2003 to 2006).

His dedication to research in the field of bioadhesion and his ambitious contributions to nature conservation will truly be missed. Jon Barnes actively devoted himself to the field and continued to pursue his fascination with tree frog adhesive systems far beyond his retirement. During the compilation of this Special Issue, he was actively working on another manuscript on the influence of surface energy on tree frog attachment, until his passing stopped him from finishing his work.

Conflicts of Interest: The authors declare no conflicts of interest.

References

1. Büscher, T.H.; Gorb, S.N. Physical constraints lead to parallel evolution of micro-and nanostructures of ani-mal adhesive pads: A review. *Beilstein J. Nanotechnol.* **2021**, *12*, 725–743. [CrossRef] [PubMed]
2. Aldred, N.; Clare, A.S. Mechanisms and principles underlying temporary adhesion, surface exploration and settlement site selection by barnacle cyprids: A short review. In *Functional Surfaces in Biology*; Gorb, S.N., Ed.; Springer: Dordrecht, The Netherlands, 2009; pp. 43–65. [CrossRef]
3. Miller, P.L. The structure and function of the genitalia in the Libellulidae (Odonata). *Zool. J. Linn. Soc.* **1991**, *102*, 43–73. [CrossRef]
4. Gorb, S.N. *Attachment Devices of Insect Cuticle*; Springer Science & Business Media: Dordrecht, The Netherlands, 2001; 305p. [CrossRef]
5. Büscher, T.H.; Gorb, S.N. Convergent Evolution of Animal Adhesive Pads. In *Convergent Evolution. Fascinating Life Sciences*; Bels, V.L., Russell, A.P., Eds.; Springer: Cham, Germany, 2023; pp. 257–287. [CrossRef]
6. Spolenak, R.; Gorb, S.N.; Arzt, E. Adhesion design maps for bio-inspired attachment systems. *Acta Biomater.* **2005**, *1*, 5–13. [CrossRef]
7. Fakhari, S.; Belleannée, C.; Charrette, S.J.; Greener, J. A microfluidic design for quantitative measurements of shear stress-dependent adhesion and motion of *Dictyostelium discoideum* cells. *Biomimetics* **2024**, *9*, 657. [CrossRef] [PubMed]
8. Seabra, S.; Zenleser, T.; Grosbusch, A.L.; Hobmayer, B.; Lengerer, B. The involvement of cell-type-specific glycans in *Hydra* temporary adhesion revealed by a lectin screen. *Biomimetics* **2022**, *7*, 166. [CrossRef]
9. Xi, P.; Qiao, Y.; Cong, Q.; Cui, Q. Experimental study on the adhesion of Abalone to surfaces with different morphologies. *Biomimetics* **2024**, *9*, 206. [CrossRef] [PubMed]
10. Thomas, J.; Gorb, S.N.; Büscher, T.H. Characterization of morphologically distinct components in the tarsal secretion of *Medauroidea extradentata* (Phasmatodea) using cryo-scanning electron microscopy. *Biomimetics* **2023**, *8*, 439. [CrossRef] [PubMed]
11. Grohmann, C.; Cohrs, A.-L.; Gorb, S.N. Underwater attachment of the water-lily leaf beetle *Galerucella nymphaeae* (Coleoptera, Chrysomelidae). *Biomimetics* **2022**, *7*, 26. [CrossRef] [PubMed]
12. Goetzke, H.H.; Burrows, M.; Federle, W. Mantises Jump from Smooth Surfaces by Pushing with "Heel" Pads of Their Hind Legs. *Biomimetics* **2025**, *10*, 69. [CrossRef] [PubMed]
13. Büscher, T.H.; Gorb, S.N. Convergent evolution of adhesive properties in leaf insect eggs and plant seeds: Cross-kingdom bioinspiration. *Biomimetics* **2022**, *7*, 173. [CrossRef] [PubMed]
14. Lyashenko, I.A.; Popov, V.L.; Borysiuk, V. Indentation and detachment in adhesive contacts between soft elastomer and rigid indenter at simultaneous motion in normal and tangential direction: Experiments and simulations. *Biomimetics* **2023**, *8*, 477. [CrossRef] [PubMed]
15. Lin, Z.; Xiao, K.; Li, L.; Zhang, Y.; Zhang, X.; Chen, D.; Xue, L. The influence of temperature on anisotropic wettability revealed by friction force measurement. *Biomimetics* **2023**, *8*, 180. [CrossRef] [PubMed]
16. Gonen, M.; Kasem, H. Effect of the mechanical properties of soft counter-faces on the adhesive capacity of mushroom-shaped biomimetic microstructures. *Biomimetics* **2023**, *8*, 327. [CrossRef] [PubMed]

17. van den Boogaart, L.M.; Langowski, J.K.A.; Amador, G.J. Studying stickiness: Methods, trade-offs, and perspectives in measuring reversible biological adhesion and friction. *Biomimetics* **2022**, *7*, 134. [CrossRef] [PubMed]
18. Kong, Y.; Ma, S.; Zhou, F. Bioinspired interfacial friction control: From chemistry to structures to mechanics. *Biomimetics* **2024**, *9*, 200. [CrossRef] [PubMed]
19. Kappl, M.; Kaveh, F.; Barnes, W.J.P. Nanoscalefriction and adhesion of tree frog toe pads. *Bioinspir. Biomim.* **2016**, *11*, 035003. [CrossRef] [PubMed]
20. Iturri, J.; Xue, L.; Kappl, M.; García-Fernández, L.; Barnes, W.J.P.; Butt, H.-J.; del Campo, A. Torrent frog-inspired adhesives: Attachment to flooded surfaces. *Adv. Funct. Mater.* **2015**, *25*, 1499–1505. [CrossRef]

biomimetics

MDPI

Article

Mantises Jump from Smooth Surfaces by Pushing with "Heel" Pads of Their Hind Legs

Hanns Hagen Goetzke, Malcolm Burrows and Walter Federle *

Department of Zoology, University of Cambridge, Cambridge CB2 3EJ, UK; hannshagen.goetzke@gmail.com (H.H.G.); mb135@cam.ac.uk (M.B.)
* Correspondence: wf222@cam.ac.uk

Abstract: Juvenile mantises can jump towards targets by rapidly extending their middle and hind legs. Here, we investigate how mantises can perform jumps from smooth surfaces such as those found on many plants. *Stagmomantis theophila* mantises possess two distinct types of attachment pads on each foot: three small proximal euplantulae ("heel pads") with microscopic cuticular ridges and one smooth large distal pair of euplantulae ("toe pad"). Microscopy showed that the surface contact of heel pads is strongly load-dependent; at low normal forces, they make only partial surface contact due to the ridges, but at higher loads they switch to larger areas in full contact. By analysing the kinematics of 64 jumps of 23 third-instar nymphs from glass surfaces and the foot contact areas of their accelerating legs, we show that heel and toe pads fulfil distinct roles. During the acceleration phase of jumps, the contact area of the hind legs' heel pads tripled, while that of the toe pad decreased strongly, and the toe pad sometimes detached completely before take-off. Although the middle legs also contribute to the jump, they showed a less consistent pattern; the contact areas of their heel and toe pads remained largely unchanged during acceleration. Our findings show that jumping mantises accelerate mainly by pushing with their hind legs and produce grip on smooth surfaces primarily with the heel pads on their proximal tarsus.

Keywords: adhesion; biomechanics; kinematics; jumping insects; take-off

Academic Editor: Thies Büscher

Received: 8 December 2024
Revised: 14 January 2025
Accepted: 19 January 2025
Published: 22 January 2025

Citation: Goetzke, H.H.; Burrows, M.; Federle, W. Mantises Jump from Smooth Surfaces by Pushing with "Heel" Pads of Their Hind Legs. *Biomimetics* **2025**, *10*, 69. https://doi.org/10.3390/biomimetics10020069

1. Introduction

Wingless juvenile mantises can jump precisely onto targets to cross gaps between twigs and leaves [1–3]. They power their jump with a rapid movement of their middle and hind legs, depressing the trochanter and femur as well as extending the tibia, while the front legs are raised off the surface [2]. Mantises regularly forage on leaves and plant stems, many of which have smooth surfaces. On such surfaces, jumping forward with a low take-off angle is potentially difficult, because the insect has to generate friction forces larger than normal forces. This is only possible if the friction coefficient $\mu > 1$. If one makes the assumption that jumping insects only rely on the classical friction of hard cuticles on the substrate, they could only make upward jumps with take-off angles > 70°, since typical friction coefficients between solids are low (e.g., claws on glass: $\mu = 0.35$, [4]). Insects must therefore improve their foot contact during the acceleration phase to generate sufficiently large friction forces. At the same time, their feet should be able to detach easily at take-off to avoid slowing down. We recently showed that leafhoppers and froghoppers have overcome this biomechanical challenge in two different ways. Leafhoppers (*Aphrodes bicinctus*/*makarovi*) produce the high friction forces required for a jump with several soft, pad-like structures (platellae) on their hind tarsi, which contact the surface only during the acceleration phase of the jump [5]. By contrast, froghoppers (*Philaenus spumarius*) produce high friction when accelerating for a jump by piercing the substrate with sharp spines of their tibia and tarsus [6]. Like froghoppers and leafhoppers, mantises are able to jump from smooth surfaces with low take-off angles. How are their legs able to produce sufficient friction for jumping?

In this study, we investigate how middle and hind legs contribute to jumps of third-instar *Stagmomantis theophila* mantises and what foot structures these insects engage in each leg pair when jumping from smooth surfaces.

2. Materials and Methods

Stagmomantis theophila (Rehn, 1904) mantises were raised from eggs of five adult males and four adult females and kept in individual boxes at room temperature. We studied the jumps of 23 third-instar nymphs (body mass: 24.8 ± 1.7 mg, mean ± S.D.).

We investigated the tarsus morphology of third- and fourth-instar nymphs using light and scanning electron microscopy. Images of front, middle, and hind feet were taken with a Canon EOS 60D digital camera attached to a Leica MZ16 stereomicroscope (Leica Microsystems GmbH, Wetzlar, Germany). For scanning electron microscopy (SEM), legs of freeze-anesthetised mantises were cut off at the femur, and immediately transferred into fixative (4% glutaraldehyde in 0.1 M PIPES buffer at pH 7.3) for 48 h at 6 °C. Legs were washed in de-ionised water and gradually dehydrated with increasing concentrations of ethanol (final concentration: 96%). Specimens were air-dried, mounted on SEM stubs, and sputter-coated with a 20 nm gold layer. Images were taken with a FEI XL30-FEG SEM (Oxford Instruments, Abingdon, UK) at 5 kV.

2.1. *Effect of Normal Force on Adhesive Pad Contact Area*

A 'see-saw' lever device was used for observing the effects of normal force on the surface contact of heel pads under the microscope (Figure S1). Two live fourth-instar mantises were mounted on their back on a light plastic sheet attached to one end of a threaded metal rod. One leg was fixed to a thin metal wire glued to the plastic sheet so that either a heel or toe pad were exposed as the highest point. The threaded rod rested on a low friction pivot, and nuts were screwed onto the opposite side of the rod to exactly balance the torque. The pads were brought into contact with a glass coverslip using a micromanipulator. Additional counterweights were then carefully attached to the rod using a micromanipulator to achieve a well-defined increase in normal load. Contact areas were imaged using a 5× or 100× oil immersion objective and monochromatic (546 nm) epi-illumination with a QICAM 10-bit monochrome camera (Qimaging, Burnaby, BC, Canada) mounted on a Leica DMR-HC microscope.

2.2. *High-Speed Contact-Area Recordings of Jumps*

Third-instar *Stagmomantis theophila* nymphs were placed on a glass coverslip on a Leica DM IRE2 inverted microscope and a paintbrush was presented as a target about two body lengths away from the glass coverslip at level height (Figure S2). Moving the paintbrush attracted the mantises' attention; they mostly walked towards the edge of the coverslip and jumped onto the paintbrush. When the insects jumped from the right position, one foot was visible from below. Only jumps in which the whole tarsus of the recorded leg was on the glass coverslip and did not protrude over the edge were included in the analysis. Two or three cameras recorded 64 jumps of 23 animals from glass coverslips. Two synchronised Phantom V7.1 high-speed cameras (Vision Research, Wayne, NJ, USA) simultaneously captured the jumps with a frame rate of 4700 frames per second, both from the side and from below, the latter using the inverted microscope with a 5× lens and epi-illumination to record contact areas. For 40 of these jumps, a third synchronised high-speed camera (Optronis CR5000x2, Optronis GmbH, Kehl, Germany) was available to capture the jump trajectory from above. Leg detachment was defined as the first frame in which the leg that was visible from below had detached from the surface and this time was defined as t = 0 ms. The first visible movement of middle or hind legs before a jump was taken as the start of the acceleration phase. In most recordings, not all legs were in focus or visible in side view, and it was therefore impossible to determine precisely when the other legs detached. On average, middle legs detached earlier than hind legs, leading to a different mean start time of the acceleration phase for recordings in which middle or hind legs were visible from below.

Contact areas and foot orientation (in the horizontal plane) were measured from the videos. The contact area of the toe pad, and the combined contact area of the three heel pads (i.e., including gaps between the cuticular ridges) were measured using a threshold algorithm in MATLAB (The Mathworks, Natick, MA, USA). The insect's take-off direction and azimuth (the horizontal angle between the take-off direction and the body orientation at the start of the acceleration phase) was measured from the dorsal view by digitising a point on the thorax at the start of the acceleration phase and when airborne. The foot orientation at the start of the acceleration phase was measured using the midline of the two most distal pads and converted into foot orientation relative to take-off direction. An estimate of the insect's take-off angle was measured from the side view of 12 jumps by digitising the position of the middle leg coxa at take-off and 2.1 ms after take-off. The digitisation was repeated three times, and the mean was taken to reduce digitising errors. To assess the force vector of middle legs at detachment, we recorded the movement direction of the detaching foot in the initial frames after take-off.

Third-instar mantises were also encouraged with a paintbrush to walk over the glass coverslip on the inverted microscope. Simultaneous contact area and side views of 17 steps of front, middle, and hind legs of six animals were recorded at 800 frames per second.

To determine the extension and acceleration distance of each leg pair during the acceleration phase and the timing of detachment, we analysed 27 jumps of nine third-instar mantises from high-density foam (Plastazote, Watkins and Doncaster, Cranbrook, UK). This dataset was recorded from side views of jumps using one Photron Fastcam SA3 high-speed camera (Photron (Europe) Ltd., West Wycombe, UK) filming at 1000 frames per second. To determine leg extension, the distance between the tibia-tarsus joint and the anterior edge of the coxa was measured at the start of the acceleration phase and in the last frame before take-off. The acceleration distance of middle and hind legs was defined as the distance that the anterior edge of the coxa travelled during the acceleration phase while the leg was in surface contact. Take-off was defined as the first frame in which all legs had detached from the surface and the animal was airborne.

The data were analysed statistically using R v3.0.2 (R Core Team, 2014). Unless specified otherwise, data are presented as means ± standard error of the mean (s.e.m.).

3. Results

3.1. Morphology

The tarsi of all legs of *Stagmomantis theophila* mantises have five segments (tarsomeres). There are attachment pads (euplantulae) ventrally at the distal end of each of the first four tarsomeres. The pads are whitish, softer to the touch, and hence less sclerotised than the surrounding cuticle (Figure 1). The four attachment pads have two distinct designs (Figure 1). The most distal pair of euplantulae ("toe pad") on the fourth tarsomere has a smooth surface structure (Figure 1). Its projected pad area was 14,920 ± 739 μm^2, more than twice the size of the pad on the third tarsomere (7179 ± 792 μm^2, N = 3 third-instars).

The three proximal pairs of euplantulae ("heel pads") increase in size from the first to the third tarsomere (ANOVA: $F_{1,7} = 29.7$, $p < 0.001$). Their surface consists of a branched pattern of ridges likely formed by the epicuticle (Figure 1E,F). The spacing between the ridges was 1.3 ± 0.1 μm (30 measurements from 10 pads of three animals). No differences of the pads were observed between front, middle, and hind legs.

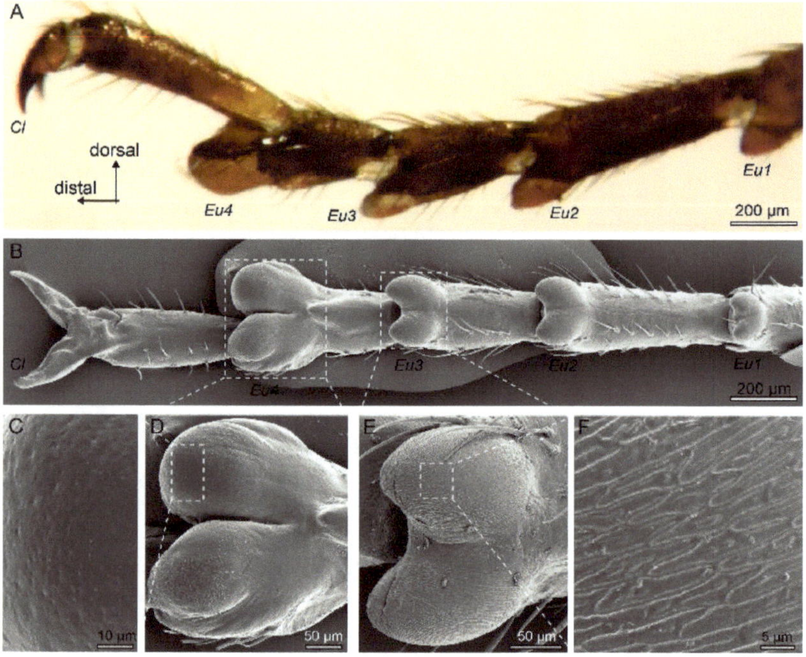

Figure 1. Light and scanning electron microscopy images of hind (**A**, lateral view) and middle legs (**B–F**, ventral view) of a fourth-instar *Stagmomantis theophila* nymph. The tarsi have five segments and euplantulae are located at the distal end of the first four tarsomeres (**A**,**B**). Euplantulae are whitish and less sclerotised than the surrounding cuticle. The most distal pair of euplantulae (Eu4, "toe" pad) have a smooth surface (**C**,**D**) and are larger than the three proximal euplantulae (Eu1-Eu3, "heel" pads), which possess a surface pattern of branching ridges (**E**,**F**). Cl: claws.

3.2. Load Dependence of Heel Pads

The contact areas of the heel pads increased with normal load (Figure 2A–D). At low normal forces, contact areas were small and pads were only in partial contact, i.e., only the ridges but not the channels in between them touched the surface (Figure 2F). At higher normal forces, contact areas increased and pads made full contact with both the ridges and the channels in between them in surface contact, aided by liquid secretion. When the normal load was decreased again, the pads initially remained in full contact. When the leg was detached, fluid droplets were left on the glass coverslip (Figure 2E). For toe pads, the contact areas also increased with load, but full contact already occurred at small normal loads.

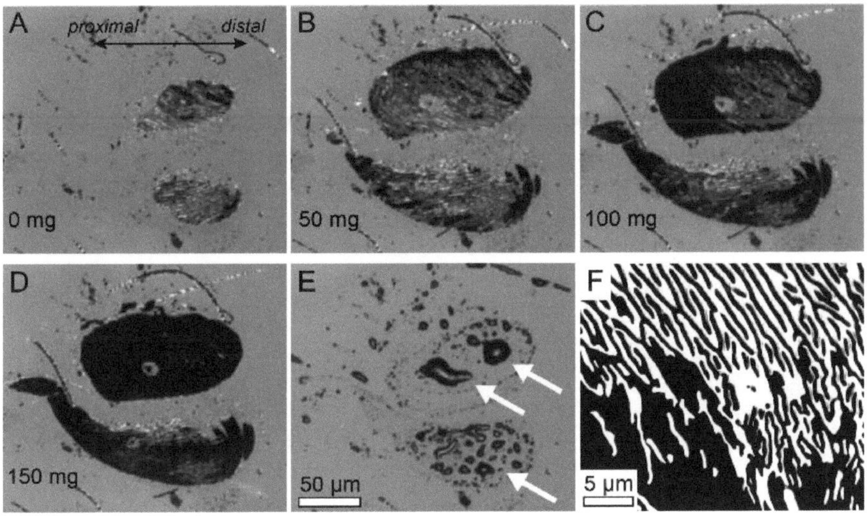

Figure 2. Contact areas of *S. theophila* heel pads (Eu2–Eu3) at different normal loads. Contact area increased with normal load and fluid accumulated in the contact zone (**A–D**) and fluid accumulated in the contact zone. Fluid droplets (white arrows) remained on the glass coverslip after removing the pad (**E**). Ridges in surface contact separated by channels filled with air or fluid secretion were visible at high magnification ((**F**), 150 mg load, contrast enhanced for clarity).

3.3. Kinematics and Tarsal Contact During Take-Off

All mantises were able to jump from the smooth glass coverslip without any slipping (n = 64 jumps of 23 insects). Mantises jumped with a mean take-off angle of 7.8 ± 3.1° (range: −10.4 to 27.8°, including downward jumps in which the target was below the glass coverslip). Since the insects jumped towards a target, this angle likely depended on the position of the target offered. At the start of the acceleration phase, the hind legs pointed backwards (mean angle 118.9 ± 2.8°), and the middle legs pointed forwards (mean angle 52.9 ± 4.3°). The front legs were never in contact during the acceleration phase (Figure 3).

During the acceleration phase, the hind legs extended significantly further than the middle legs (increase in coxa–tarsus distance: hind legs 9.52 ± 0.33 mm, middle legs 3.42 ± 0.24 mm, 27 jumps by nine third-instar nymphs; Welch's *t*-test: $t_{47.1} = 14.9$, $p < 0.001$; Figure 4). At take-off, the middle and hind legs were fully extended. While the hind legs mainly extended in the direction of the jump, the middle legs changed from a forward to a backward orientation by a rotation in the coxa. On average, the middle legs detached earlier than the hind legs (paired *t*-test: $t_{26} = 3.13$, $p = 0.004$, Figure S3A). The total acceleration distance, i.e., the distance the coxa travelled while the foot was in contact with the surface, was therefore significantly larger for the hind legs than for the middle legs (hind legs 10.30 ± 0.38 mm, middle legs 9.18 ± 0.41 mm; Welch's *t*-test: $t_{51.7} = 2.0$, $p = 0.0498$; Figure S3B).

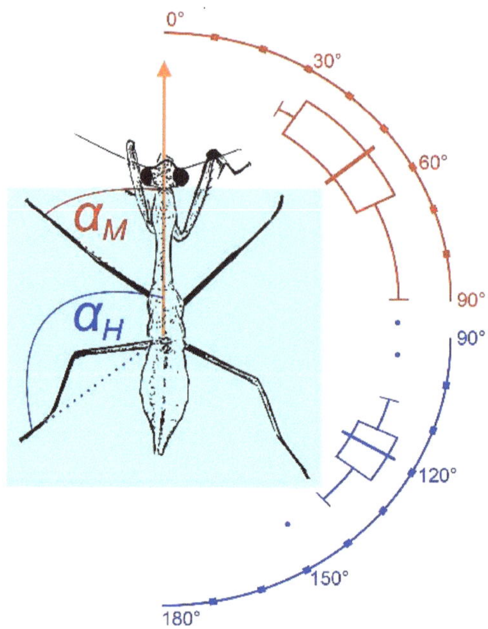

Figure 3. Foot orientation of *S. theophila* hind and middle legs (α_H, α_M) relative to the take-off direction at the start of the acceleration phase. When preparing for the jump, mantises placed their forward-pointing middle legs close to the edge of the glass coverslip (light blue) while their hind legs were pointing laterally backward. Front legs are not in contact with the surface during the acceleration phase.

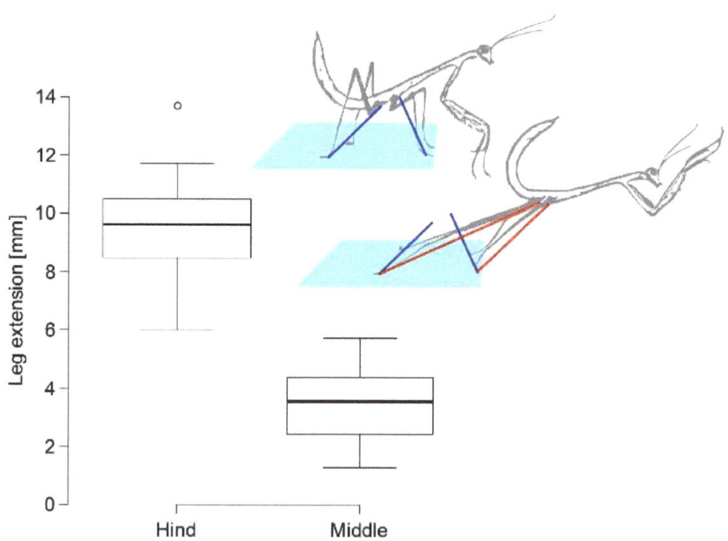

Figure 4. Leg extension of *S. theophila* middle and hind legs during the acceleration phase of a jump. Leg extension was measured as the change in tarsus–coxa distance from the start (blue lines) to the end (red lines) of the acceleration phase.

We observed a characteristic foot movement and change in surface contact for all hind legs during the jumps (38 jumps by 23 mantises; Figures 5 and 6, Videos S1–S4). While in most jumps, both heel and toe pads came into surface contact, the last two tarsal segments were raised during the acceleration phase and the contact area of the toe pad gradually decreased. In contrast, the projected contact area of the heel pads (both Eu3 and Eu2 in all 38 jumps; in addition, Eu1 came into contact with the surface during only one of the 38 jumps) increased threefold at the start of the acceleration phase and reached a plateau. Although the resolution of the contact area recordings was not sufficient to see the cuticular ridges, the contact zones of the heel pads appeared mostly grey and lighter than those of the toe pads (Figure 5B, Video S2), indicating that they were in partial contact. During the acceleration phase, the contact area of the heel pads often became locally darker, indicating that they made full contact in these regions. In the jumps where both pad types of the hind leg were in contact, the toe pad detached on average earlier than the heel pads (median difference 0.1 ms, mean difference 2.3 ms; Wilcoxon signed rank test: W = 191.5, n= 36, $p = 0.044$; Figure 6). In freely walking mantises, no such pattern for hind leg contacts was observed. Walking mantises used all three leg pairs and both heel and toe pads (Figure S4).

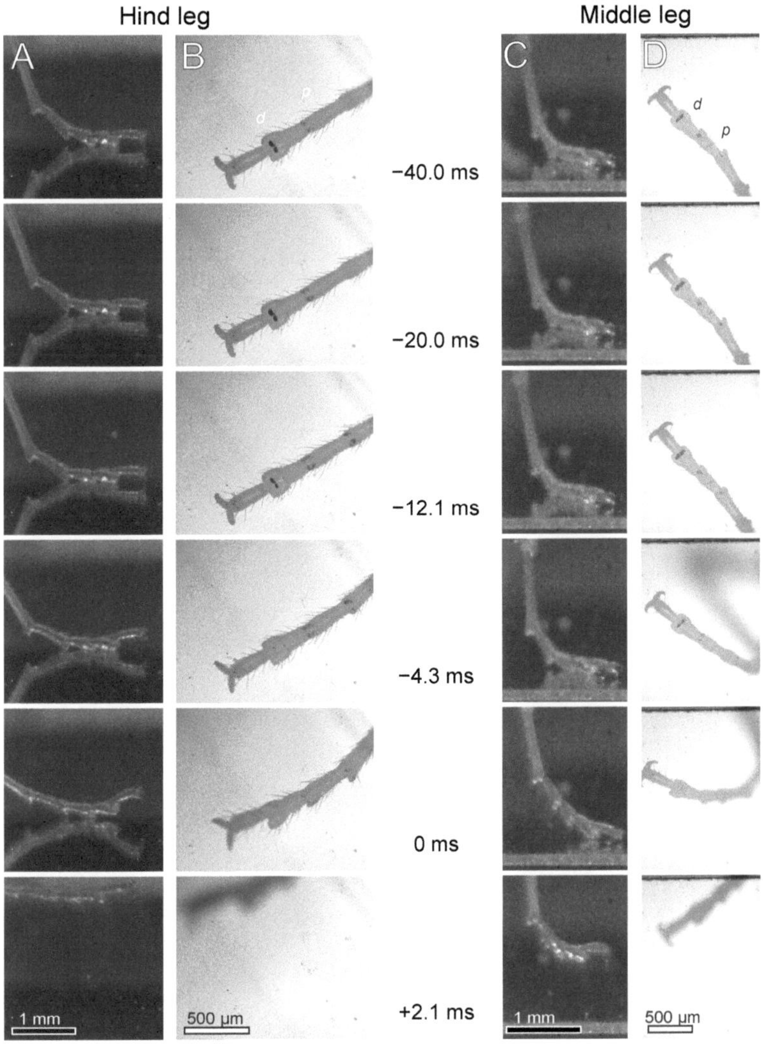

Figure 5. *Cont.*

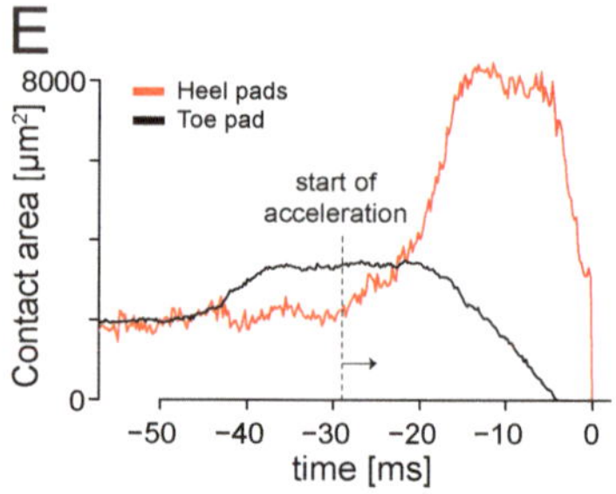

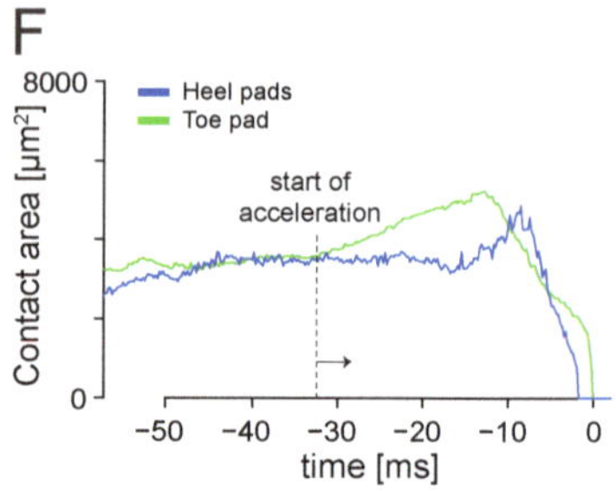

Figure 5. High-speed image sequences of the acceleration phase of two jumps of *S. theophila*, showing hind (**A**,**B**) and middle legs (**C**,**D**) from the side and below, as well as the adhesive contact areas of their heel and toe pads (**E**,**F**). Contact area images are rotated so that the jump direction is toward the top of the page. During the acceleration phase of the jump shown in (**A**,**B**,**E**), which lasted 29.1 ms, the contact area of the hind legs' heel pads increased fourfold while their toe pads decreased in contact area and detached first. During the acceleration phase of the jump shown in (**C**,**D**,**F**), which lasted 32.3 ms, the contact area of the middle legs' heel pads increased only slightly and their toe pad did not detach before the heel pads. The first frame in which a propulsive movement of the leg was visible in the side view was defined as the start of the acceleration phase. The first frame without any adhesive contact was defined as detachment and set to 0 ms.

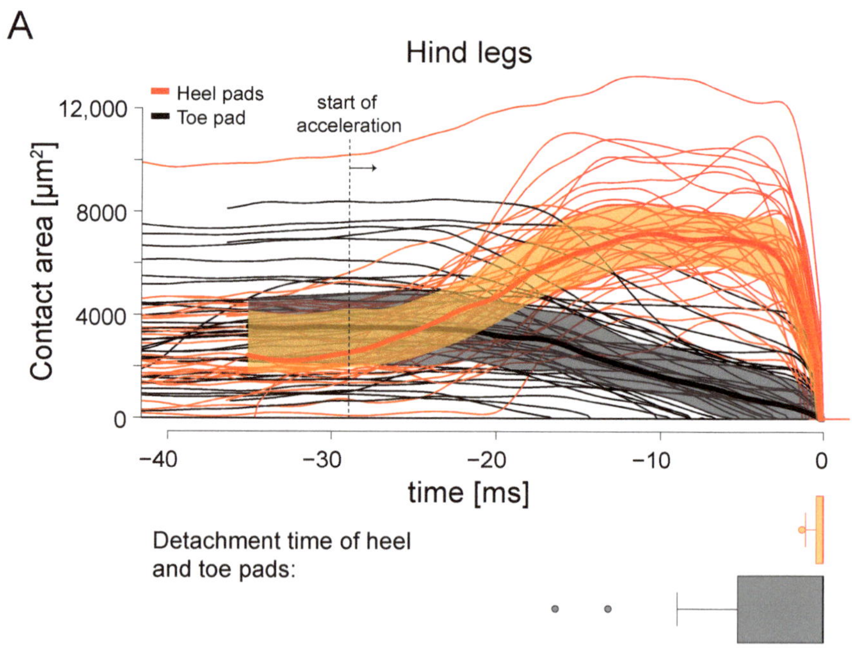

Figure 6. *Cont.*

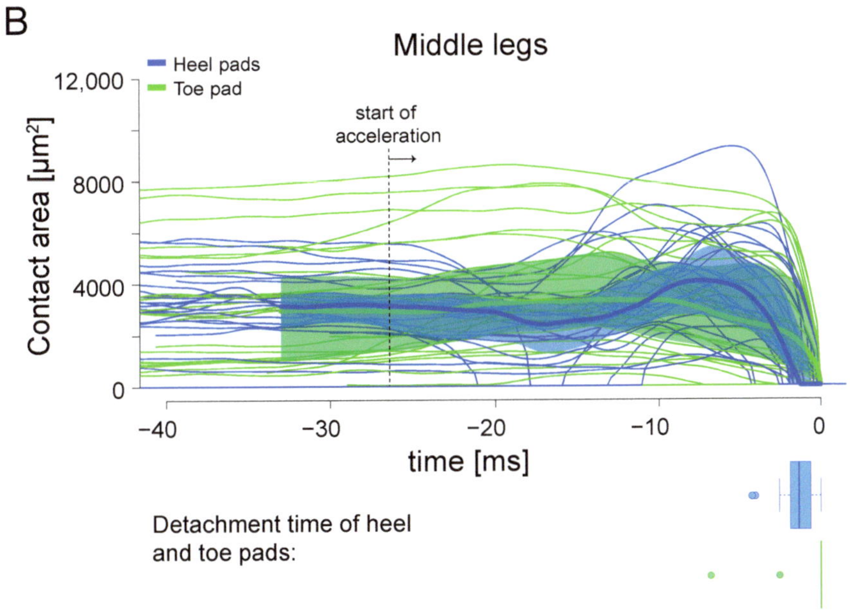

Figure 6. Contact areas for heel and toe pads of hind legs (**A**) and middle legs (**B**) during the acceleration phase of *S. theophila* jumps. Raw data was filtered using a low-pass Butterworth filter with a cut-off frequency of 470 Hz. A median curve (bold line) and interquartile range (shaded) was calculated from 38 jumps by 23 animals for hind legs and 26 jumps by 15 animals for middle legs and plotted for the range of the shortest recorded jump. Horizontal boxplots below the contact area curves indicate the detachment times of heel and toe pads (0 ms is defined as the last pad detaching at take-off). In the hind legs, the changes in contact area differed significantly between heel and toe pads, whereas no clear pattern was visible for the middle legs. The mean start time of the acceleration phase for hind and middle legs is indicated by the dotted line; the time at which the leg detached was set to 0 ms.

In contrast to the situation in hind legs, we did not observe any lifting of the last two tarsal segments in the middle legs, or any decrease in the contact area of the toe pad or increase in area of the heel pads (26 jumps by 15 mantises; Figures 5 and 6, Videos S3 and S4). The heel pads (Eu3 in all 26 jumps; Eu2 in 17 out of 26 jumps) detached first and lost contact up to 8.1 ms before the toe pad (median difference 1.5 ms, mean difference 1.1 ms; Wilcoxon signed rank test: W(25) = 255.5, $p = 0.003$; boxplots in Figure 6). The contact areas of middle legs varied strongly between jumps. Before the start of the acceleration phase, contact areas of toe and heel pads were similar in both the middle and hind legs ($F_{3,122} = 0.8$, $p = 0.481$). During the acceleration phase, the contact area of heel pads exceeded that of toe pads only in the hind legs but not in the middle legs. The middle legs often rotated around their foot contact during the acceleration phase and heel pads sometimes detached and re-attached again. Some of the variation in contact area in middle leg toe pads may be explained by the foot orientation prior to the jump: when middle legs were oriented forward, parallel to the jump direction, the maximum contact area of the toe pad was larger. This indicates that the middle legs contributed to the jump acceleration by pulling, and that this increased the contact area most strongly when the legs pointed in the direction of the jump; however, the effect was small ($F_{1,16} = 5.3$, $p = 0.035$, $R^2 = 0.201$).

4. Discussion

Our results show that juvenile mantises are able to perform jumps from smooth surfaces without slipping. Contact area recordings during the acceleration phase revealed that they mainly engage the heel pads of their hind legs, suggesting a division of labour between heel and toe pads similar to that found in other climbing insects [7–10]. Our results suggest that middle and hind legs perform different functions during the acceleration phase. While the hind legs likely generate most of the propulsion, the middle legs can be used to control the jump trajectory.

4.1. Division of Labour Between Attachment Pads During the Acceleration Phase of Jumps

As the tarsi of *Stagmomantis theophila* mantises have the same number of attachment pads on the front, middle, and hind legs, and pads are similar in morphology and size their foot attachment structures are probably not particularly specialised for jumping. Like other mantises, they possess two distinct types of attachment pads: one large pair of euplantulae with a smooth surface on the fourth tarsomere (toe pad) and three smaller pairs of euplantulae with a cuticular ridge pattern on the first three tarsomeres (heel pads). Our observations suggest that the cuticular ridges on the heel pads enable load-dependent control of contact area and thus high friction coefficients combined with low detachment forces, similar to the function of "nubby" stick insect tarsal friction pads with conical cuticular outgrowths [8,11]. The cuticular ridges on the mantises' heel pads are similar to those reported for *Nauphoeta cinerea* cockroaches, where they have been found to increase friction on rough surfaces [12]. It is possible that the ridges on the heel pads also aid mantises in interlocking on rough surfaces.

Like other mantises [13,14], *Stagmomantis theophila* lack an adhesive pad on the pretarsus, and the distal pair of euplantulae may have taken over the function of the adhesive pad. A similar arrangement and division of labour between two distinct tarsal pad types has been described for *Tettigonia viridissima* and *Acanthoproctus diadematus* bush-crickets [9,15]. For various insects it has been shown that toe and heel pads perform different tasks during climbing; toe pads are mainly used to generate adhesion and friction forces in the pulling direction, whereas heel pads are used to generate friction forces under compression in the pushing direction [7–10,16]. Jumping mantises used their toe and heel pads in accordance with this division of labour. To enable forward jumps on smooth surfaces, hind legs must generate large friction forces when pushing. As we did not observe any slipping even when mantises jumped from glass with low take-off angles, their pads indeed produced friction forces much greater than the normal forces. As the contact area of the hind legs' heel pads increased threefold during the acceleration phase, it is likely that these pads are mainly responsible for the high friction. In contrast, the contact area of the hind legs' toe pads decreased during the acceleration phase, and on average they detached earlier than the heel pads. This suggests that toe pads did not contribute much to the hind legs' pushing forces, consistent with findings for distal adhesive pads in other insects, which typically detach when pushed [8,10,17,18]. When the toe pads did not detach before the heel pads, their contact areas were very small just before the jump, so that their adhesion could hardly slow down the jump. The middle legs pointed in the jump direction or were oriented laterally to it, allowing them to contribute to the jump by pulling. In contrast to the hind legs, pad contact areas of the middle legs varied strongly between different animals and jumps, but on average did not change much during the acceleration phase. When the tarsi of middle legs were aligned with the jump direction, the maximum contact area of the toe pads was larger, indicating that the middle legs contribute to the jump by pulling.

4.2. Contribution of Middle and Hind Legs to the Jump

While the fastest jumping insects only use their hind legs to power their jumps [19–21], some insects and jumping spiders accelerate with two leg pairs [22–32]. Three hypotheses have been suggested as to why the use of an additional leg pair might be advantageous [24]: First, spreading the forces over four legs would reduce the force on each individual foot, thereby allowing jumps from softer substrates [24,25]. Second, species with thin legs might require two leg pairs to produce sufficient power to jump [26,28]. Third, two leg pairs might enable the animal to control pitch more easily [24]. Why is it beneficial for mantises to use four legs for jumping?

Juvenile mantises can jump precisely onto targets [1,2]. Similarly, jumping spiders catch prey by jumping with their third and fourth leg pairs while the anterior two leg pairs are lifted off the surface [31]. The spiders' third and fourth legs can vary considerably in length between species, and thus in their contribution to acceleration [30,33]. If more than two legs contribute to the jump, the different legs could take on different tasks during the acceleration phase. Our results indicate that this is the case for mantises: the different contact area progressions of pads in middle and hind legs, and the larger acceleration distance for hind legs, suggest that hind legs provide most of the thrust for the jump. The contribution of each leg to the kinetic energy of the jump is the integral of that leg's ground reaction force over the acceleration distance. This distance was significantly larger for the hind legs, as the middle legs usually detached first. If middle and hind legs contributed equally to the energy of the jump, the forces of the middle legs would therefore have to exceed those of the hind legs. However, larger normal forces would probably result in larger contact areas of the heel pads, just as we observed in the hind legs during the acceleration phase. As the contact areas of middle legs were smaller and did not increase much during the acceleration phase, the middle legs may contribute only little to the acceleration of the jump.

What then is the function of the middle legs for the jump? It is possible that they are mainly used to control the take-off angle, pitch, and azimuth during the acceleration phase of the jump. The control of these parameters has been studied in insects that jump only with hind legs. Locusts adjust the position of their hind legs before accelerating to control their take-off angle [34] and froghoppers adjust the lateral position of their hind legs to control azimuth [35]. Rapidly jumping small insects employ powerful catapult mechanisms, e.g., [19,20,36,37], where the acceleration lasts no longer than a few milliseconds and is therefore too short for neuronal feedback. The jumps of these insects mainly serve a quick escape and are less optimised for a precise landing. In contrast, mantises take much longer to accelerate (juvenile mantises: >20 ms [2,3]), which would in principle allow neuronal feedback to adjust the take-off angle during the acceleration phase. Our observations indicate that middle legs are indeed involved in the control of azimuth: in one jump with an unusually large (right) azimuth, the weighting of heel and toe pads was strongly asymmetrical in the laterally oriented left middle leg and similar to that of the hind legs, indicating that this middle leg pushed sideways to correct the azimuth (Figure 7, Videos S5 and S6). For the targeted jumps of mantises, accurate landing is probably more important than power. Therefore, the additional control by the middle legs and the longer acceleration time may be required for the precise control of take-off to land in the right spot.

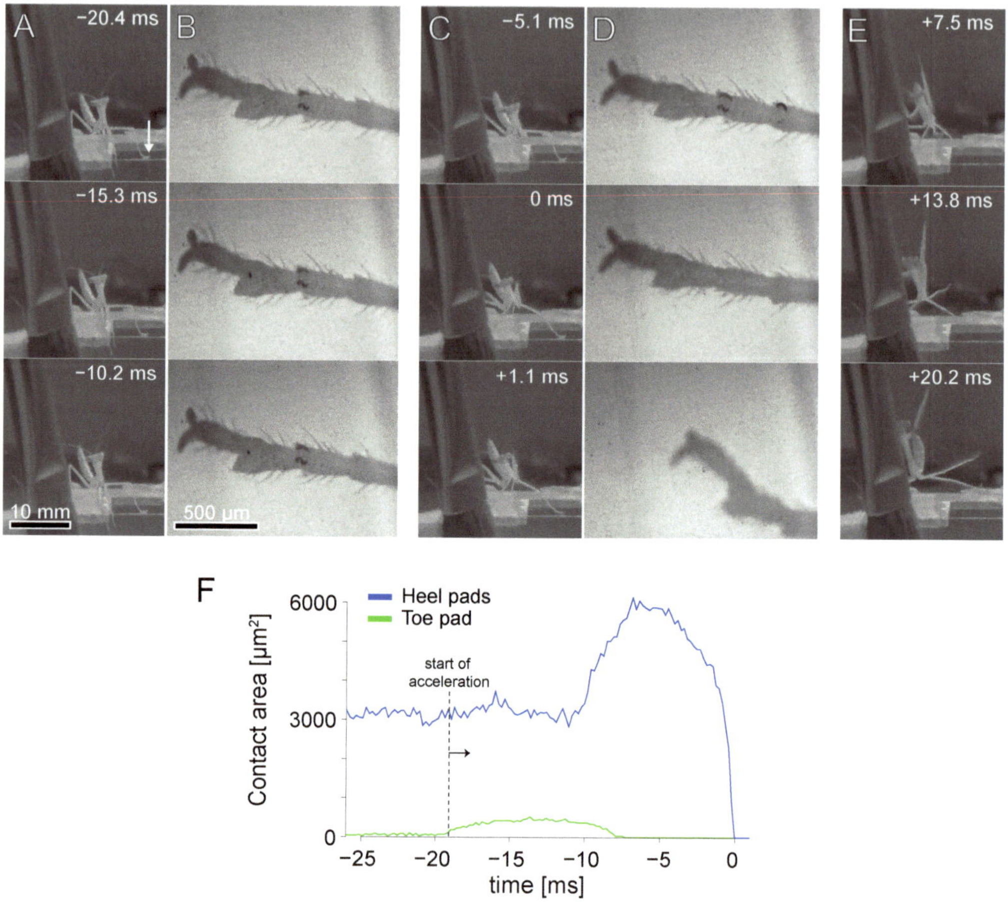

Figure 7. Image sequence of a *S. theophila* jump with a large azimuth angle of 42.1° from the side (**A**,**C**,**E**) and below, showing the left middle leg (**B**,**D**). The adhesive contact area of toe and heel pads of the left middle leg is plotted below (**F**). In (**A**), a white arrow points at left middle leg in the first image. The contact area images are rotated so that the direction of the jump points to the top of the page. In contrast to most other recordings of middle legs, the contact area of the toe pad was very small, and the toe pad detached before the heel pads.

Supplementary Materials: The following supporting information can be downloaded at https://www.mdpi.com/article/10.3390/biomimetics10020069/s1. Figure S1: 'See-saw' setup to visualise the effect of normal load. Sketch of 'see-saw' setup used to visualise the effect of normal load on the contact area of mantis pads. Mantises were mounted on a plastic sheet and immobilised using Parafilm. One leg was attached using dental wax to a metal wire fixed to the plastic sheet so that the ventral side of one tarsal attachment pad was the highest point. The plastic sheet was attached to a threaded metal cylinder resting on a low friction pivot and was balanced by a nut that acted as a counterweight. Additional weights were placed to achieve defined normal loads. Contact areas were recorded through a microscope using reflected light. Figure S2: Setup to record contact areas. Setup used to record contact areas of mantis attachment pads during the acceleration phase of jumps. Mantises were motivated to jump from a glass coverslip towards a target. The glass coverslip was positioned on an inverted microscope and contact areas were recorded with a high-speed camera using reflected light. The movement of the foot and whole-body movements were filmed from the side and, for some jumps, from above. Figure S3: Relative detachment time of hind vs. middle legs. (A) When jumping, the middle legs detached on average before the hind legs. (B) Acceleration distance (distance travelled by the coxa while the foot was in surface contact) in middle and hind legs. Figure S4: Contact area progression during walking steps. contact area progression during steps of front, middle, and hind legs in walking mantises (front legs: three steps of three insects; middle legs: seven steps of six insects; hind legs: six steps of six insects). Video S1: Mantis jump hind leg side view. High-speed video of a jump of *S. theophila* from a glass surface, showing the left hind leg during the acceleration phase from the side (frames of this video are shown in Figure 5A). The video was recorded at 4700 fps, and the width of the field of view is 9.43 mm. Video S2: Mantis jump hind leg contact area. High-speed video of the same jump of *S. theophila* as in Video S1, showing the left hind leg during the acceleration phase from below (frames of this video are shown in Figure 5B). The video was recorded at 4700 fps, and the width of the field of view is 1.66 mm. Video S3: Mantis jump middle leg side view. High-speed video of a jump of *S. theophila* from a glass surface, showing the left middle leg during the acceleration phase from the side (frames of this video are shown in Figure 5C). The video was recorded at 4700 fps, and the width of the field of view is 10.6 mm. Video S4: Mantis jump middle leg contact area. High-speed video of the same jump of *S. theophila* as in Video S3, showing the left middle leg during the acceleration phase from below (frames of this video are shown in Figure 5D). The video was recorded at 4700 fps, and the width of the field of view is 1.61 mm. Video S5: Mantis jump large azimuth middle leg side view. High-speed video of a jump of *S. theophila* with a large azimuth angle, showing the left middle leg during the acceleration phase from the side (frames of this video are shown in Figure 7A,C,E). The video was recorded at 4700 fps, and the width of the field of view is 21.3 mm. Video S6: Mantis jump large azimuth middle leg contact area. High-speed video of the same jump of *S. theophila* as in Video S5, showing the left middle leg during the acceleration phase from below (frames of this video are shown in Figure 7B,D). The video was recorded at 4700 fps, and the width of the field of view is 1.56 mm.

Author Contributions: Conceptualisation, H.H.G., M.B. and W.F.; methodology, H.H.G., M.B. and W.F.; validation, H.H.G. and W.F.; formal analysis, H.H.G.; investigation, H.H.G.; resources, H.H.G., M.B. and W.F.; data curation, H.H.G.; writing—original draft preparation, H.H.G.; writing—review and editing, H.H.G., M.B. and W.F.; visualisation, H.H.G.; supervision, W.F.; project administration, H.H.G. and W.F.; funding acquisition, H.H.G. and W.F. All authors have read and agreed to the published version of the manuscript.

Funding: This research was funded by a Gates Cambridge Scholarship and the Balfour Fund of the Department of Zoology to H.H.G., and a grant from the Human Frontier Science Programme (RGP0034/2012).

Institutional Review Board Statement: Not applicable.

Data Availability Statement: All data of the recorded contact areas and jump kinematics underlying Figures 3–7, S3 and S4 have been uploaded to the Dryad digital repository (https://doi.org/10.5061/dryad.x69p8czvh, accessed on 14 January 2025).

Acknowledgments: The authors gratefully acknowledge the EPSRC Engineering Instrument Pool for multiple loans of the Phantom high-speed camera system. We thank Darron Cullen for sharing study insects and helpful advice, and the members of the Insect Biomechanics Group in Cambridge for helpful discussions.

Conflicts of Interest: The authors declare no conflicts of interest.

References

1. Walcher, F.; Kral, K. Visual deprivation and distance estimation in the praying mantis larva. *Physiol. Entomol.* **1994**, *19*, 230–240. [CrossRef]
2. Burrows, M.; Cullen, D.A.; Dorosenko, M.; Sutton, G.P. Mantises exchange angular momentum between three rotating body parts to jump precisely to targets. *Curr. Biol.* **2015**, *25*, 786–789. [CrossRef] [PubMed]

3. Sutton, G.P.; Doroshenko, M.; Cullen, D.A.; Burrows, M. Take-off speed in jumping mantises depends on body size and a power-limited mechanism. *J. Exp. Biol.* **2016**, *219*, 2127–2136. [CrossRef]
4. Dai, Z.; Gorb, S.N.; Schwarz, U. Roughness-dependent friction force of the tarsal claw system in the beetle *Pachnoda marginata* (Coleoptera, Scarabaeidae). *J. Exp. Biol.* **2002**, *205*, 2479–2488. [CrossRef] [PubMed]
5. Clemente, C.J.; Goetzke, H.H.; Bullock, J.M.R.; Sutton, G.P.; Burrows, M.; Federle, W. Jumping without slipping: Leafhoppers (Hemiptera: Cicadellidae) possess special tarsal structures for jumping from smooth surfaces. *J. R. Soc. Interface* **2017**, *14*, 20170022. [CrossRef]
6. Goetzke, H.H.; Pattrick, J.G.; Federle, W. Froghoppers jump from smooth plant surfaces by piercing them with sharp spines. *Proc. Natl. Acad. Sci. USA* **2019**, *116*, 3012–3017. [CrossRef]
7. Bullock, J.M.R.; Federle, W. Division of labour and sex differences between fibrillar, tarsal adhesive pads in beetles: Effective elastic modulus and attachment performance. *J. Exp. Biol.* **2009**, *212*, 1876–1888. [CrossRef] [PubMed]
8. Labonte, D.; Federle, W. Functionally different pads on the same foot allow control of attachment: Stick insects have load-sensitive "heel" pads for friction and shear-sensitive "toe" pads for adhesion. *PLoS ONE* **2013**, *8*, e81943. [CrossRef] [PubMed]
9. Grohmann, C.; Henze, M.J.; Nørgaard, T.; Gorb, S.N. Two functional types of attachment pads on a single foot in the Namibia bush cricket *Acanthoproctus diadematus* (Orthoptera: Tettigoniidae). *Proc. R. Soc. B* **2015**, *282*, 20142976. [CrossRef] [PubMed]
10. Clemente, C.J.; Federle, W. Pushing versus pulling: Division of labour between tarsal attachment pads in cockroaches. *Proc. R. Soc. B* **2008**, *275*, 1329–1336. [CrossRef]
11. Labonte, D.; Williams, J.A.; Federle, W. Surface contact and design of fibrillar 'friction pads' in stick insects (*Carausius morosus*): Mechanisms for large friction coefficients and negligible adhesion. *J. R. Soc. Interface* **2014**, *11*, 20140034. [CrossRef]
12. Clemente, C.J.; Dirks, J.-H.; Barbero, D.R.; Steiner, U.; Federle, W. Friction ridges in cockroach climbing pads: Anisotropy of shear stress measured on transparent, microstructured substrates. *J. Comp. Physiol. A* **2009**, *195*, 805–814. [CrossRef] [PubMed]
13. Beutel, R.G.; Gorb, S.N. Ultrastructure of attachment specializations of hexapods (Arthropoda): Evolutionary patterns inferred from a revised ordinal phylogeny. *J. Zool. Syst. Evol. Res.* **2001**, *39*, 177–207. [CrossRef]
14. Beutel, R.G.; Gorb, S.N. A revised interpretation of attachment structures in Hexapoda with special emphasis on Mantophasmatodea. *Arthrop. Syst. Phyl.* **2006**, *64*, 3–25. [CrossRef]
15. Gorb, S.; Jiao, Y.; Scherge, M. Ultrastructural architecture and mechanical properties of attachment pads in *Tettigonia viridissima* (Orthoptera Tettigoniidae). *J. Comp. Physiol. A* **2000**, *186*, 821–831. [CrossRef] [PubMed]
16. Büscher, T.H.; Gorb, S.N. Complementary effect of attachment devices in stick insects (Phasmatodea). *J. Exp. Biol.* **2019**, *222*, jeb209833. [CrossRef] [PubMed]
17. Niederegger, S.; Gorb, S. Tarsal movements in flies during leg attachment and detachment on a smooth substrate. *J. Insect Physiol.* **2003**, *49*, 611–620. [CrossRef]
18. Bullock, J.; Drechsler, P.; Federle, W. Comparison of smooth and hairy attachment pads in insects: Friction, adhesion and mechanisms for direction-dependence. *J. Exp. Biol.* **2008**, *211*, 3333–3343. [CrossRef] [PubMed]
19. Burrows, M. Jumping performance of froghopper insects. *J. Exp. Biol.* **2006**, *209*, 4607–4621. [CrossRef] [PubMed]
20. Burrows, M. Jumping performance of planthoppers (Hemiptera, Issidae). *J. Exp. Biol.* **2009**, *212*, 2844–2855. [CrossRef]
21. Burrows, M. Jumping mechanisms in dictyopharid planthoppers (Hemiptera, Dicytyopharidae). *J. Exp. Biol.* **2014**, *217*, 402. [CrossRef] [PubMed]
22. Tautz, J.; Hölldobler, B.; Danker, T. The ants that jump: Different techniques to take off. *Zoology* **1994**, *98*, 1–6.
23. Baroni-Urbani, C.; Boyan, G.S.; Blarer, A.; Billen, J.; Musthak Ali, T.M. A novel mechanism for jumping in the indian ant *Harpegnathos saltator* (Jerdon) (Formicidae, Ponerinae). *Experientia* **1994**, *50*, 63–71. [CrossRef]
24. Burrows, M. Jumping mechanisms and performance of snow fleas (Mecoptera, Boreidae). *J. Exp. Biol.* **2011**, *214*, 2362–2374. [CrossRef]
25. Burrows, M. Jumping from the surface of water by the long-legged fly Hydrophorus (Diptera, Dolichopodidae). *J. Exp. Biol.* **2013**, *216*, 1973–1981. [CrossRef] [PubMed]
26. Burrows, M.; Dorosenko, M. Jumping mechanisms in lacewings (Neuroptera, Chrysopidae and Hemerobiidae). *J. Exp. Biol.* **2014**, *217*, 4252–4261. [CrossRef] [PubMed]
27. Burrows, M.; Morris, O. Jumping in a winged stick insect. *J. Exp. Biol.* **2002**, *205*, 2399–2412. [CrossRef] [PubMed]
28. Burrows, M.; Dorosenko, M. Jumping mechanisms and strategies in moths (Lepidoptera). *J. Exp. Biol.* **2015**, *218*, 1655. [CrossRef] [PubMed]
29. Koh, J.S.; Yang, E.; Jung, G.P.; Jung, S.P.; Son, J.H.; Lee, S.I.; Jablonski, P.G.; Wood, R.J.; Kim, H.Y.; Cho, K.J. Jumping on water: Surface tension-dominated jumping of water striders and robotic insects. *Science* **2015**, *349*, 517–521. [CrossRef] [PubMed]
30. Goetzke, H.-H.; Federle, W. Role of legs and foot adhesion in salticid spiders jumping from smooth surfaces. *J. Comp. Physiol. A* **2021**, *207*, 165–177. [CrossRef]
31. Parry, D.A.; Brown, R.H.J. The jumping mechanism of Salticid spiders. *J. Exp. Biol.* **1959**, *36*, 654–664. [CrossRef]

32. Weihmann, T.; Karner, M.; Full, R.; Blickhan, R. Jumping kinematics in the wandering spider *Cupiennius salei*. *J. Comp. Physiol. A* **2010**, *196*, 421–438. [CrossRef] [PubMed]
33. Ehlers, M. Untersuchungen über Formen aktiver Lokomotion bei Spinnen. *Zool. Jb. Syst.* **1939**, *72*, 373–499.
34. Sutton, G.; Burrows, M. The mechanics of elevation control in locust jumping. *J. Comp. Physiol. A* **2008**, *194*, 557–563. [CrossRef]
35. Sutton, G.P.; Burrows, M. The mechanics of azimuth control in jumping by froghopper insects. *J. Exp. Biol.* **2010**, *213*, 1406–1416. [CrossRef]
36. Sutton, G.P.; Mendoza, E.; Azizi, E.; Longo, S.J.; Olberding, J.P.; Ilton, M.; Patek, S.N. Why do large animals never actuate their jumps with latch-mediated springs? Because they can jump higher without them. *Integr. Comp. Biol.* **2019**, *59*, 1609–1618. [CrossRef] [PubMed]
37. Burrows, M. Kinematics of jumping in leafhopper insects (Hemiptera, Auchenorrhyncha, Cicadellidae). *J. Exp. Biol.* **2007**, *210*, 3579–3589. [CrossRef]

biomimetics

MDPI

Article

A Microfluidic Design for Quantitative Measurements of Shear Stress-Dependent Adhesion and Motion of *Dictyostelium discoideum* Cells

Sepideh Fakhari [1,2,3], Clémence Belleannée [2,3], Steve J. Charrette [4] and Jesse Greener [1,2,*]

1 Department of Chemistry, Faculty of Science and Engineering, Université Laval, Québec City, QC G1V 0A6, Canada
2 Centre de Recherche du Centre Hospitalier Universitaire de Québec, Université Laval, Québec City, QC G1V 0A6, Canada
3 Centre de Recherche en Reproduction, Développement et Santé Intergénérationnelle, Department of Obstetrics, Gynecology, and Reproduction, Faculty of Medicine, Université Laval, Québec City, QC G1V 0A6, Canada
4 Department of Biochemistry, Microbiology and Bioinformatics, Faculty of Science and Engineering, Université Laval, Québec City, QC G1V 0A6, Canada; steve.charette@bcm.ulaval.ca
* Correspondence: jesse.greener@chm.ulaval.ca

Abstract: Shear stress plays a crucial role in modulating cell adhesion and signaling. We present a microfluidic shear stress generator used to investigate the adhesion dynamics of *Dictyostelium discoideum*, an amoeba cell model organism with well-characterized adhesion properties. We applied shear stress and tracked cell adhesion, motility, and detachment using time-lapse videomicroscopy. In the precise shear conditions generated on-chip, our results show cell migration patterns are influenced by shear stress, with cells displaying an adaptive response to shear forces as they alter their adhesion and motility behavior. Additionally, we observed that DH1-10 wild-type *D. discoideum* cells exhibit stronger adhesion and resistance to shear-induced detachment compared to *phg2* adhesion-defective mutant cells. We also highlight the influence of cell density on detachment kinetics.

Keywords: cell adhesion; cell motility; shear stress; microfluidic devices; amoeba; *Dictyostelium discoideum*

Citation: Fakhari, S.; Belleannée, C.; Charrette, S.J.; Greener, J. A Microfluidic Design for Quantitative Measurements of Shear Stress-Dependent Adhesion and Motion of *Dictyostelium discoideum* Cells. *Biomimetics* **2024**, *9*, 657. https://doi.org/10.3390/biomimetics9110657

Academic Editors: Stanislav N. Gorb and Thies Büscher

Received: 10 October 2024
Revised: 23 October 2024
Accepted: 25 October 2024
Published: 27 October 2024

1. Introduction

Shear stress is an external mechanical force that can influence cell adhesion, motility, and other biological processes. This includes regulating cells' interaction with, and attachment to, various surfaces based on mediation through specialized protein complexes [1]. The mechanical aspects of cell adhesion and motility are under intensive ongoing theoretical and experimental investigation [2,3] due to their crucial relation to the behavior of many microorganisms. Being a fundamental property of natural and synthetic cellular biosystems, these are also hot topics in biomimetics. To move this field forward, new platforms to accurately study cellular adhesion are required.

Amoebae are excellent model microorganisms for further biomimetic development due to their natural affinity to surfaces [4] from artificial cells [5] and their applications to soft robotics [6]. *Dictyostelium discoideum*, a social amoeba that naturally inhabits soil, has been extensively studied for cell adhesion, motility, and development, and other biological processes [7]. *D. discoideum* undergoes two primary life stages: the first is a unicellular vegetative stage, and the second involves multicellular development. In the first stage, the cells employ phagocytosis for nutrition. This process involves extending pseudopods—temporary cytoplasmic projections—used to engulf and internalize prey [8]. Successful phagocytosis depends on the amoeba's ability to adhere to surfaces and move effectively, even under shear stress conditions. When the food source is exhausted, *D. discoideum* cells switch to the second life stage, which starts with the accumulation of hundreds of thousands of cells into dense colonies (pre-aggregation phase) before aggregating into three-dimensional structures and undergoing significant changes via gene expression, leading to the formation of mature fruiting bodies bearing spores [9]. During the pre-aggregation phase, the cell density increases beyond 100 cells mm^{-2}, reaching at least 1000 cells mm^{-2} [10].

Previous research on *D. discoideum* cell adhesion under shear stress has elucidated several critical mechanisms that govern cellular adhesion and detachment. For example, the number of adhered cells exhibits a sigmoidal decrease with increasing shear stress, suggesting normally distributed adhesion across the cell population [11]. The initial reversible adhesion is typically followed by a stabilization phase, which highlights the importance of temporal factors in adhesion specificity [12]. Adhesion force measurements have revealed the role of surface hydrophobicity [13], and the presence of biomolecules at the attachment surface [14]. Other studies have focused on the role of the hydrodynamic environment in cell motility and directionality, determining that shear flow acts as a mechanical cue that influences cellular behavior [15].

Most studies of cell shear stress adhesion, including those on *D. discoideum*, are conducted using macroscale setups, which involve application of external forces originating from spinning disks [11], centrifugation [12,16], or flow chambers (radial [17] or rectangular [14,18]). However, these conventional methods have limitations, including low throughput, complex assembly, restricted range and limited precision in the application of detachment forces, often due in part to non-uniform velocity distributions and turbulence.

Microfluidic devices have become accepted tools for precise measurements pertaining to microorganisms and their multicellular constructs [19–21]. The low material consumption and high surface-area-to-volume ratios in microchannels can significantly decrease experimental costs and increase cell–surface interactions. Most relevant for the present work, the specific ability to accurately manipulate and control fluids at the micro-/nanoscale offers the potential for detailed studies into cell adhesion and motility due to the combination of highly controllable shear forces and the compatibility of transparent planar microchannels with high-quality microscopy [22]. Various microfluidic setups have been used to study the effect of shear forces applied to different cell types, including, but not limited to, neutrophils [23–26], endothelial cells [22,27–29], bacterial cells [30–34], and, as we have recently shown, their biofilms [35–40].

Microfluidic studies of *D. discoideum* have included single-cell studies [41–43] as well as chemotaxis studies [44–47]. Other studies used high-resolution time-lapse microscopy to monitor the real-time interactions between *D. discoideum* and bacterial pathogens [43]. Though less frequent, microfluidic adhesion studies on *D. discoideum* have also been demonstrated. For example, tapered channels in which shear stress was correlated to downstream position (based on the corresponding local channel width) were used to measure detachment kinetics and shear stress-dependent motion under different applied shear stresses [48]. The results noted a lognormal distribution of the threshold stresses for detachments and first-order kinetics. However, while innovative, the tapered channel design also could also result in artifacts from wall effects due to the coupling of applied shear forces with changes to position during cell motility studies. Tarantola et al. employed a microfluidic device to study surface adhesion of vegetative *D. discoideum*, in which they reported a ten-fold decrease in substratum adhesion [49]. The highly engineered devices included several channel branch points that admitted liquid to various attachment chambers. Based on the complex interconnected flow paths and chamber sizes, different shear stresses were developed in the various chambers [50]. Drawbacks included non-obvious flow properties due to the multitude of branchpoints and complex channel structures. Advantages included large rectangular cell adhesion chambers in which constant shear stresses could be applied while minimizing the confounding influence of cell-wall interactions. Although it was not exploited in the study by Tarantola et al., the large cell adhesion chambers have the potential for motility studies in parallel with adhesion studies.

In this study, our primary goal was the design and implementation of a microfluidic shear stress generator coupled with time-lapse video microscopy. This device was used to study both the motility and adhesion of native cells. We designed the system to prioritize precision and reliability, with a focus on controlled single-condition experiments rather than automated or high-throughput methods. By using the same chamber and sequentially applying different shear stresses via flow rate adjustments, this approach allowed for precise real-time observation of cell detachment. Our simplified yet effective system enabled accurate quantification of shear stress impacts and initial cell density measurements. Using this platform, we analyzed the detachment and movement of the adhesion-competent wild-type *D. discoideum* strain (DH1-10) and the adhesion-deficient mutant strain (*phg2*). By comparing the behavior of these cells, we validated the ability of our microfluidic shear stress generator to quantitatively assess cell adhesion and investigate the role of cell type and density.

2. Materials and Methods

2.1. Cell Culture

D. discoideum DH1-10 [51] and *phg2* cells [52] were used in this study. DH1-10 cells are a spontaneously isolated clone derived from the DH1 strain. The DH1 strain itself was created by deleting the pyr5-6 locus from the AX3 cell line [53]. The *phg2* mutant cells were subsequently generated from the DH1-10 line through restriction enzyme-mediated integration (REMI) [52]. Therefore, DH1-10 and *phg2* mutant cells are genetically identical, with the sole difference being the inactivation of the *phg2* gene in the mutant cells. The *phg2* gene encodes the Phg2 protein, a putative serine/threonine kinase essential for phagocytosis, and cell adhesion and mutations in the *phg2* gene impair cell adhesion [52,54,55], which also emphasizes the significance of *phg2* in coordinating adhesion-related signaling pathways and actin cytoskeleton reorganization [56]. As previously described [51], these cells were cultured at 21 °C in an HL5 medium supplemented with 15 μg mL^{-1} tetracycline, following the method outlined by Mercanti et al. [57]. The cells were subcultured twice per week in a fresh medium to prevent them from reaching confluence.

2.2. Microfluidic Device Fabrication

A microfluidic device with a total volume of approximately 2 μL was designed using standard photolithography techniques [58], including CAD software, and a photomask was fabricated by photoplotter (FPS25000, Fortex Engineering Ltd., Lincoln, UK) for use in photolithography. A mold for casting the device was created from a single dry photoresist film (SY300 film, Fortex Engineering Ltd., UK), which was then laminated onto a glass slide (12–550C, Fisher Scientific, Ottawa, ON, Canada) using a lamination system (FL-0304-01, Fortex Engineering Ltd., UK). The height of the laminated photoresist film determined the final channel height (50 μm). The shadow mask and the photoresist-coated glass slide were exposed to UV light in a UV exposure system (AY-315, Fortex Engineering Ltd., UK) to selectively crosslink portions of the photoresist. Subsequent immersion in developer and rinse baths (SY300 Developer/Rinse, Fortex Engineering Ltd., UK) removed the uncrosslinked portions of the photoresist, resulting in formation of the final mold. A polydimethylsiloxane (PDMS) and crosslinker solution were mixed at a 10:1 ratio, cast on top of the mold, and cured at 70 °C overnight. Two inlet holes and one outlet hole were punched into the PDMS at each channel termination point. The two inlets were included in the design for the administration of liquid without the need to disconnect the system during multiple switches between culture media flow and cell flow, thereby preventing the introduction of air bubbles. Air plasma activation (PCD001, Harrick Plasma, Ithaca, NY, USA) was used to bond and seal the PDMS device onto a glass slide. The bonded device underwent brief annealing at 70 °C to enhance the bond strength. The final device was fully transparent, enabling real-time optical measurements while selected flow rates were provided by syringe pumps (PhD 2000, Harvard Apparatus, Holliston, MA, USA).

2.3. *Videomicroscopy and Image Analysis*

In situ imaging of cells was conducted using an inverted microscope (IX73, Olympus, Richmond Hill, ON, Canada) equipped with a 10× Olympus Plan Fluorite objective lens with a 0.30 numerical aperture. Sequential time-lapse imaging was conducted with an 8 s delay for 40 min using a CCD camera (Lumenera Infinity 31 U, Ottawa, ON, Canada). Image sequences were then transferred to a computer for frame-by-frame analysis using the Fiji distribution of ImageJ software (version 1.54f, NIH, Bethesda, MD, USA). Image treatment was used to facilitate cell identification. This included Fiji background subtraction using a rolling-ball technique of 0.5 pixels and image conversion to binary. Following this, the number of detached cells was deduced based on a manual count of the number of remaining cells at each time point. From this, the number of detached cells was normalized by the initial population count at t = 0 to determine the ratio of remaining cells.

2.4. *Cell* Seeding and Application of Shear Stress

D. discoideum cells (DH1-10 or *phg2*) were cultured in HL5 medium in 10 cm plastic culture dishes (Falcon) until they covered 60% to 80% of the dish surface. The cells were pelleted by centrifugation, and the resulting cell pellet was resuspended in fresh medium at one-tenth of the initial volume. These suspended cells (the inoculum) were then introduced into a sterilized microfluidic chip, which was pre-filled with HL5 medium. The inoculum was injected via a syringe pump set at a near-zero flow rate (0.1 mL h^{-1}) until initial cell densities between 100 and 1000 cells mm^{-2} were achieved. Static conditions (zero flow rate) were never applied in order to avoid nutrient depletion, which can otherwise occur rapidly in the small volumes on-chip. Subsequently, the inoculum was replaced with a culture medium, which was first introduced at a low flow rate (0.1 mL h^{-1}) until all culture medium was washed out. Under hydrodynamic conditions of the sterile culture medium, unattached cells were flushed out of the device, leaving only the attached cells in the channels. In separate experiments, we confirmed that under a flow rate of 0.1 mL h^{-1}, attached cells persisted on the attachment surface for 40 min. Therefore, we considered this as a baseline flow rate, instead of using 0 mL h^{-1} as a control experiment. Then, volumetric flow rates of 2, 5, and 10 mL h^{-1} were applied to probe the effect of fluid shear stresses on the cells. The hydraulic retention times were 60, 24, and 12 min (1, 2.5, and 5 medium recharges per hour) for flow rates of 2, 5, and 10 mL h^{-1}, respectively. A new device was used for each cell seeding experiment.

2.5. *Statistical Analysis*

Statistical analyses were conducted using GraphPad Prism version 10.1.2. Data are expressed as the mean ± standard error (σ/n, where σ is the standard deviation and n is the number of samples) of the mean from at least three experiments per condition. Cell densities ranged from 100 to 1000 cells mm^{-2}, representing cell densities similar to those during the pre-aggregation phase. We avoided cell densities higher than 1000 cells mm^{-2} to minimize cell-to-cell contact, which can significantly influence adhesion, detachment, and motility. Statistical analyses were conducted to ensure significance between data sets that were run with different conditions (e.g., cell types, flow conditions). Differences between any two data sets were evaluated using unpaired *t*-tests, whereas comparisons among more than two data sets were performed using two-way ANOVA followed by Tukey's post hoc test. Statistical significance was defined as a *p*-value less than 0.05. The levels of significance are denoted in figures as $p < 0.05$ (*), $p < 0.01$ (**), $p < 0.001$ (***), and not significant for $p > 0.05$ (n.s.).

2.6. *Computer Simulation*

In this study, computational fluid dynamics (CFD) was used to simulate the hydrodynamic flow conditions within a cell adhesion chamber, with a specific focus on calculating the shear stress exerted on immobilized amoeba cells. The simulation was performed using simulation software (COMSOL Multiphysics 4.2a, Stockholm, Sweden). The simulation employed fluid–

structure interaction to concurrently analyze the effect of fluid flow across the amoeba, under simplified cellular adhesion patterns on the solid surface. The model's geometrical features the microfluidic system, including two inlets and one outlet, with amoebae being represented as solid semi-spherical objects with radii of 5 μm based on approximations from microscope imaging. We fixed the field of view at the center of the cell adhesion chamber (Figure 1) to avoid wall effects. A selection of important model properties is presented in Table 1.

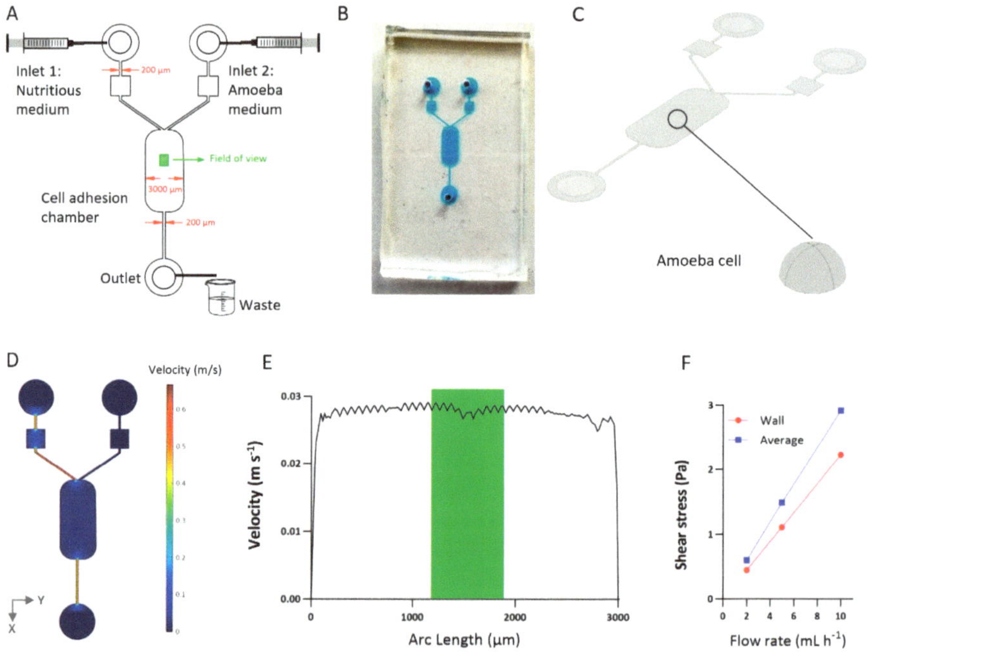

Figure 1. Microfluidic device design. (**A**) CAD model of the microfluidic array, featuring 2 inlets, 1 outlet, and a cell adhesion chamber, designed to ensure controlled and uniform shear stress, provide ample space for cell mobility, and enable short-term and long-term experiments. (**B**) Microfluidic chip fabricated based on the CAD design in (**A**). (**C**) Three-dimensional numerical model of the device with a semi-spherical amoeba model fixed in the middle of the cell adhesion chamber. The shape of the cell model is shown zoomed in for clarity. (**D**) Velocity magnitude along the device at a flow rate of 5 mL h^{-1} at inlet 1, showing a uniform distribution of velocity across the device. (**E**) Velocity magnitude along a line through the cell adhesion chamber width (Y-direction), highlighting the uniformity of shear stress distribution; the green box shows the width of the field of view in (**A**). (**F**) Difference between the analytically calculated wall shear stress and the numerically calculated average shear stress applied to the amoeba cell surface.

The pressure-driven flow of an incompressible liquid through the microfluidic channel can be described using the Navier–Stokes and the continuity equation. To simulate flow through the microfluidic shear stress generator, we used the Poiseuille model to analytically investigate the wall shear stress in a cell adhesion chamber. The shear stress (τ) is a function of shear rate (γ), which is determined by the volumetric flow rate (Q), the dimensions of the channel height and width (h and w, respectively), and the liquid viscosity (μ):

$$\tau = \mu \times \gamma = \mu \frac{6Q}{h^2 w} \quad (1)$$

A mesh was developed to define spatial locations where hydrodynamic values and forces were simulated by CFD. A mesh refinement step was conducted iteratively, with enhanced mesh density, until variations in the computed shear stress on the amoeba cell were negligible. A "finer" mesh, resulting in a simulation error of less than 1%, was determined to be optimal (Table 2).

Table 1. Properties of the microfluidic device model used in CFD simulation.

Amoeba radius	5 μm
Fluid delivery channel width	200 μm
Cell adhesion chamber length (l)	6.6 mm
Cell adhesion chamber width (w)	3 mm
Channel height (h)	50 μm
Fluid density	1000 kg m^{-3}
Fluid viscosity (μ)	1 mPa s

Table 2. Summary of mesh independence analysis results.

Element Size	Number of Mesh Elements	Shear Stress (Pa)	Relative Error
Coarse	59,442	1.87	36%
Normal	83,001	2.59	12%
Fine	119,616	2.91	1.2%
Finer	434,694	2.92	0.8%
Extra fine	804,595	2.95	-

3. Results

3.1. Device Design and Shear Stress Simulation

The design of the final microfluidic shear stress generator, with the design schematic and image shown in Figure 1A,B, aimed to meet several key experimental objectives for cell adhesion studies: ensuring a controlled and uniform shear stress across the cell adhesion chamber, providing ample space for cell motility while minimizing interferences such as the wall effect, and enabling measurements at time scales ranging from minutes to more than 12 h. In this work, all experiments were run for 40 min. The region of interest (900 × 673 μm) was positioned 1000 μm downstream from the cell adhesion chamber entry and centered within the cell adhesion chamber so that the visualized amoeba was not influenced by the vertical sidewalls. We validated this assumption using numerical simulations (Figure 1C), demonstrating that along the Y-axis (perpendicular to the flow direction), over 95% of the channel maintained a uniform velocity pattern throughout the channel width, with the exception of locations directly beside the walls (Figure 1D,E). This uniformity is due to the low aspect ratio design of the channel (channel height divided by channel width), which minimized the influence of the sidewalls. The low aspect ratio also provided adequate space for cell adhesion, movement, and growth.

According to Equation (1), the wall shear stresses in the cell adhesion chamber at flow rates of 2, 5, and 10 mL h^{-1} are calculated as 0.45, 1.11, and 2.23 Pa, respectively. However, due to the microscale flow disturbances around the cells, the actual shear stress experienced by the cells may differ from the wall shear stress. To assess this effect, we modeled the presence of an idealized cell as a semi-sphere with a radius of 5 μm (Figure 1C). The model indicated that the average shear stresses on the cell surface at flow rates of 2, 5, and 10 mL h^{-1} were 0.60, 1.49, and 2.92 Pa, respectively. These values represent 35.5%, 33.4% and 33.13% higher shear stresses experienced by the cell at 2, 5, and 10 mL h^{-1}, respectively, compared to the shear stress applied to the channel wall (Figure 1F). It is important to note that our simulations did not account for dynamic changes in cell shape. In the future, higher-resolution imaging combined with automated image analysis could be applied to obtain more detailed information about the dimensions on a cell-by-cell basis during the different stages of the experiments and then used to calculate the precise shear stress.

3.2. Cell Tracking and Migration

The cells exhibited normal migration before detachment. In advance of the shear-induced detachment, the cells initially extended several forward and lateral pseudopods. Eventually, they ceased migration, became more rounded, with less contact to the surface, and then finally detached (Figure 2A). This dynamic response indicates an adaptive behavior to shear stress conditions.

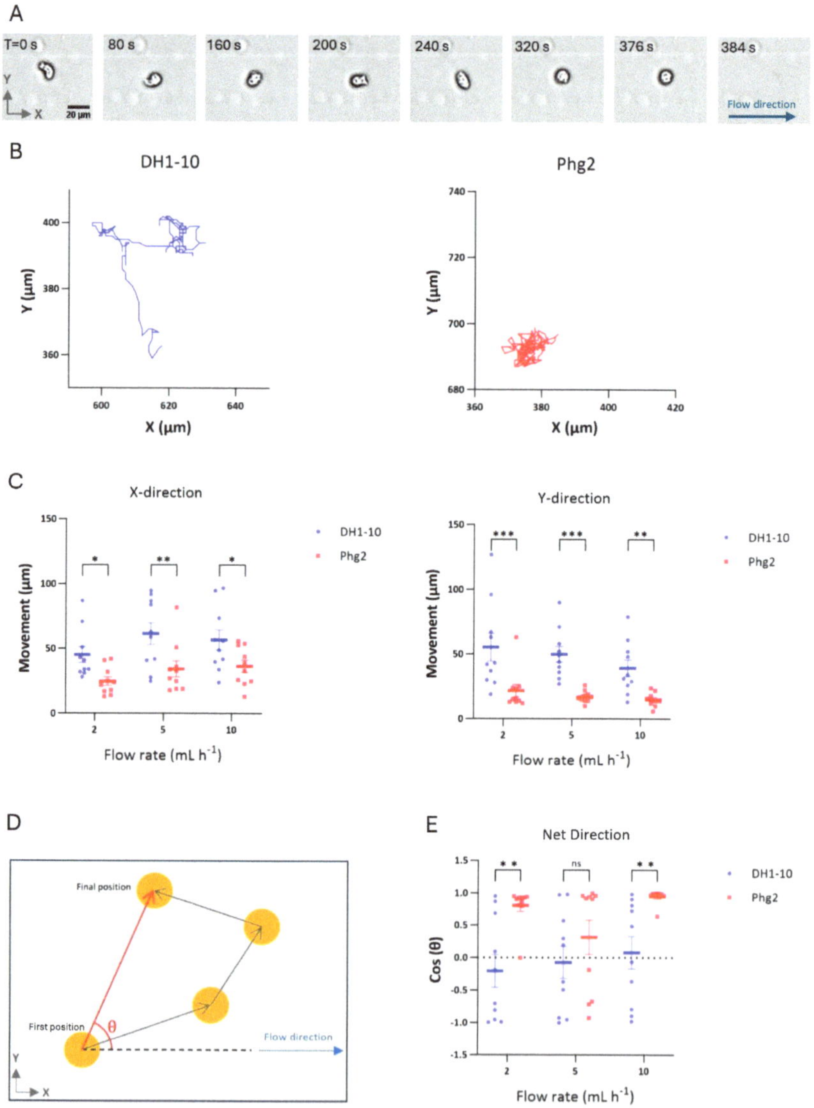

Figure 2. Analysis of *D. discoideum* cell motility under shear stress. (**A**) Detachment of a single DH1-10 cell at 10× magnification, at 10 mL h^{-1}; the arrow in the last frame indicates the flow direction. (**B**) Migration

of a single DH1-10 and single *phg2* cell under fluid flow over a period of 40 min. (**C**) Cell migration of 10 cells under shear stress. (**D**,**E**) Mean directionality of cell movement as a function of applied shear stress, indicating to what extent migration is aligned with flow direction. Directionality is defined as the angle between the flow direction and the cell movement direction over a period of 40 min. Therefore, $\cos(\theta) = 1$ indicates fully biased cell movement in the flow direction, $\cos(\theta) = 0$ indicates cell movement perpendicular to fluid flow, and $\cos(\theta) = -1$ indicates cell movement opposite to fluid flow direction. Error bars represent the standard error. Significance levels in the figures are represented as follows: $p < 0.05$ (*), $p < 0.01$ (**), $p < 0.001$ (***), and not significant (ns) for $p > 0.05$.

The motility of the attached cells was quantified by measuring the maximum absolute displacement in the X-direction (parallel to flow) and Y-direction (perpendicular to flow) at each flow rate (Figure 2B,C). DH1-10 cells showed greater movement in both directions compared to *phg2* mutant cells, suggesting higher cellular activity and response to shear stress. When evaluating the effect of flow on motility within the same cell class, the trends indicated that motility in the direction of flow increased and motility perpendicular to flow decreased. However, the statistical differences were lower than our identified threshold for significance.

The directionality of net cell movement was further evaluated by measuring the angle between the initial (t = 0 min) and final cell positions (t = 40 min), as illustrated in Figure 2D. DH1-10 cells exhibited diffusive behavior, characterized by random movement within their environment, and maintained this diffusive pattern even under high shear stress (Figure 2E). In contrast, *phg2* mutant cells demonstrated a movement predominantly aligned with the flow direction (X-direction), as indicated by $\cos(\theta)$ values close to 1 (Figure 2E) indicating a more pronounced response to the flow conditions. In contrast, movement distance in the X-direction was approximately equivalent for movement upstream compared to movement downstream for DH1-10 cells, indicating a motility that was nearly independent of the flow rate.

3.3. Cell Detachment Under Shear Flow

We investigated the adhesion and detachment dynamics of *D. discoideum* cells under varying shear stress conditions using the microfluidic shear stress generator. By comparing the detachment and movement of *D. discoideum* wild-type cells and adhesion-defective mutant cells, we aimed to validate the ability of our microfluidic shear stress generator to quantitatively assess comparative cell adhesion.

Various shear stresses were applied by adjusting the inlet flow rate, based on Equation (1). The flow rate was carefully selected to ensure that both cell movement and detachment could be observed for *D. discoideum* DH1-10 and *phg2* cells. A low flow rate of $Q = 2$ mL h^{-1} was identified in preliminary assays as the minimum required to study the cell detachment percentage in DH1-10. Below this flow rate, the number of detached cells was found to be insignificant. On the other hand, a high flow rate of $Q = 10$ mL h^{-1} was sufficient to study cell detachment in DH1-10. This approach allowed us to assess the effects of varying shear stress on cell adhesion and detachment for both cell types. In this study, we were interested in the average results over a range of cell densities from 100 to 1000 cells mm^{-2}, a range relevant to the initial densification during the pre-aggregation phase. According to our work, pre-aggregation cell densities can reach 3000 cells mm^{-2} but result in complications related to cell-to-cell contact, which we sought to avoid in this work.

Primary detachment data are shown in Figure 3A–C. Generally, the detachment was initially rapid and slowed until it reached a pseudo-plateau where changes to the number of cells were nearly constant over the time scale of the 40-minute experiment. Cell detachment analysis was conducted, with the results presented as curves in Figure 3D,E for DH1-10 and *phg2* amoeba, respectively. The DH1-10 cell type exhibited a gradual increase in the detachment percentage and a lower pseudo-plateau in detachment levels across all flow rates compared to that of *phg2*. Based on the two-way ANOVA test, all data were statistically significant (for example, for DH1-10 detachment at 5 and 10 mL h^{-1}). The initial rate of cell detachment and the total percentage of detached cells after reaching a plateau increased with the flow rate but, again, were notably less affected compared to *phg2* cells, which were generally prone to detachment at all flow rates. In both cases, the detachment process did not reach 100% before the plateau region, indicating that strains contained a population subset with stronger attachment properties. This strongly attached population was larger for the DH1-10 cells than for the *phg2* cells. Overall, *phg2* cells have weaker adhesion, including a heightened sensitivity to shear forces.

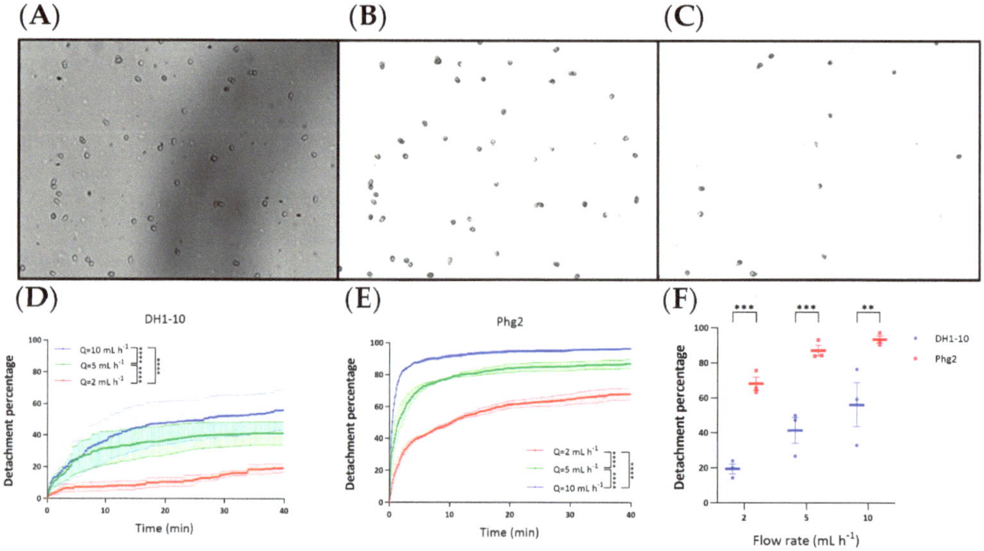

Figure 3. Analysis of cell detachment under various shear stress conditions. (**A**) Primary data showing the microscope raw image at time, t = 0 for DH1-10 cells. (**B**) Same t = 0 data after image treatment, including background subtraction and conversion to a binary image. (**C**) Treated image of shear stress chambe r after t = 40 min. Flow for images (**A–C**) was 10 mL h^{-1}. (**D–F**) Comparative analysis of cell detachments for all flow rates and cell types. (D) Cell detachment curves for *Dictyostelium discoideum* DH1-10 in medium at three flow rates of Q = 10 mL h^{-1} (blue), Q = 5 mL h^{-1} (green), and Q = 2 mL h^{-1} (red), illustrating increased detachment with higher shear stresses from increasing fluid flow rates. (**E**) Cell detachment curves for the *phg2* adhesion-defective mutant in medium at three flow rates of Q = 10 mL h^{-1} (blue), Q = 5 mL h^{-1} (green), and Q = 2 mL h^{-1} (red), showing a rapid increase in detachment levels. (**F**) Differential response of DH1-10 and *phg2* final detachment percentages (after 40 min) for flow rates of 2, 5, and 10 mL h^{-1}. Figures denote significance as $p < 0.01$ (**), $p < 0.001$ (***), $p < 0.0001$ (****), and $p > 0.05$ (ns).

We further analyzed the final detachment percentages of DH1-10 and *phg2* mutants after arriving at the pseudo-plateau (Figure 3F); it is evident that *phg2* cells are less adherent and detach more readily in response to fluid flow, indicative of weaker adhesion Conversely, DH1-10 cells display more robust adhesion, demonstrating resistance to shear-induced detachment. These outcomes were anticipated and likely reflect the intrinsic adhesion properties of each cell type [52]. This differentiation not only highlights the utility of our microfluidic design in assessing cell adhesion under dynamic conditions but also provides valuable insights into the cellular mechanisms governing adhesion, with potential implications for understanding various biological processes and diseases.

3.4. Effect of Cell Density on Detachment

Next, we deepened the analysis by evaluating the effect of initial cell density (number of initial cells per unit area) on the detachment process to test our hypothesis that this may affect the accumulation of cells during the pre-aggregation phase. To begin, we ran simulations to determine how the shear stress was modified on a cell in the presence of a single upstream cell. Using modeling, we accounted for the effect of a single upstream amoeba on the applied shear stress on a (second) neighboring amoeba (downstream). When the intercellular separation distances were large (over 25 μm), the shear stress on the downstream amoeba was the same as that reported in Figure 1F (τ = 0.6, 1.49, and 2.92 Pa at 2, 5, and 10 mL h^{-1}, respectively). When the separation distances were reduced, our simulation found that the downstream amoeba was partially shielded, resulting in a lower applied shear stress. The results indicated that, at all three flow rates, the presence of an upstream amoeba at an edge-to-edge distance of 10 μm (twice the amoeba's radius) resulted in a reduction in shear stress on the downstream amoeba of 10%, whereas a 16% reduction in shear stress was observed when the distance was reduced to 5 μm. Next, we extended this simulation to determine the effect of several upstream amoebae at a cell density that mirrored that of our experiments. Based on the initial cell densities in our experiments, we estimated that the average intercellular distance varied significantly, from 7 μm (at 1068 cells mm^{-2}) to 55 μm (at 76 cells mm^{-2}). We re-ran the simulation with 0, 2, 4, and 7 upstream amoebae, each separated by an edge-to-edge distance d (Figure 4A). We conducted these simulations at the three flow rates used in this study and changed the inter-cellular distances to either d = 7 or 55 μm to represent the minimum and maximum average cell-to-cell distances based on the range of cell densities in our experiments. As seen in Figure 4B, the effect of more upstream amoebae serves to further reduce the applied shear stress on the final amoeba. From these simulations, we see that the shear stress on the most downstream amoeba reaches a stable value after approximately four upstream cells for the highest flow rate used. Based on a close analysis, this stability is reached earlier at lower flow rates. Therefore, we are confident that the results with four or more upstream amoebae represent an accurate average applied shear stress applied on most of the cells in the experiments. For dense colonies (d = 7 μm), the applied shear stress on the most downstream cell was 2.25 Pa at 10 mL h^{-1}, marking a reduction of nearly 25% in the applied shear stress compared to that of a single cell. In the case of the lowest cell density (d = 55 μm), the absolute shear stress values were nearly unchanged, with only a small reduction from 0.6 Pa to approximately 0.5 Pa after the flow passed by four upstream cells with d = 7 μm.

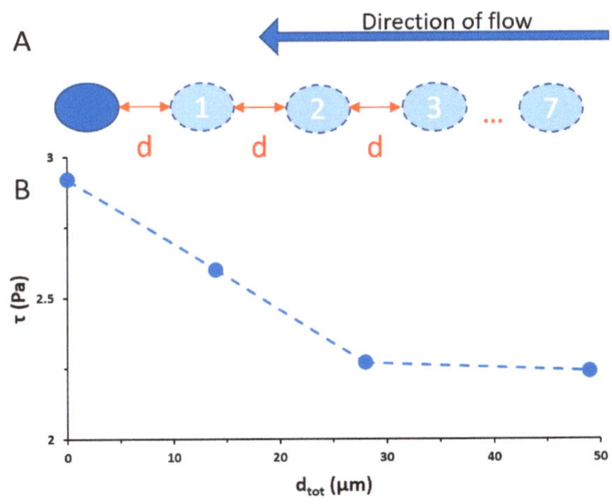

Figure 4. Simulation of shear stresses on an amoeba and accounting for the influence of upstream cells. (**A**) Schematic of the simulation showing a test cell (dark blue), from which the shear stresses are obtained, and up to 7 upstream cells (light blue) that are separated by distance d. (**B**) Shear stress (τ) as a function of the total distance (d_{tot}) to the most distant amoeba with data points for the number of cells equal to 0, 2, 4, and 7.

Upon investigating the data presented in Figure 5, we concluded that cell density can indeed affect detachment, but that this effect is dependent on cell type and flow rate. For example, DH1-10 wild-type cells (Figure 5A) did show a lower sensitivity to cell detachment when the initial cell densities were high. Unfortunately, the highest cell density for the 2 mL h^{-1} experiment was lower than for the others, so the data have a more pronounced appearance for higher flow rates. In contrast, for *phg2* mutant cells, regardless of the initial cell density, the detachment behavior remained nearly consistent at 2 mL h^{-1}, 5 mL h^{-1}, and 10 mL h^{-1} (Figure 5B). This suggests that the reductions in applied shear stresses at high cell densities were not significant enough to impact the very weak adhesion arising from the *phg2* mutation.

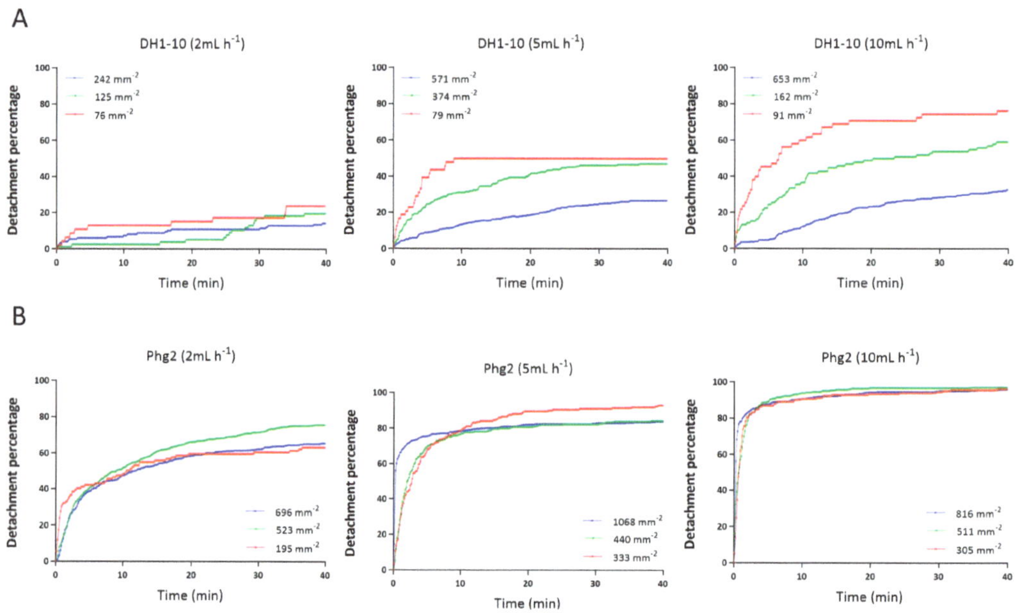

Figure 5. Detachment percentages of (**A**) DH1-10 wild-type cells and (**B**) *phg2* mutant cells at different initial cell densities under flow rates of 2 mL h^{-1}, 5 mL h^{-1}, and 10 mL h^{-1}.

4. Discussion

Our study employed a custom-designed microfluidic shear stress generator device to analyze the adhesion and detachment of *D. discoideum* DH1-10 and *phg2* mutant cells under various shear stress conditions. The results revealed distinct detachment behaviors between the wild-type DH1-10 and *phg2* mutant cells. Wild-type DH1-10 cells exhibited a gradual increase in detachment rates with increasing shear stress, whereas *phg2* cells demonstrated a rapid and substantial detachment response. This observation confirms that *phg2* cells possess inherently weaker adhesion properties or an increased sensitivity to shear stress, likely due to the disrupted function of the Phg2 protein. This observation aligns with previous studies, highlighting the crucial role of Phg2 in organizing the actin cytoskeleton, regulating adhesion molecules, and enabling signal transduction for mechanical stimuli response [48,59]. In *phg2* mutant cells, these processes are disrupted, leading to disorganized actin filaments and altered adhesion properties, resulting in increased detachment under shear stress, as our study confirms. Our results on the role of the Phg2 protein in cell adhesion align with those of studies on other adhesion-related proteins such as talin [17,60] and myosin II [61,62]. For example, mutants lacking talin have been shown to exhibit weakened adhesion under shear stress, similar to the *phg2* mutant in our study [60].

In comparison with the literature, our study provides insights into the detachment and cell motility under shear stress conditions. First, our results compare well with those of Décavé et al., who reported cell motility and directionality at higher shear flow rates. Specifically, our observations on the directionality and distance of the cell movement along the flow direction for both DH1-10 and *phg2* cells are coherent with the literature [17].

A major deviation from the literature, however, is our results showing that cell density plays a measurable effect on the detachment rate of wild type DH1-10 amoeba due to hydrodynamic shielding from neighboring cells, as confirmed by simulations. This is in contrast to the known loss of adhesion that occurs at high cell densities. However, such studies usually have investigated the highest surface coverage (e.g., up to 3000 cell mm^{-2}), which result in the destabilizing effects of cell-to-cell contact. In this work, we largely avoided this effect by limiting cell densities to less than 1000 cell mm^{-2}.

The use of microfluidic devices in studying *D. discoideum* adhesion under controlled shear stress provides precise control and real-time observation of cell detachment, offering significant advantages over traditional methods. Our approach using large culture areas offers key strengths, including facility in fabrication, accurate application of uniform shear stress over large distances and observation over long duration. Consequently, this design is optimal for investigating dynamic cellular responses and the mechanisms of adhesion and detachment over time. Furthermore, the design enables cells to freely move and thrive within the device, further supporting motility and cell density experiments. Supporting simulations complement the cell density experiments by quantifying the effect of neighbouring cells on applied shear stress, thereby revealing a mutual shielding function that appears to help maintain surface contact in the lead up to pre-aggregation densification process. Future testing should further develop the model to account for two-dimensional cell clusters and more realistic cell shapes, and future experiments should be run under the low nutrient conditions that are usually responsible for triggering aggregation. The versatility of microfluidic fabrication, including various surface functionalizations, architectures, and embedded sensors for surface sensitive spectroscopy for sensing microorganisms [63,64], makes our approach adaptable for future studies exploring environmental effects on adhesion, motility, aggregation, and cellular surface chemistry.

5. Conclusions

The presented microfluidic shear stress generator can be used to effectively assess cell adhesion under varying shear stresses. We used *Dictyostelium discoideum* as a model organism due to its known adhesion properties. Our device measures cell adhesion more accurately than other designs, with minimal wall effects. We compared wild-type DH1-10 cells to adhesion-defective mutant *phg2* cells, finding that the generator distinguishes between cell types based on adhesion. Mutant cells showed weaker adhesion and greater sensitivity to shear forces. The device also allows for detailed analysis of cell adhesion and migration under uniform shear stress, with potential applications in studying mechanotransduction and cell behavior in response to mechanical stimuli.

Author Contributions: S.F.: Software, formal analysis, investigation, writing—original draft; C.B.: Supervision, funding acquisition; S.J.C.: Conceptualization, methodology, resources, writing—review and editing, funding acquisition; J.G.: Conceptualization, methodology, resources, writing—original draft, writing—review and editing, supervision, project administration, funding acquisition. All authors have read and agreed to the published version of the manuscript.

Funding: This research was funded by the Natural Sciences and Engineering Research Council (NSERC) for funding under grant number RGPIN-2020-03603 and the Fonds de Recherche du Québec (FRQ) under the programme Audace.

Data Availability Statement: Raw data is available upon request.

Acknowledgments: The authors also acknowledge Molly K. Gregas for technical edits.

Conflicts of Interest: The authors declare no conflicts of interest.

References

1. Bachir, A.I.; Horwitz, A.R.; Nelson, W.J.; Bianchini, J.M. Actin-based adhesion modules mediate cell interactions with the extracellular matrix and neighboring cells. *Cold Spring Harb. Perspect. Biol.* **2017**, *9*, a023234. [CrossRef] [PubMed]
2. Hammer, D.A. Adhesive dynamics. *J. Biomech. Eng.* **2014**, *136*, 021006. [CrossRef] [PubMed]
3. Zhao, X.; Jin, L.; Shi, H.; Tong, W.; Gorin, D.; Kotelevtsev, Y.; Mao, Z. Recent advances of designing dynamic surfaces to regulate cell adhesion. *Colloid Interface Sci. Commun.* **2020**, *35*, 100249. [CrossRef]
4. Tsompanas, M.-A.I.; Sirakoulis, G.C. Modeling and hardware implementation of an amoeba-like cellular automaton. *Bioinspiration Biomim.* **2012**, *7*, 036013. [CrossRef] [PubMed]
5. Yi, J.; Schmidt, J.; Chien, A.; Montemagno, C.D. Engineering an artificial amoeba propelled by nanoparticle-triggered actin polymerization. *Nanotechnology* **2009**, *20*, 085101. [CrossRef]
6. Leon-Rodriguez, H.; Le, V.H.; Ko, S.Y.; Park, J.-O.; Park, S. Ferrofluid soft-robot bio-inspired by Amoeba locomotion. In Proceedings of the 2015 15th International Conference on Control, Automation and Systems (ICCAS), Busan, Republic of Korea, 13–16 October 2015; IEEE: Piscataway, NJ, USA; pp. 1833–1838.
7. Bozzaro, S. The past, present and future of Dictyostelium as a model system. *Int. J. Dev. Biol.* **2019**, *63*, 321–331. [CrossRef]
8. Pears, C.J.; Gross, J.D. Microbe Profile: *Dictyostelium discoideum*: Model system for development, chemotaxis and biomedical research. *Microbiology* **2021**, *167*, 001040. [CrossRef]
9. Bozzaro, S.; Eichinger, L. The professional phagocyte *Dictyostelium discoideum* as a model host for bacterial pathogens. *Curr. Drug Targets* **2011**, *12*, 942–954. [CrossRef]
10. Sodeman, W.A., Jr. Intestinal protozoa: Amebas. In *Medical Microbiology*, 4th ed.; University of Texas Medical Branch at Galveston: Galveston, TX, USA, 1996.
11. García, A.J.; Ducheyne, P.; Boettiger, D. Quantification of cell adhesion using a spinning disc device and application to surface-reactive materials. *Biomaterials* **1997**, *18*, 1091–1098. [CrossRef]
12. McClay, D.R.; Wessel, G.M.; Marchase, R.B. Intercellular recognition: Quantitation of initial binding events. *Proc. Natl. Acad. Sci. USA* **1981**, *78*, 4975–4979. [CrossRef]
13. Francis, G.W.; Fisher, L.R.; Gamble, R.A.; Gingell, D. Direct measurement of cell detachment force on single cells using a new electromechanical method. *J. Cell Sci.* **1987**, *87*, 519–523. [CrossRef] [PubMed]
14. Kapur, R.; Rudolph, A.S. Cellular and cytoskeleton morphology and strength of adhesion of cells on self-assembled monolayers of organosilanes. *Exp. Cell Res.* **1998**, *244*, 275–285. [CrossRef] [PubMed]
15. Décavé, E.; Rieu, D.; Dalous, J.; Fache, S.; Bréchet, Y.; Fourcade, B.; Satre, M.; Bruckert, F. Shear flow-induced motility of *Dictyostelium discoideum* cells on solid substrate. *J. Cell Sci.* **2003**, *116*, 4331–4343. [CrossRef] [PubMed]
16. Chu, L.; Tempelman, L.A.; Miller, C.; Hammer, D.A. Centrifugation assay of IgE-mediated cell adhesion to antigen-coated gels. *AIChE J.* **1994**, *40*, 692–703. [CrossRef]
17. Décavé, E.; Garrivier, D.; Bréchet, Y.; Fourcade, B.; Bruckert, F. Shear flow-induced detachment kinetics of *Dictyostelium discoideum* cells from solid substrate. *Biophys. J.* **2002**, *82*, 2383–2395. [CrossRef] [PubMed]
18. Boura, C.; Muller, S.; Vautier, D.; Dumas, D.; Schaaf, P.; Voegel, J.C.; Stoltz, J.F.; Menu, P. Endothelial cell—Interactions with polyelectrolyte multilayer films. *Biomaterials* **2005**, *26*, 4568–4575. [CrossRef]
19. Zhang, D.; Bi, H.; Liu, B.; Qiao, L. Detection of pathogenic microorganisms by microfluidics based analytical methods. *Anal. Chem.* **2018**, *90*, 5512–5520. [CrossRef]
20. Qi, P.; Lv, J.; Yan, X.; Bai, L.; Zhang, L. Microfluidics: Insights into Intestinal Microorganisms. *Microorganisms* **2023**, *11*, 1134. [CrossRef]
21. Ugolini, G.S.; Wang, M.; Secchi, E.; Pioli, R.; Ackermann, M.; Stocker, R. Microfluidic approaches in microbial ecology. *Lab Chip* **2024**, *24*, 1394–1418. [CrossRef]
22. Varma, S.; Voldman, J. A cell-based sensor of fluid shear stress for microfluidics. *Lab Chip* **2015**, *15*, 1563–1573. [CrossRef]
23. Jannat, R.A.; Dembo, M.; Hammer, D.A. Traction forces of neutrophils migrating on compliant substrates. *Biophys. J.* **2011**, *101*, 575–584. [CrossRef] [PubMed]
24. Jones, C.N.; Hoang, A.N.; Dimisko, L.; Hamza, B.; Martel, J.; Irimia, D. Microfluidic platform for measuring neutrophil chemotaxis from unprocessed whole blood. *JoVE (J. Vis. Exp.)* **2014**, *88*, 51215.
25. Yang, K.; Peretz-Soroka, H.; Wu, J.; Zhu, L.; Cui, X.; Zhang, M.; Rigatto, C.; Liu, Y.; Lin, F. Fibroblast growth factor 23 weakens chemotaxis of human blood neutrophils in microfluidic devices. *Sci. Rep.* **2017**, *7*, 3100. [CrossRef] [PubMed]
26. Zhao, W.; Zhao, H.; Li, M.; Huang, C. Microfluidic devices for neutrophil chemotaxis studies. *J. Transl. Med.* **2020**, *18*, 168. [CrossRef]
27. Chau, L.; Doran, M.; Cooper-White, J. A novel multishear microdevice for studying cell mechanics. *Lab Chip* **2009**, *9*, 1897–1902. [CrossRef]
28. Rossi, M.; Lindken, R.; Hierck, B.P.; Westerweel, J. Tapered microfluidic chip for the study of biochemical and mechanical response at subcellular level of endothelial cells to shear flow. *Lab Chip* **2009**, *9*, 1403–1411. [CrossRef]
29. Morimoto, Y.; Nagata, S.; Matsumoto, M.; Sugahara, K.; Miura, S.; Takeuchi, S. Microfluidic system for applying shear flow to endothelial cells on culture insert with collagen vitrigel membrane. *Sens. Actuators B Chem.* **2021**, *348*, 130675. [CrossRef]
30. Park, A.; Jeong, H.-H.; Lee, J.; Kim, K.P.; Lee, C.-S. Effect of shear stress on the formation of bacterial biofilm in a microfluidic channel. *Biochip. J.* **2011**, *5*, 236–241. [CrossRef]

31. Mosier, A.P.; Kaloyeros, A.E.; Cady, N.C. A novel microfluidic device for the in situ optical and mechanical analysis of bacterial biofilms. *J. Microbiol. Methods* **2012**, *91*, 198–204. [CrossRef]
32. Straub, H.; Eberl, L.; Zinn, M.; Rossi, R.M.; Maniura-Weber, K.; Ren, Q. A microfluidic platform for in situ investigation of biofilm formation and its treatment under controlled conditions. *J. Nanobiotechnol.* **2020**, *18*, 166. [CrossRef]
33. Rusconi, R.; Lecuyer, S.; Autrusson, N.; Guglielmini, L.; Stone, H.A. Secondary flow as a mechanism for the formation of biofilm streamers. *Biophys. J.* **2011**, *100*, 1392–1399. [CrossRef] [PubMed]
34. Bruchmann, J.; Sachsenheimer, K.; Rapp, B.E.; Schwartz, T. Multi-channel microfluidic biosensor platform applied for online monitoring and screening of biofilm formation and activity. *PLoS ONE* **2015**, *10*, e0117300. [CrossRef] [PubMed]
35. Parvinzadeh Gashti, M.; Bellavance, J.; Kroukamp, O.; Wolfaardt, G.; Taghavi, S.M.; Greener, J. Live-streaming: Time-lapse video evidence of novel streamer formation mechanism and varying viscosity. *Biomicrofluidics* **2015**, *9*, 041101. [CrossRef] [PubMed]
36. Greener, J.; Parvinzadeh Gashti, M.; Eslami, A.; Zarabadi, M.P.; Taghavi, S.M. A microfluidic method and custom model for continuous, non-intrusive biofilm viscosity measurements under different nutrient conditions. *Biomicrofluidics* **2016**, *10*, 064107. [CrossRef] [PubMed]
37. Greener, J.; Harvey, W.Y.; Gagné-Thivierge, C.; Fakhari, S.; Taghavi, S.M.; Barbeau, J.; Charette, S.J. Critical shear stresses of *Pseudomonas aeruginosa* biofilms from dental unit waterlines studied using microfluidics and additional magnesium ions. *Phys. Fluids* **2022**, *34*, 021902. [CrossRef]
38. Paquet-Mercier, F.; Gashti, M.P.; Bellavance, J.; Taghavi, S.M.; Greener, J. Through thick and thin: A microfluidic approach for continuous measurements of biofilm viscosity and the effect of ionic strength. *Lab Chip* **2016**, *16*, 4710–4717. [CrossRef]
39. Zarabadi, M.P.; Paquet-Mercier, F.; Charette, S.J.; Greener, J. Hydrodynamic effects on biofilms at the biointerface using a microfluidic electrochemical cell: Case study of *Pseudomonas* sp. *Langmuir* **2017**, *33*, 2041–2049. [CrossRef]
40. Aznaveh, N.B.; Safdar, M.; Wolfaardt, G.; Greener, J. Micropatterned biofilm formations by laminar flow-templating. *Lab Chip* **2014**, *14*, 2666–2672. [CrossRef]
41. Rodogiannis, K.; Duong, J.T.; Kovarik, M.L. Microfluidic single-cell analysis of oxidative stress in *Dictyostelium discoideum*. *Analyst* **2018**, *143*, 3643–3650. [CrossRef]
42. Khodaparastasgarabad, N.; Couture, M.; Greener, J. A microfluidic study of acetate conversion kinetics in a microbial electrolysis cell: The role of age, concentration and flow on biofilm permeability. *Sens Actuators B Chem* **2024**, *412*, 135779. [CrossRef]
43. Delincé, M.J.; Bureau, J.B.; López-Jiménez, A.T.; Cosson, P.; Soldati, T.; McKinney, J.D. A microfluidic cell-trapping device for single-cell tracking of host–microbe interactions. *Lab Chip* **2016**, *16*, 3276–3285. [CrossRef] [PubMed]
44. Wu, C.; Asokan, S.B.; Berginski, M.E.; Haynes, E.M.; Sharpless, N.E.; Griffith, J.D.; Gomez, S.M.; Bear, J.E. Arp2/3 is critical for lamellipodia and response to extracellular matrix cues but is dispensable for chemotaxis. *Cell* **2012**, *148*, 973–987. [CrossRef] [PubMed]
45. Scherer, A.; Kuhl, S.; Wessels, D.; Lusche, D.F.; Raisley, B.; Soll, D.R. Ca^{2+} chemotaxis in *Dictyostelium discoideum*. *J Cell Sci.* **2010**, *123*, 3756–3767. [CrossRef] [PubMed]
46. Fache, S.; Dalous, J.; Engelund, M.; Hansen, C.; Chamaraux, F.; Fourcade, B.; Satre, M.; Devreotes, P.; Bruckert, F. Calcium mobilization stimulates *Dictyostelium discoideum* shear-flow-induced cell motility. *J. Cell Sci.* **2005**, *118*, 3445–3458. [CrossRef] [PubMed]
47. Fisher, P.R.; Merkl, R.; Gerisch, G. Quantitative analysis of cell motility and chemotaxis in *Dictyostelium discoideum* by using an image processing system and a novel chemotaxis chamber providing stationary chemical gradients. *J. Cell Biol.* **1989**, *108*, 973–984. [CrossRef]
48. Rupprecht, P.; Golé, L.; Rieu, J.P.; Vézy, C.; Ferrigno, R.; Mertani, H.C.; Riviere, C. A tapered channel microfluidic device for comprehensive cell adhesion analysis, using measurements of detachment kinetics and shear stress-dependent motion. *Biomicrofluidics* **2012**, *6*, 014107. [CrossRef]
49. Tarantola, M.; Bae, A.; Fuller, D.; Bodenschatz, E.; Rappel, W.J.; Loomis, W.F. Cell substratum adhesion during early development of *Dictyostelium discoideum*. *PLoS ONE* **2014**, *9*, e106574. [CrossRef]
50. Loomis, W.F.; Fuller, D.; Gutierrez, E.; Groisman, A.; Rappel, W.-J. Innate non-specific cell substratum adhesion. *PLoS ONE* **2012**, *7*, e42033. [CrossRef]
51. Cornillon, S.; Pech, E.; Benghezal, M.; Ravanel, K.; Gaynor, E.; Letourneur, F.; Brückert, F.; Cosson, P. Phg1p is a nine-transmembrane protein superfamily member involved in *dictyostelium* adhesion and phagocytosis. *J. Biol. Chem.* **2000**, *275*, 34287–34292. [CrossRef]
52. Gebbie, L.; Benghezal, M.; Cornillon, S.; Froquet, R.; Cherix, N.; Malbouyres, M.; Lefkir, Y.; Grangeasse, C.; Fache, S.; Dalous, J.; et al. Phg2, a kinase involved in adhesion and focal site modeling in *Dictyostelium*. *Mol. Biol. Cell* **2004**, *15*, 3915–3925. [CrossRef]
53. Caterina, M.J.; Milne, J.L.; Devreotes, P.N. Mutation of the third intracellular loop of the cAMP receptor, cAR1, of *Dictyostelium* yields mutants impaired in multiple signaling pathways. *J. Biol. Chem.* **1994**, *269*, 1523–1532. [CrossRef] [PubMed]
54. Kortholt, A.; Rehmann, H.; Kae, H.; Bosgraaf, L.; Keizer-Gunnink, I.; Weeks, G.; Wittinghofer, A.; Van Haastert, P.J. Characterization of the GbpD-activated Rap1 pathway regulating adhesion and cell polarity in *Dictyostelium discoideum*. *J. Biol. Chem.* **2006**, *281*, 23367–23376. [CrossRef] [PubMed]
55. Kortholt, A.; van Haastert, P.J.M. Highlighting the role of Ras and Rap during *Dictyostelium* chemotaxis. *Cell. Signal.* **2008**, *20*, 1415–1422. [CrossRef] [PubMed]

56. Cherix, N.; Froquet, R.; Charette, S.J.; Blanc, C.; Letourneur, F.; Cosson, P. A Phg2-Adrm1 pathway participates in the nutrient-controlled developmental response in *Dictyostelium*. *Mol. Biol. Cell* **2006**, *17*, 4982–4987. [CrossRef] [PubMed]
57. Mercanti, V.; Charette, S.J.; Bennett, N.; Ryckewaert, J.J.; Letourneur, F.; Cosson, P. Selective membrane exclusion in phagocytic and macropinocytic cups. *J. Cell Sci.* **2006**, *119*, 4079–4087. [CrossRef]
58. Wegner, G.; Allard, N.; Al Shboul, A.; Auger, M.; Beaulieu, A.M.; Bélanger, D.; Bénard, P.; Bilem, I.; Byad, M.; Byette, F.; et al. *Functional Materials: For Energy, Sustainable Development and Biomedical Sciences*; Walter de Gruyter GmbH & Co KG: Berlin, Germany, 2014.
59. Cornillon, S.; Gebbie, L.; Benghezal, M.; Nair, P.; Keller, S.; Wehrle-Haller, B.; Charette, S.J.; Brückert, F.; Letourneur, F.; Cosson, P. An adhesion molecule in free-living *Dictyostelium* amoebae with integrin β features. *EMBO Rep.* **2006**, *7*, 617–621. [CrossRef]
60. Tsujioka, M.; Yoshida, K.; Inouye, K. Talin B is required for force transmission in morphogenesis of *Dictyostelium*. *EMBO J.* **2004**, *23*, 2216–2225. [CrossRef]
61. Uyeda, T.Q.P.; Nagasaki, A. Variations on a theme: The many modes of cytokinesis. *Curr. Opin. Cell Biol.* **2004**, *16*, 55–60. [CrossRef]
62. Pasternak, C.; Spudich, J.A.; Elson, E.L. Capping of surface receptors and concomitant cortical tension are generated by conventional myosin. *Nature* **1989**, *341*, 549–551. [CrossRef]
63. Jia, N.; Daignault-Bouchard, A.; Deng, T.; Mayerhöfer, T.G.; Bégin-Drolet, A.; Greener, J. SpectIR-fluidics: Completely customizable microfluidic cartridges for high sensitivity on-chip infrared spectroscopy with point-of-application studies on bacterial biofilms. *Lab Chip* **2023**, *23*, 3561–3570. [CrossRef]
64. Pousti, M.; Zarabadi, M.P.; Amirdehi, M.A.; Paquet-Mercier, F.; Greener, J. Microfluidic bioanalytical flow cells for biofilm studies: A review. *Analyst* **2019**, *144*, 68–86. [CrossRef] [PubMed]

biomimetics

MDPI

Article

Experimental Study on the Adhesion of Abalone to Surfaces with Different Morphologies

Peng Xi [1,2], Yanqi Qiao [1,2], Qian Cong [3,4] and Qingliang Cui [1,2,*]

1 College of Agricultural Engineering, Shanxi Agricultural University, Jinzhong 030801, China; yydljxp2000@126.com (P.X.); 20233053@stu.sxau.edu.cn (Y.Q.)
2 Dryland Farm Machinery Key Technology and Equipment Key Laboratory of Shanxi Province, Shanxi Agricultural University, Jinzhong 030801, China
3 College of Biological and Agricultural Engineering, Jilin University, Changchun 130022, China; congqian@jlu.edu.cn
4 Key Laboratory of Bionic Engineering Ministry of Education, Jilin University, Changchun 130022, China
* Correspondence: 13029127775@163.com; Tel.: +86-13029127775

Abstract: To date, research on abalone adhesion has primarily analyzed the organism's adhesion to smooth surfaces, with few studies on adhesion to non-smooth surfaces. The present study examined the surface morphology of the abalone's abdominal foot, followed by measuring the adhesive force of the abalone on a smooth force measuring plate and five force measuring plates with different surface morphologies. Next, the adhesion mechanism of the abdominal foot was analyzed. The findings indicated that the abdominal foot of the abalone features numerous stripe-shaped folds on its surface. The adhesion of the abalone to a fine frosted glass plate, a coarse frosted glass plate, and a quadrangular conical glass plate was not significantly different from that on a smooth glass plate. However, the organism's adhesion to a small lattice pit glass plate and block pattern glass plate was significantly different. The abalone could effectively adhere to the surface of the block pattern glass plate using the elasticity of its abdominal foot during adhesion but experienced difficulty in completely adhering to the surface of the quadrangular conical glass plate. The abdominal foot used its elasticity to form an independent sucker system with each small lattice pit, significantly improving adhesion to the small lattice pit glass plate. The elasticity of the abalone's abdominal foot created difficulty in handling slight morphological size changes in roughness, resulting in no significant differences in its adhesion to the smooth glass plate.

Keywords: abalone; abdominal foot; adhesion; non-smooth surface; force measuring plate

Citation: Xi, P.; Qiao, Y.; Cong, Q.; Cui, Q. Experimental Study on the Adhesion of Abalone to Surfaces with Different Morphologies. *Biomimetics* **2024**, *9*, 206. https://doi.org/10.3390/biomimetics9040206

Academic Editors: Stanislav N. Gorb and Thies Büscher

Received: 19 February 2024
Revised: 26 March 2024
Accepted: 27 March 2024
Published: 29 March 2024

1. Introduction

Over a long period of evolution, organisms have developed unique and exceptional adaptations to thrive in their natural environments. For example, many animals in nature possess adhesive capabilities [1–3]. Animals are able to firmly adhere to different surfaces in their environment using their adhesive abilities, helping such organisms with fundamental survival tasks like crawling, hunting, grabbing, and fleeing [4–8]. Adhesion is not only used by many animal varieties in nature but also plays an important role in human production and life. The most typical applications of this organism involve vacuum suckers, which are widely used in industrial production and people's daily lives through the adhesion of different pressures inside and outside the sucker [9–12]. However, vacuum suckers have high requirements for the adhesion surface and offer good adhesion only on smooth surfaces. The adhesion effect on non-smooth surfaces is poor or absent. At the same time, this adhesion effect is prone to leakage failure, resulting in accidents [13–15]. However, organisms with adhesion capabilities not only adhere to smooth surfaces but also yield good adhesion effects on non-smooth surfaces. The ability to produce strong adhesion effects on both smooth and non-smooth surfaces has attracted great interest from relevant researchers. These researchers have selected organisms with adhesion capabilities and observed the structures of their biological suckers in detail. The adhesion capabilities of biological suckers onto different morphological surfaces have also been measured using experimental methods. This study seeks to improve the poor adhesion capabilities of vacuum suckers onto non-smooth surfaces using the method of engineering bionics, thus facilitating the development of vacuum suckers.

To date, researchers have studied and achieved results related to the adhesion capabilities of some common adhesive organisms such as the octopus, leech, remora, tree frog, and Northern clingfish. Tramacere et al. studied octopus suckers and found that each sucker was divided into upper and lower chambers, with the two chambers connected through an orifice in the center. When the sucker is adsorbed, the lower chamber first adheres to the surface and then gradually flattens to increase the adhesion area. Finally, the sucker forms a sealing structure with the surface. The upper cavity of the sucker has a protruding structure covered by a large number of fibers. The water in the lower chambers of the sucker is extruded through the orifice into the upper cavity via the gradual extrusion of the lower chambers of the sucker. At the same time, the protruding structure of the upper cavity and the orifice form an effective seal, ultimately producing vacuum pressure that enables the sucker to adhere to the object's surface. The fiber structure on the surface of the protrusion can improve the sealing ability between the protrusion and the orifice [16–18]. Ditsche et al. observed the abdominal suckers of Northern clingfish, which have adhesion capabilities, and found that the suckers' surfaces were composed of numerous micron-sized fibers with different size grades. When Northern clingfish adhere to a surface using their suckers, these fibers ensure that a sealed structure is produced regardless of the roughness of the surface, enabling the adhesion of Northern clingfish to different surfaces in nature. Adhesion experiments showed that the Northern clingfish has good adhesion (30–40 kPa) and adaptability to non-smooth surfaces [19,20]. Chuang et al. observed the abdominal sucker of the Pulin river loach (*Sinogastromyzon puliensis*) with adhesion capabilities and found that its surface was composed of many radial fins. The surfaces of the radial fins also had micron-sized fiber structures. When the Pulin river loach adhere to a surface, the radial fins and fiber structure of the abdominal sucker enable the adhesion of the organism to non-smooth surfaces [21]. Kampowski et al. observed the suckers of leeches with adhesion capabilities with scanning electron microscopy and found that a large number of pores were distributed on the surfaces of the suckers in front of and behind the leech. When the leech engages in adsorption, the small holes on the sucker can secrete mucus to fill the unevenness of the adhesion surface, improving the sealing performance of the leech sucker when the surface roughness is larger. This mechanism increases the adhesion force of the sucker. Abalone is an adhesive organism in the ocean whose abdominal foot has strong adhesion capabilities [22–24]. According to reports, an abalone with a body length of about 15 cm has an adhesion force of up to 200 kg, highlighting the organism's strong adhesion force [25]. Due to the strong adhesive properties of abalone, researchers have conducted extensive studies on the adhesion of the organism's muscular foot. Lin et al. studied the American red abalone and found that its abdominal foot surface is composed of fibers with two sizes. This multi-level fiber structure enables the abdominal foot sucker to form an interlocking structure on surfaces with a variety of roughness types, effectively improving the adaptability of abalone to different adhesion surfaces [26]. Li et al. tested the adhesion force of abalone in both water and air using various force measuring plates. The authors found that the adhesion force of the abalone's abdominal foot primarily comes from vacuum adhesion force, van der Waals force, and capillary force [27]. Xi analyzed the measurement results of abalone's adhesion force on different force measuring plates and determined that the vacuum adhesion force plays a significant role in the total adhesion force of abalone [28].

To date, research on abalone has mainly focused on the adhesion capabilities of the organism's abdominal foot on smooth surfaces, as well as the composition of the adhesion force and the surface structure of abalone's abdominal foot. However, there are few studies on the adhesion capabilities of abalone's abdominal foot on non-smooth surfaces. To fill this gap, the present study offers a new direction for the bionic design of vacuum suckers by studying the abalone's adhesion capabilities and modes of action on non-smooth surfaces. The specific research contents are as follows. Firstly, abalone samples of basically the same mass and size were selected for feeding, and then the surface morphology of the abalone's abdominal foot was observed macroscopically and microscopically. Force measuring plates with different surface morphologies were then selected, and the adhesion force of the abalone's abdominal foot on force measuring plates with different surface morphologies was measured via tensile testing. The corresponding adhesion stress was obtained according to the area of the abalone's abdominal foot. The effects of force measuring plates with different surface morphologies on the adhesion of the abalone's abdominal foot were compared, and the mode of action between them was explored. This paper provides a reference for studying the adhesion capabilities of other organisms with adhesion capabilities on non-smooth surfaces and the interactions between them.

2. Materials and Methods

2.1. Observation of the Abalone's Abdominal Foot

2.1.1. Abalone Sample Preparation

The abalone used in this experiment was *Haliotis discus hannai*, which was acquired from an aquatic market and promptly transferred to a laboratory aquarium for feeding. The aquarium measured 1500 × 1000 × 600 mm and was equipped with a filtration system and water circulation system. The water temperature in the tank was maintained between 15 and 20 °C, with a salinity of 30% and a water depth of 0.5 m. The abalone samples were nourished with wakame to ensure their survival in the aquarium [29,30]. The abalone samples weighed between 50 and 65 g and were acclimated in the aquarium for a minimum of 10 days before the experiment to mitigate errors stemming from individual variations.

2.1.2. Observations of the Abalone's Abdominal Foot Surface Morphology

The main structure of abalone is shown in Figure 1a,b and is composed of a hard shell with soft abdominal feet. Figure 1c presents the positional relationship between the abdominal foot and the shell. The abdominal foot serves as the primary organ responsible for abalone's adhesion and crawling. To facilitate further research on abalone adhesion, the surface morphology of the abdominal foot was examined using a stereomicroscope (Stemi 2000-C, ZEISS, Oberkochen, Germany). The surface morphology of an abalone's abdominal foot is shown in Figure 1d, with the abdominal foot surface segmented into three layers: the outer layer, the middle layer, and the inner layer. The inner layer, which encompasses the majority of the abdominal foot area, displays numerous striped folds on its surface, as shown in Figure 1e. The abdominal foot has a certain degree of elasticity and stretch through the different areas of the abdominal foot's striped folds, generated by driving forward movement.

2.2. Adhesion Test

2.2.1. Preparation of the Force Measuring Plate

Using a tensile test, we measured the adhesion of the abalone's abdominal foot to force measuring plates with different surface morphologies. Due to the good adhesion and adaptability of the abalone's abdominal foot to the glass plate, this surface enabled us to observe changes in the abalone's abdominal foot. Thus, the glass plate was selected as the force measuring plate for the tensile test. In this paper, six types of glass plates with different surface morphologies were selected: (1) a smooth glass plate; (2) a fine frosted glass plate, roughness Ra = 0.86 μm; (3) a coarse frosted glass plate, roughness Ra = 480 μm; (4) a quadrangular conical glass plate with a side length of 1.5~5 mm and a height of 1 mm, as shown in Figure 2b; (5) a block pattern glass plate with a block side length of 10~20 mm and height of 0.5 mm, as shown in Figure 2c; and (6) a small lattice pit glass plate with a pit length of 0.8 mm, as shown in Figure 2d. The surface morphologies of the six force measuring plates are shown in Figure 2a.

Figure 1. (**a**) Abalone shell; (**b**) abalone abdominal foot (crouching state); (**c**) the positional relationship between the abdominal foot and the shell; (**d**) abalone abdominal foot surface; (**e**) three layers and striped folds of the abalone's abdominal foot.

2.2.2. Design and Processing of the Hook

To measure the adhesion force of the abalone, we had to detach the abalone in its adhesion state from the force measuring plate, as the adhesion force of the abalone's abdominal foot is notably strong. Moreover, abalone shell shapes present certain differences between individuals. For this purpose, we designed a type of hook that could hook the shell without affecting the adhesion of the abalone. The 3D design model of the hook and relevant design details are shown in Figure 3A, as follows: (a) a hole diameter of 5 mm; (b) a concave design to avoid contact with the abalone shell; (c) a certain radian designed to adapt to the abalone shell; (d) a shallow groove design conveniently able to catch the abalone shell; (e) a chamber structure designed to prevent scratching the abalone; and (f) a wedge structure designed for easy insertion into the abalone shell while increasing strength Then, the hook was processed via 3D printing (UP! three-dimensional printer) with PLA used as the printing material. The 3D-printing process of the hook is shown in Figure 3B and the solid machined hook is shown in Figure 3C.

2.2.3. Abalone Abdominal Foot Adhesion Test

The universal testing machine (WSM-500N) used in the tensile test was controlled with a computer. Before the test, six force measuring plates were placed at the bottom of each leaching basin. Then, one abalone (purchased and kept in the aquarium, as shown in Section 2.1.1) was placed on each force measuring plate. Finally, the leaching basin was placed at the bottom of the aquarium together with the abalone samples on the force measuring plates, which were left to slowly adhere. In each test, the force measuring plate and abalone adhering to its surface were placed together on the testing machine for the tensile test. In this test, the force measuring plate was fixed first, and then the left and right sides of the abalone shell were hooked with two self-made hooks. Next, the tensile test was carried out. Figure 4 presents a schematic diagram of the tensile test. In this test, the lifting speed of the tensile testing machine was 100 mm/min. As the abalone became subjected to increasing upward tension, the abdominal foot was gradually separated from the force measuring plate. The test ended when the abalone's abdominal foot was completely detached from the force measuring plate. Then, the maximum tensile force on the abalone during the whole tensile process was recorded and taken as the adhesion force of the abalone's abdominal foot. The time interval between each tensile test was 24 h to ensure that the adhesion force of the abdominal foot did not increase with extended time. Five tests were performed for each force measuring plate.

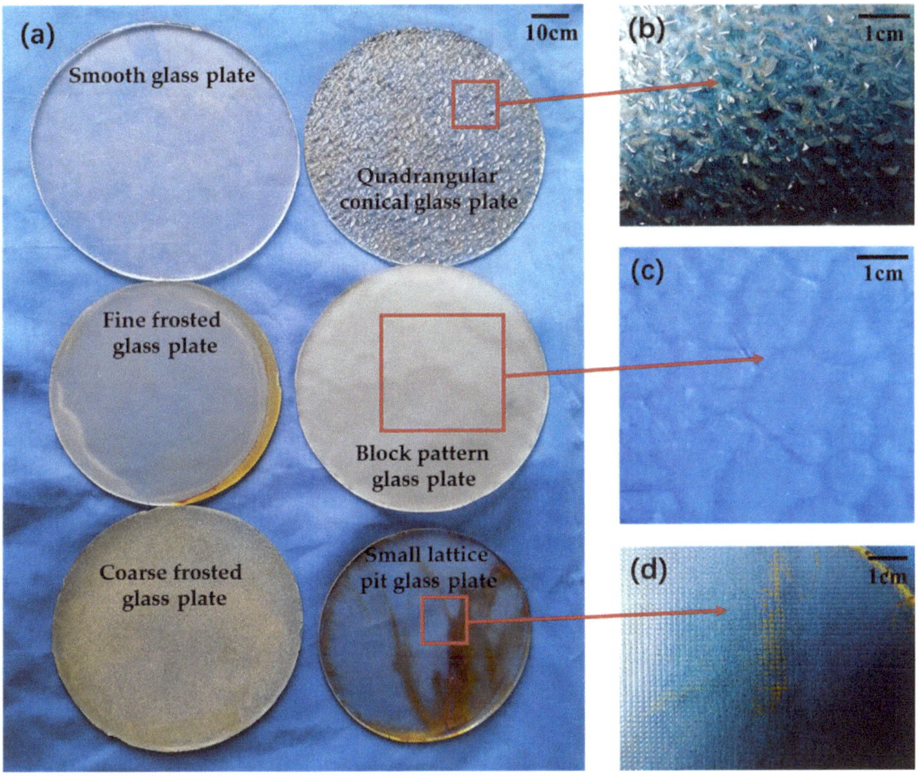

Figure 2. (**a**) The six force measuring plates used for the tensile test; (**b**) the quadrangular conical glass plate and the specific quadrangular conical morphology; (**c**) the block pattern glass plate and block pattern's specific morphology; (**d**) the small lattice pit glass plate and the pit's specific morphology.

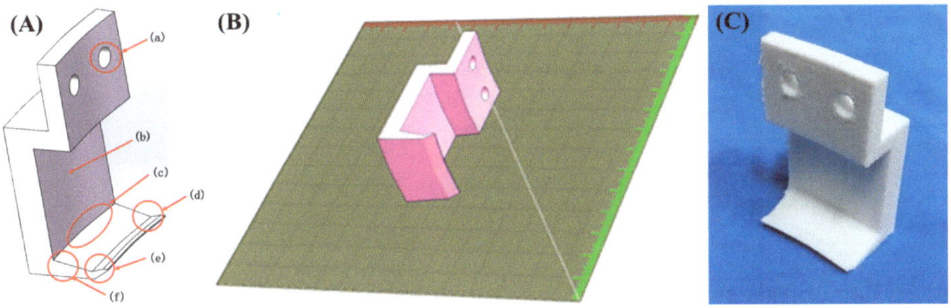

Figure 3. (**A**) Hook three-dimensional model and design details; (**B**) 3D-printing process of the hook; (**C**) 3D-printing hook entity.

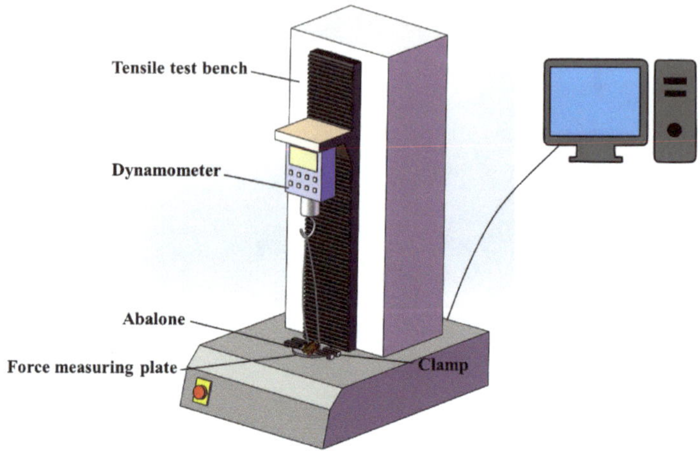

Figure 4. Tensile test schematic diagram.

3. Results and Discussion

3.1. Calculation and Analysis of Test Results

Table 1 presents the adhesion force of the abalone on the six force measuring plates along with the corresponding mass values. To analyze the ability of the abalone's abdominal foot to adhere to a surface, the relevant adhesion stress (f) must be determined. The adhesion stress (f) is defined as

$$f = F/A \tag{1}$$

where F represents the adhesion force of the abalone on the force measuring plate, and A is the corresponding abalone's abdominal foot area. Since an abalone is a living creature, it was not possible to artificially unfold the abalone's abdominal foot and force it to adhere onto the force measuring plate. Indeed, each abalone's abdominal foot was generally curled up, as shown in Figure 1b, making it difficult to measure the area of the abdominal foot when adhered onto the force measuring plate immediately after each tensile test. Due to the challenges in directly measuring the abdominal foot area during adhesion in this experiment, we instead measured the mass of the abalone and calculated the corresponding abdominal foot area. Ten abalones used in the experiment were randomly selected. The mass (g) of each abalone and the corresponding area (mm^2) of its abdominal foot were measured separately. The ratio (S) of the abdominal foot area (mm^2) to mass (g) was calculated separately for each abalone based on measurements, and then the average of ratio (S) for 10 abalones was calculated as S_{AVG}. The average S_{AVG} value was 43.15. Thus, the abdominal foot area (A) of the abalone was calculated in this experiment by measuring the abalone's mass and multiplying the ratio, S_{AVG}.

Table 1 presents the tensile test results and corresponding abalone mass. The calculated corresponding adhesion stress value (f) of the abalone's abdominal foot on the six different force measuring plates is shown in Table 2.

Figure 5 presents a box plot of the average adhesion stress (f) of the abalone's abdominal foot on different force measuring plates. It can be seen from Figure 5 that the average values of adhesion stress (f) on the smooth glass plate, fine frosted glass plate, coarse frosted glass plate, and quadrangular conical glass plate were basically the same, whereas the stress value on the block pattern glass plate was slightly larger. The stress value on the small lattice pit glass plate was largest.

Table 1. Adhesion force and corresponding mass of the abalone on the six force measuring plates.

Maximum Adhesion Force/N	Type of Force Measuring Plate					
Test Times	Smooth Glass Plate	Fine Frosted Glass Plate	Coarse Frosted Glass Plate	Small Lattice Pit Glass Plate	Quadrangular Conical Glass Plate	Block Pattern Glass Plate
1	80.5	110.9	105.2	176.1	61.17	126.6
Abalone mass/g	49.1	67.5	54.9	56	57.4	60.2
2	89.13	95.92	89.46	149.5	175.6	125.1
Abalone mass/g	48.3	67	51.5	57.2	65.3	60.3
3	116.9	142.9	103.2	204	108.2	113.3
Abalone mass/g	59.7	58.1	56.9	56	56	63.6
4	96.9	114.6	112.7	144.9	94.7	129.4
Abalone mass/g	57.4	60.3	54.7	56	59.4	51.5
5	101.6	102.5	92.64	191.6	114.5	116.1
Abalone mass/g	60.1	60.3	53.8	58.5	58	51.5

Table 2. Adhesion stress (f) of abalone on the six force measuring plates.

Adhesion Stress/kPa	Type of Force Measuring Plate					
Test Times	Smooth Glass Plate	Fine Frosted Glass Plate	Coarse Frosted Glass Plate	Small Lattice Pit Glass Plate	Quadrangular Conical Glass Plate	Block Pattern Glass Plate
1	38.00	38.08	44.41	72.88	24.70	48.74
2	42.77	33.18	40.26	60.57	62.32	48.08
3	45.38	57.00	42.03	84.42	44.78	41.28
4	39.12	44.04	47.75	59.97	36.95	58.23
5	39.18	39.39	39.91	75.90	45.75	52.24
Average value	40.89	42.34	42.87	70.75	42.90	49.72

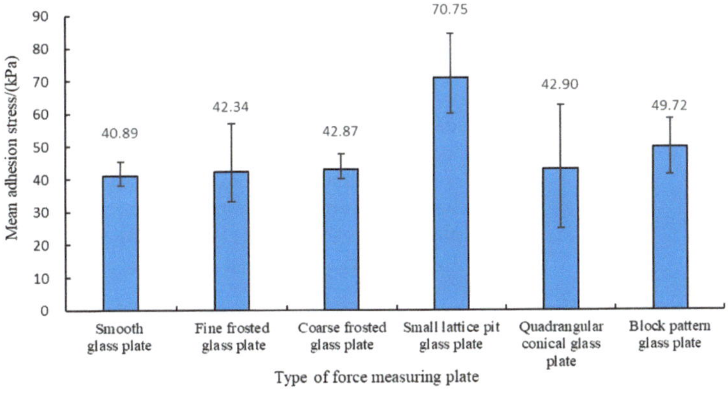

Figure 5. The average adhesion stress of the abalone's abdominal foot on the six different force measuring plates.

3.2. Adhesion Mechanism Analysis

The adhesion stresses of the abalone on five force measuring plates with different roughness types and surface morphologies (Table 2) were analyzed for significance (a significance level of 0.05) to comparatively analyze the effects of the plates on the adhesion of the abalone's abdominal foot. A rank sum test was used to compare the values of five abalone adhesion stresses on a fine frosted glass plate, a coarse frosted glass plate, a small lattice pit glass plate, a quadrangular conical glass plate, and a block pattern glass plate with those on a smooth glass plate, respectively. The results are shown in Table 3.

Table 3. The significance analysis results of the abalone's adhesion stress on the six different force measuring plates.

Type of Force Measuring Plate	p Value	Explanation
Fine frosted glass plate	0.917	Comparison with smooth glass plate
Coarse frosted glass plate	0.251	Comparison with smooth glass plate
Small lattice pit glass plate	0.009	Comparison with smooth glass plate
Quadrangular conical glass plate	0.754	Comparison with smooth glass plate
Block pattern glass plate	0.028	Comparison with smooth glass plate

The significance analysis results in Table 3 show that the adhesion stress (f) of abalone on the fine frosted glass plate, coarse frosted glass plate, and quadrangular conical glass plate was not significantly different from that on the smooth glass plate ($p > 0.05$). The adhesion stress on the glass plate with small lattice pits and that on the block pattern glass plate were significantly different, indicating that the adhesion stress (f) of the abdominal foot on these two force measuring plates was significantly different from that on the smooth glass plate ($p < 0.05$).

The adhesion stress of abalone on the block pattern glass plate was significantly different from that on the smooth glass plate, primarily because the surface morphology of the block pattern glass plate changed slowly, with blunt corners. This slow rate of change better enabled the abalone's abdominal foot to exert its stretching capabilities. The abalone abdominal foot can completely adhere to the morphology surface of the block pattern force measuring plate, as shown in Figure 6a. The adhesion area of the abalone was larger than the area of the abdominal foot upon complete attachment to the morphological surface of the block pattern glass plate, thereby increasing the adhesion force and the adhesion stress of the abalone on its surface, as shown in Figure 6b. The shape of the quadrangular conical glass plate did not have a significant impact on the adhesion of the abalone's abdominal foot, primarily because the quadrangular cone itself in the quadrilateral conical glass plate and the shape between the quadrangular cone changed rapidly; i.e., the rotation angle was sharp, and the ridges were too numerous. As a result, it was difficult for the abalone's abdominal foot to exert its stretching characteristics and completely adhere to the morphological surface of the quadrilateral conical glass plate, as shown in Figure 7a As can be seen from Figure 7a, there are not many clear quadrangular conical imprints on the surface of the abalone abdominal foot, indicating that the elasticity and stretching capabilities of the abalone abdominal foot does not allow it to adhere well to the surface of the quadrangular conical glass plate. The adhesion area of the abalone to the force measuring plate was essentially the same as the size of the organism's abdominal foot area Thus, the adhesion stress of the abalone on the plate's surface did not change significantly, as shown in Figure 7b.

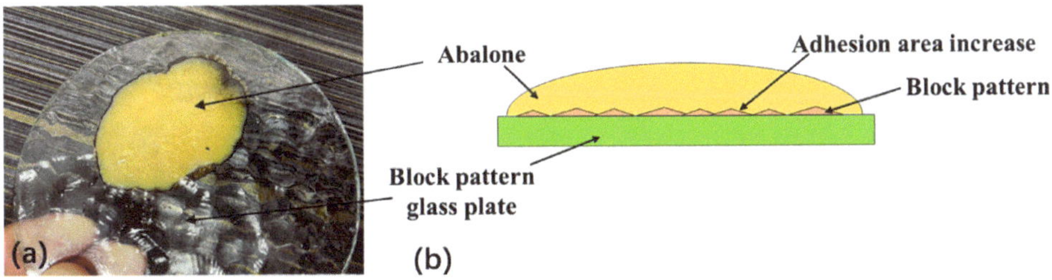

Figure 6. (**a**) The adhesion of the abalone's abdominal foot to the block pattern glass plate; (**b**) The adhesion area of the abalone was larger than the area of the abdominal foot upon complete attachment to the morphological surface of the block pattern glass plate.

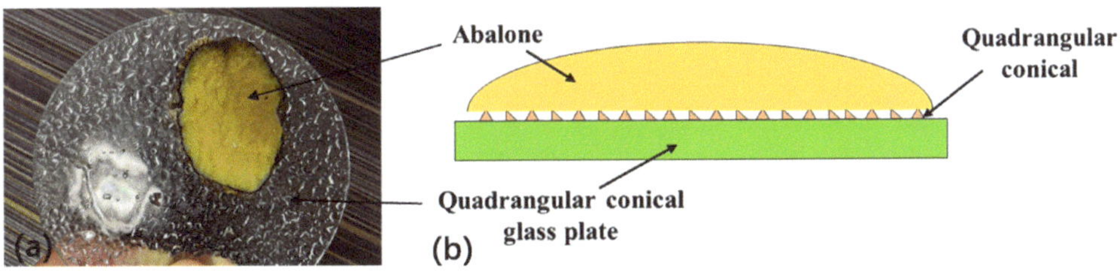

Figure 7. (**a**) The adhesion of the abalone's abdominal foot to the quadrangular conical glass plate; (**b**) The adhesion area of the abalone to the quadrangular conical glass plate was essentially the same as the size of the abdominal foot area.

The adhesion stress of abalone on the small lattice pit glass plate was significantly higher than that on the smooth glass plate because, under normal circumstances, the abalone formed a sucker structure on the smooth force measuring plate, as shown in Figure 8a. Because an abalone is a living creature, according to the experiment, the degree of the vacuum between the abdominal foot and the force measuring plate was far less than 100% (about 40%) when adhering to the smooth force measuring plate. When the abalone adhered to the force measuring plate with small, shallow lattice pits, the abdominal foot, which has certain elasticity, wrapped around the small lattice pit and excluded some of the gas in the pit. Figure 8b shows the adhesion state of the abalone abdominal foot on the small lattice pit glass plate, and it can be seen that there are a lot of small squares on the surface of the abdominal foot, which is formed by the abalone abdominal foot due to the elastic deformation squeezed into the small pits. In this way, each small lattice pit formed a separate sucker structure, as shown in Figure 8c. As a result, the vacuum degree of the whole abdominal foot on the force measuring plate with small lattice pits was significantly increased, so the adhesion stress of the abalone on the force measuring plate with small lattice pits was greater than that on the smooth force measuring plate.

The results on the various measuring plates (Table 2) along with the significance analysis results (Table 3) showed that the adhesion stress of the abalone's abdominal foot did not change significantly with an increase in the roughness of the force measuring plate (i.e., the smooth glass plate, fine frosted glass plate (Ra = 0.86 μm), and coarse frosted glass plate (Ra = 480 μm)). This result shows that the elasticity and stretch of the abdominal foot sucker create difficulties in adhering to small surface morphologies of roughness dimension levels (micron-sized). The adhesion area of the abalone on the force measuring plate was essentially the same as the size of the abdominal foot. Thus, the increase in the roughness of the force measuring plate did not yield an increase in the adhesion of the abdominal foot.

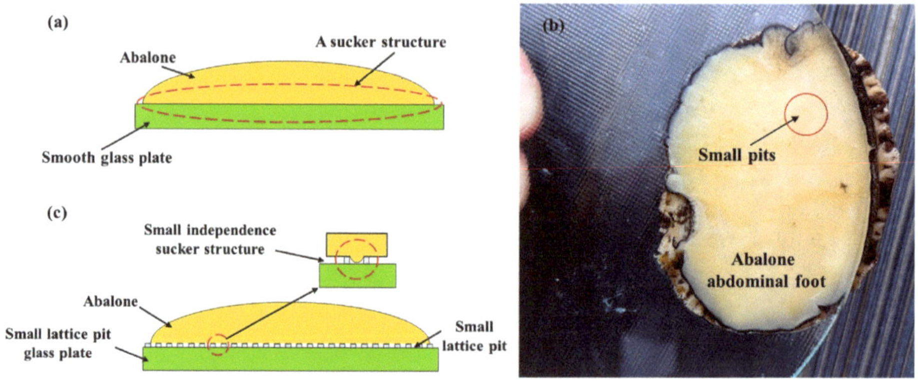

Figure 8. (**a**) The abalone formed a sucker structure on the smooth force measuring plate; (**b**) The adhesion of the abalone's abdominal foot to the small lattice pit glass plate; (**c**) Each small lattice pit of the small lattice pit glass plate formed a independence sucker structure with the abalone's abdominal foot.

In this paper, the mechanism analysis of abalone adhesion only considered the effect between the elasticity of the abdominal foot and the surface morphology of different force measuring plates, and other factors were not considered. Therefore, the analysis of the adhesion mechanism of abalone on different force measuring plates is only a hypothesis, which needs further experiments to confirm.

4. Conclusions

Through observations and tensile tests, we studied the effects of using force measuring plates with different surface morphologies on the adhesion of the abalone's abdominal foot, and corresponding conclusions were drawn.

(1) There was no significant difference in the adhesion of the abalone to the fine frosted glass plate, coarse frosted glass plate, quadrangular conical glass plate, or smooth glass plate. However, adhesion to the small lattice pit glass plate and block pattern glass plate was significantly different.
(2) The quadrangular conical shape in the quadrangular conical glass plate changed rapidly, making it difficult for the abalone's abdominal foot to fully adhere to the morphological surface of this plate. Conversely, the surface morphology of the block pattern glass plate changed slowly, enabling the abalone's abdominal foot to fully adhere to this plate's surface. When the abalone adhered to the small lattice pit glass plate, each small lattice pit was enclosed, excluded some of the gas in the pit, forming an independent sucker system due to the stretching characteristics of the abdominal foot and resulting in a significant increase in the adhesion of the abdominal foot.
(3) Changes in the stretching of the abdominal foot created difficulties in achieving small morphological size changes based on the roughness, leading to no significant differences in the adhesion of abalone to force measuring plates with different types of roughness.

Author Contributions: Conceptualization, P.X. and Q.C. (Qian Cong); methodology, P.X. and Q.C. (Qingliang Cui); software, P.X. and Y.Q.; writing—original draft preparation, P.X.; writing—review and editing, P.X. and Y.Q. All authors have read and agreed to the published version of the manuscript.

Funding: This research was funded by the National Natural Science Foundation of China (grant no. 51775234), Scientific and Technological Innovation Programs of Higher Education Institutions in Shanxi (2021L162), the Doctoral Scientific Research Launch Project of Shanxi Agricultural University (2021BQ35), the Scientific Research Project of Reward Fund for Doctoral Graduates in Shanxi Province

(SXBYKY2021076), and The 2022 Shanxi Provincial Science and Technology Major Special Project Plan "Leading the Charge with Open Competition" Project (202201140601023).

Institutional Review Board Statement: Not applicable.

Data Availability Statement: Data are contained within the article.

Conflicts of Interest: The authors declare no conflicts of interest.

References

1. Peng, X.; Ma, C.; Ji, J.; Li, J. Underwater adhesion mechanisms and biomimetic study of Marine Life. *Tribology* **2020**, *40*, 816–830. [CrossRef]
2. Maie, T.; Blob, R.W. Adhesive force and endurance of the pelvic sucker across different modes of waterfall-climbing in gobiid fishes: Contrasting climbing mechanisms share aspects of ontogenetic change. *Zoology* **2021**, *149*, 125969. [CrossRef]
3. Palecek, A.M.; Schoenfuss, H.L.; Blob, R.W. Sucker shapes, skeletons, and bioinspiration: How hard and soft tissue morphology generates adhesive performance in waterfall climbing goby fishes. *Integr. Comp. Biol.* **2022**, *62*, 934–944. [CrossRef] [PubMed]
4. Wang, S.H.; Luo, H.Y.; Linghu, C.H.; Song, J. Elastic energy storage enabled magnetically actuated, octopus-inspired smart adhesive. *Adv. Funct. Mater.* **2021**, *31*, 2009217.1–2009217.9. [CrossRef]
5. Baik, S.; Hwang, G.W.; Jang, S.; Kim, K.H.; Yang, T.H.; Pang, C.H. Bioinspired microsphere-embedded adhesive architectures for an electrothermally actuating transport device of dry/wet pliable surfaces. *ACS Appl. Mater. Interfaces* **2021**, *13*, 6930–6940. [CrossRef]
6. Huie, J.M.; Summers, A.P. The effects of soft and rough substrates on suction-based adhesion. *J. Exp. Biol.* **2022**, *225*, jeb243773. [CrossRef]
7. Kazuma, T.; Yasutaka, M.; Masatsugu, S.; Yuji, H. A new concept for an adhesive material inspired by clingfish sucker nanofilaments. *Langmuir ACS J. Surf. Colloids* **2022**, *38*, 1215–1222. [CrossRef]
8. Popov, V.L.; Filippov, A.E.; Gorb, S.N. Biological microstructures with high adhesion and friction. Numerical approach. *Phys. Uspekhi* **2016**, *59*, 829–845. [CrossRef]
9. Bai, L.; Song, Z.; Wang, P. Simulation analysis of internal flow field of vacuum sucker under different leakage conditions. *Mech. Eng.* **2018**, *7*, 33–35+38. [CrossRef]
10. Qin, J.; Deng, C.; Wang, D.; Li, W. Influence of different layouts based on finite element on vacuum adhesion device. *J. Guilin Univ. Technol.* **2017**, *37*, 713–717.
11. Zhen, J.; Yang, Z.; Wang, X.; Song, H. Design of an intelligent vacuum sucker device. *J. Vac. Sci. Technol.* **2017**, *37*, 1038–1043.
12. Zhang, J.; Xue, W.; Liang, Y.; Yang, M. Vacuum adhesion technology and its application in construction machinery assembly. *Constr. Mach. Maint.* **2015**, *S1*, 302–305.
13. Nie, J.; Wang, T.; Xu, Y.; Chen, Z.; Ma, Q. Vacuum adhesion performance test of flexible sucker. *Hydraul. Pneum.* **2020**, 131–137. [CrossRef]
14. Chen, L.; Yun, Z.; Jiang, Y. Design and performance analysis of bionic sucker based on remora adhesion principle. *J. Beijing Univ. Chem. Technol. Nat. Sci. Ed.* **2018**, *45*, 100–105.
15. Li, J.; Liang, D.; Lu, B.; Li, G. Research on non-contact handling sucker system. *J. Instrum.* **2018**, *39*, 108–116. [CrossRef]
16. Francesca, T.; Esther, A.; Barbara, M.; Stanislav, N.G. Hairy suckers: The surface microstructure and its possible functional significance in the *Octopus vulgaris* sucker Francesca. *Beilstein J. Nanotechnol.* **2014**, *5*, 561–565. [CrossRef]
17. Francesca, T.; Nicola, M.P.; Michael, J.K.; Barbara, M. Unveiling the morphology of the acetabulum in octopus suckers and its role in attachment. *Interface Focus* **2014**, *5*, 20140050. [CrossRef]
18. Francesca, T.; Lucia, B.; Michael, K.; Alessandro, G.; Angelo, B.; Barbara, M. The Morphology and Adhesion Mechanism of Octopus vulgaris Suckers. *PLoS ONE* **2013**, *8*, e65074. [CrossRef]
19. Ditsche, P.; Wainwright, D.K.; Summers, A.P. Attachment to challenging substrates—Fouling, roughness and limits of adhesion in the northern clingfish (*Gobiesox maeandricus*). *J. Exp. Biol.* **2014**, *217*, 2548–2554. [CrossRef]
20. Ditsche, P.; Summers, A.P. Learning from Northern clingfish (*Gobiesox maeandricus*): Bioinspired suction cups attach to rough surfaces. *Philos. Trans. R. Soc. B Biol. Sci.* **2019**, *374*, 20190204. [CrossRef]
21. Chuang, Y.C.; Chang, H.K.; Liu, G.L.; Chen, P.Y. Climbing upstream: Multi-scale structural characterization and underwater adhesion of the Pulin river loach (*Sinogastromyzon puliensis*). *J. Mech. Behav. Biomed. Mater.* **2017**, *73*, 76–85. [CrossRef] [PubMed]
22. Zhang, Y.; Li, S.; Zuo, P.; Xu, X.; Liu, J. Adhesion Behaviors of Abalone Under the Action of Water Flow. *Front. Mech. Eng.* **2021**, *7*, 659468. [CrossRef]
23. Zhang, Y.; Li, S.; Zuo, P.; Li, J.; Liu, J. A mechanics study on the self-righting of abalone from the substrate. *Appl. Bionics Biomech.* **2020**, *2020*, 8825451. [CrossRef] [PubMed]
24. Li, J.; Ma, C.; Liu, J.; Dong, X.; Liu, J. The co-effect of microstructures and mucus on the adhesion of abalone from a mechanical perspective. *Biosurf. Biotribol.* **2021**, *7*, 180–186. [CrossRef]
25. Liu, X. Biological characteristics and artificial breeding technology of abalone. *Anhui Agric. Sci.* **2009**, *37*, 5872–5874.
26. Lin, A.; Brunner, R.; Chen, P.; Talke, F.; Meyers, M. Underwater adhesion of abalone: The role of van der Waals and capillary forces. *Acta Mater.* **2009**, *57*, 4178–4185. [CrossRef]

27. Li, J.; Zhang, Y.; Liu, S.; Liu, J. Insights into adhesion of abalone: A mechanical approach. *J. Mech. Behav. Biomed. Mater.* **2018**, *77* 331–336. [CrossRef] [PubMed]
28. Xi, P.; Ye, S.; Cong, Q. Abalone adhesion: The role of various adhesion forces and their proportion to total adhesion force. *PLoS ONE* **2023**, *18*, e0286567. [CrossRef]
29. Cheng, Y.; Du, M. Abalone aquaculture technology. *Agric. Eng. Technol.* **2016**, *36*, 72.
30. Sun, L. Study on high adhesion performance of abalone and design of bionic sucker. Master's Thesis, Jilin University, Changchun China, 2017.

biomimetics

MDPI

Review

Bioinspired Interfacial Friction Control: From Chemistry to Structures to Mechanics

Yunsong Kong [1,2], Shuanhong Ma [1,*] and Feng Zhou [1]

1 State Key Laboratory of Solid Lubrication, Lanzhou Institute of Chemical Physics, Chinese Academy of Sciences, Lanzhou 730000, China; yunsongkong@licp.cas.cn (Y.K.); zhouf@licp.cas.cn (F.Z.)
2 College of Materials Science and Opto-Electronic Technology, University of Chinese Academy of Sciences, Beijing 100049, China
* Correspondence: mashuanhong@licp.cas.cn; Tel.: +86-931-4968079

Abstract: Organisms in nature have evolved a variety of surfaces with different tribological properties to adapt to the environment. By studying, understanding, and summarizing the friction and lubrication regulation phenomena of typical surfaces in nature, researchers have proposed various biomimetic friction regulation theories and methods to guide the development of new lubrication materials and lubrication systems. The design strategies for biomimetic friction/lubrication materials and systems mainly include the chemistry, surface structure, and mechanics. With the deepening understanding of the mechanism of biomimetic lubrication and the increasing application requirements, the design strategy of multi-strategy coupling has gradually become the center of attention for researchers. This paper focuses on the interfacial chemistry, surface structure, and surface mechanics of a single regulatory strategy and multi-strategy coupling approach. Based on the common biological friction regulation mechanism in nature, this paper reviews the research progress on biomimetic friction/lubrication materials in recent years, discusses and analyzes the single and coupled design strategies as well as their advantages and disadvantages, and describes the design concepts, working mechanisms, application prospects, and current problems of such materials. Finally, the development direction of biomimetic friction lubrication materials is prospected.

Keywords: friction control; lubrication regulation; chemistry; surface structure; mechanics

Citation: Kong, Y.; Ma, S.; Zhou, F. Bioinspired Interfacial Friction Control: From Chemistry to Structures to Mechanics. *Biomimetics* **2024**, *9*, 200. https://doi.org/10.3390/biomimetics9040200

Academic Editors: Stanislav N. Gorb and Thies Büscher

Received: 23 February 2024
Revised: 22 March 2024
Accepted: 22 March 2024
Published: 27 March 2024

1. Introduction

Friction, the process of energy dissipation when two surfaces slide relative to each other, can be found everywhere in daily life and industrial manufacturing. On the one hand, friction plays a vital role in everyday life and production; on the other hand, friction causes severe wear and tear phenomena and requires colossal energy consumption. Therefore, it is necessary to develop various lubricant materials to regulate interfacial friction in specific situations. People have studied friction extensively and intensively for sustaining industrial production, conserving energy, and improving the quality of life, which has continued to drive the development of mechanical and materials science. However, with the improvement of people's quality of life and rapid technological innovation, traditional lubrication materials sometimes find it challenging to meet the requirements of specific friction systems, which requires us to propose new strategies for controlling interfacial friction.

The diversity of life in nature showcases the beauty and functionality of matching form and purpose across all scales. The unique structures that have evolved in organisms due to common materials or specific physiological processes can inspire us to design materials, devices, or processes with desirable functions, which is the fundamental concept behind "bionics." Over 3.8 billion years, a wide range of natural organisms have evolved organs and structures that can be adapted to complex operating conditions, including a wide range of ingenious friction and lubrication systems. Many of these organisms realize a wide range of tribological properties through different interfacial chemistry, surface structures at various scales, and mechanical properties of the biological structures to achieve the desired lubrication effect in a long-lasting and efficient manner. By further understanding how the complex functionalization and modulation of biological structures can be achieved, we can optimize the performance and realize the intellectualization of materials.

Humans have long noticed the excellent tribological properties of various organisms internal and external physiological structures and have conducted a series of related studies. From the perspective of solid surface lubrication, organisms in nature exhibit three main types of friction regulation strategies (Figure 1). One is the particular chemical nature of the surface, which realizes lubrication through the macromolecular layer on the surface of organisms with unique functions or the secretion of chemical substances with lubricating effects, such as the mucus secreted by the plant [1] and the synovial fluid and cartilage layer of mammals [2]. The second is the formation of structures on surfaces at various scales, such as the arrays of gecko feet [3] and the grooves on the surface of shark skin [4]. The third is to change the mechanical properties of the surface or subsurface to drastically alter the friction state at the interface, such as the hardening of the dermis of the sea cucumber and the contraction of fish muscles [5,6]. In the face of complex environmental conditions, it is often challenging to design biomimetic lubrication materials based on a single strategy to cope with the wide range of influencing factors in real situations, so researchers usually need to couple multiple strategies for material development.

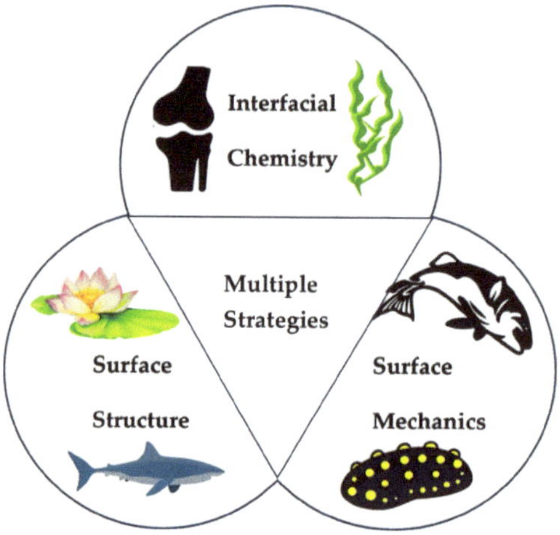

Figure 1. The schematic shows the three biomimetic strategies for achieving friction control.

This paper introduces the common forms of bio-lubrication modulation in nature and the corresponding application of biomimetic materials in friction systems from the standard lubrication systems in nature. This paper introduces the mechanisms of biomodulation of interfacial friction from three perspectives, namely, interfacial chemistry, surface structure, and surface mechanics, respectively, and analyzes the advantages and disadvantages of various biomimetic strategies, discusses the possibilities and superiority of multi-strategy coupling, and looks forward to the direction of the development of biomimetic interfacial friction modulation and the prospects for its application.

2. Surface Chemistry-Dominated Friction

It has long been noted that many plants and animals in nature can achieve lubrication effects through good hydration of their secretions or soft tissue surfaces. Jacob Klein, a famous tribologist, proposed the concept of hydration lubrication, described the role of the hydration layer in water lubrication, and explained the principles of many biological lubrication systems [7]. The water molecule appears to be electrically neutral. However, due to the dipoles caused by residual charges on the hydrogen and oxygen atoms, the water molecules will form a hydration layer around the polar groups (Figure 2). The hydration charges will repel each other when they are close, making it difficult for the hydration layer to overlap [8,9]. During aqueous lubrication, the charged groups at the interface can immobilize the oppositely charged hydrated groups during sliding via strong electrostatic interactions, meaning that the hydrated layer also reduces interfacial friction under high normal pressures, which is consistent with the working conditions in many cases in living organisms [10,11].

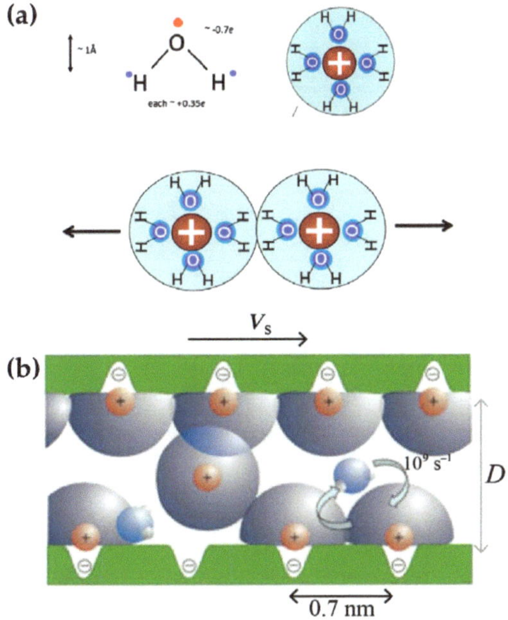

Figure 2. (**a**) The large dipole of water and the formation of hydration shells about charges. (**b**) The mechanism of hydration lubrication between charged surfaces across trapped hydrated ions [7]. Copyright Permission from Springer Nature, 2013.

Researchers have studied biological structures with good lubricating properties and found that the mucus or surface with lubricating functions in plants and animals usually has a special chemical composition. For example, the components of plant secretions that play a lubricating role mainly include well-hydrated macromolecules such as polysaccharides and cellulose. One of the strategies for developing new lubricants is to analyze the mucus by extracting specific components or designing based on its composition. The mucilage in aloe vera is a suitable polysaccharide water-based bio-lubricant. Aloe leaves are rich in mucilage, whose main component is polysaccharides. Xu et al. [12] investigated the tribological properties of aloe mucilage and found that the mucilage can exhibit friction consistent with thin-film lubrication. Hakala et al. [13] extracted mucilage with a lubricating effect from fresh papaya fruit (Figure 3a–c), and the combination of nanofibers and water-soluble polysaccharides can form a gel-like structure. Arad et al. [14] evaluated the sulfated polysaccharide obtained from the red microalga Porphyridium sp., which showed good lubrication properties in rheological studies. Li et al. [15] reported the excellent lubricating properties of Brasenia schreberi mucilage (Figure 3d–f), in which there are a large number of polysaccharide cross-linked nanosheets, which can be combined into a solid polysaccharide layer on the glass surface through hydrogen bonding and the adsorption of a large number of water molecules during the lubrication process, and they form a hydration layer between the layers in order to effectively reduce the friction. The plant secretions mentioned in this paragraph and their tribological properties have been summarised in Table 1.

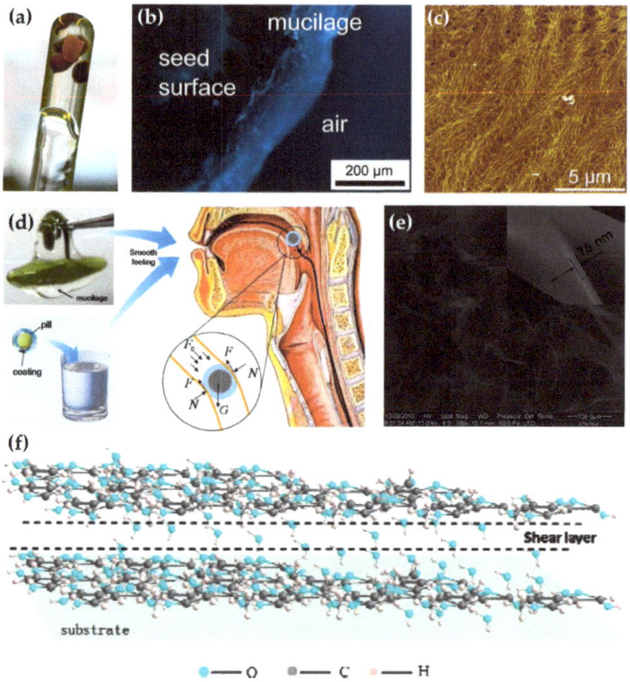

Figure 3. (**a**) Photograph of gel-like mucus obtained from papaya seeds, (**b**) gel-like layer formed on the surface of the seeds after being dissolved for 20 min using calcium fluoride solution, and (**c**) AFM morphology image of fresh papaya mucus aggregated on a mica sheet [13]. Copyright Permission from Elsevier, 2014. (**d**) Brasenia schreberi mucilage and its lubrication, (**e**) SEM image of Brasenia schreberi mucilage after treatment by the vacuum freeze-drying method, and (**f**) schematic of polysaccharide nanosheets in mucilage during lubrication [15]. Copyright Permission from ACS, 2012.

Table 1. Friction-reducing properties of the secretions of natural plants.

Creature/Tissue	Friction Substitutes and Velocity	COF	Reference
Aloe mucilage	WC ball/DLC flat; 150 mm·s^{-1}	0.04	[12]
Papaya seed mucilage	Polyethylene flat/stainless steel flat; 100 mm·s^{-1}	0.03	[13]
Red microalga secretion	Si_3N_4 ball/alumina flat; 0.2 mm·s^{-1}	0.003	[14]
Brasenia mucilage	Glass flat/glass flat; 0.01 mm·s^{-1}	0.005	[15]

Compared to the limited lubricating properties of plant mucus, the lubrication system in animals usually maintains a lower COF and efficient lubrication under more complex and demanding conditions, as required for the proper functioning of various functions. In the human body, biological lubrication plays a role in almost every organ and tissue in the body all the time, such as the blinking lubrication by the tear fluid between the cornea and the eyelids [16], the lubrication of the esophagus by mucus containing biomolecules when swallowing food [17], the boundary lubricant film formed by salivary proteins in the oral cavity [18], and the synergistic lubrication of synovial fluid and cartilage in the joints [19] Among them, the human joint lubrication system has been widely studied because of its close correlation with people's quality of life and its excellent lubrication performance which can work normally with a very low COF under high load conditions and shows excellent lubricating and anti-wear properties [20]. The synergistic effect of synovial fluid and structurally specialized cartilage in the joint system contributes to the excellent and stable lubricating properties.

The main components of synovial fluid include hyaluronic acid (HA), polyproteoglycans, and lubricin (Figure 4) [21]. HA is a high-molecular-weight linear polysaccharide that can bind many water molecules and separate the cartilage on both sides of the joint during sliding, which is essential for increasing synovial fluid viscosity [22]. At the same time, HA binds to phospholipids to anchor to the vesicle surface, and the combination of the two dramatically improves the hydration properties of synovial fluid. Polyproteoglycans have a natural hierarchical bottle-brush structure, with a backbone capable of forming interconnections or adsorbing onto the cartilage surface and hydrophilic glycan side chains capable of binding to water molecules [23]. Lubricin is also a glycoprotein with a bottle-brush structure that can act as a protective agent for chondrocytes. Klein et al. [24,25] explained the mechanism in detail for the specific form of action in polymer brush joint lubrication. Hydrophilic macromolecules contract in the dry state, ionize to form high osmotic pressures when hydrated, and maintain a stretched and swollen morphology, which prevents interfacial contact and resists applied loads [26,27]. At the same time, the hydration of polymer brushes causes them to aggregate at the sliding interface to form a boundary lubrication layer, further reducing friction [28].

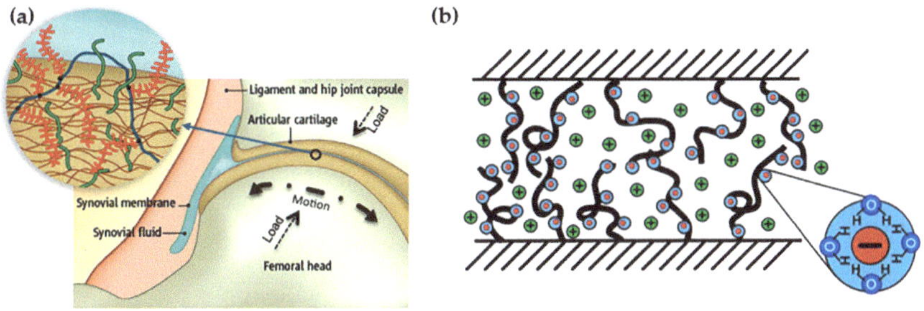

Figure 4. (**a**) Illustration of the natural articular cartilage system and the functional biomolecules in it: HA (blue), polyproteoglycan (red bottle-brush molecule), and lubricin (green) [21]. Copyright Permission from AAAS, 2009. (**b**) Highly hydrated phosphorylcholine groups are a highly effective lubricating element, and the figure illustrates the hydrated phosphorylcholine headgroups exposed on the surface of the liposomes as they slide relative to one another [27]. Copyright Permission from ACS, 2015.

Researchers have discovered or synthesized many macromolecular bio-lubricants with excellent properties based on understanding the lubrication mechanism of synovial fluid. Natural chitosan is a naturally available cationic glycan that functions similarly to HA and can act as a bio-lubricant for treating arthritis. The clinical lubrication properties of KiOmedine® CM-chitosan, a non-animal carboxymethyl chitosan, have been evaluated by Vandeweerd et al. [29]. In vitro tribological experiments showed that this chitosan significantly reduced the COF due to the lubricating ability of the cross-linked HA formulations. In addition, chondroitin sulfate with glucosamine has also been used as a biological lubricant, which is commonly used clinically for arthritis relief and treatment [30,31]. Synthetic bio-lubricants have also shown good performance in terms of the lubrication and therapeutic effects. Through the ring-opening disproportionation polymerization of methyl 5-oxonorbornene-2-carboxylate, Wathier et al. [32] synthesized a polyanionic bio-lubricant (Figure 5). Friction experiments have shown that the polymers with low molecular weight showed a lower COF and significantly enhanced the viscosity of synovial fluid compared to saline, Synvisc, and bovine synovial fluid (BSF). Inspired by the bottle-brush structure possessed by biomolecules, Hartung et al. [33,34] prepared a series of brush lubricants with poly-L-lysine (PLL) or polyallylamine (PAAm) as the main chain and flexible PEG as the side chain, and their lubricating properties were also related to the length of the PEG chain and the grafting density. The PLL or PAAm can be bonded to negatively charged surfaces by electrostatic interactions to form a boundary lubrication layer [35]. Pettersson et al. [36] copolymerized PEO_{45}MEMA with methacryloxyethyl trimethyl ammonium chloride to obtain a new type of bio-lubricant, which can also form a boundary lubrication layer on the substrate surface through electrostatic interaction, and the lubrication performance is mainly determined by its chain density.

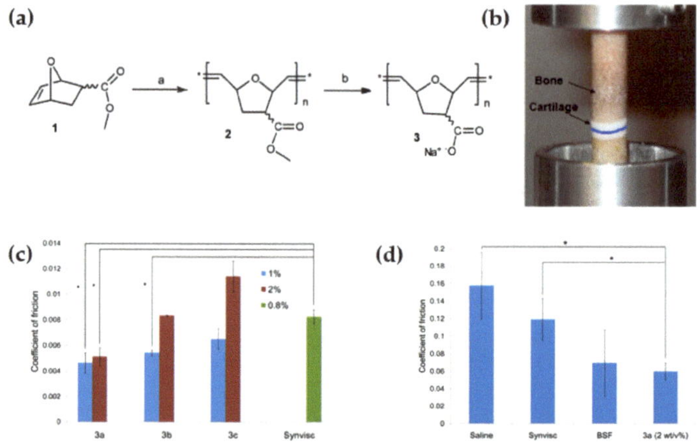

Figure 5. (**a**) Structural formula for Poly(7-oxanorbornene-2carboxylate) and (**b**) schematic diagram of its lubrication test model. (**c**) Polymer 3a with the lowest molecular weight has the lowest COF, (**d**) and exhibits lubricating properties superior to those of BSF [32]. Copyright Permission from ACS, 2013.

While synovial fluid provides good lubrication as a fluid environment, the articular cartilage plays a more critical role in lubrication. The synovial joints of the human body are covered with a thin layer of articular cartilage (1–3 mm thick), which has a sponge-like macromolecular network structure. The synovial fluid's water will penetrate the network during the sliding process, while charged water-soluble biomolecules can be assembled onto the cartilage surface to realize boundary lubrication [37]. The surface of the cartilage is also covered with HA, polyproteoglycans, and lubricin. The size of these macromolecules creates a site-barrier effect, and their strong hydration capacity allows them to freely extend into the solution and form a hydration layer [38]. This stable and dense layer has good adhesion and hydrated fluidity, allowing it to withstand high loads while maintaining a low friction factor. Inspired by the human joint lubrication mechanism, polymer brushes have been utilized to obtain superior tribological properties and good biocompatibility by grafting them onto desired surfaces to achieve functional mimicry of the joint lubrication system. Surface-grafted biomimetic polymer brushes mainly refer to the grafting of polymers from or onto surfaces by physical adsorption or covalent bonding, with the hydrophilic portion at the other end having no or only weak forces with the substrate. When the polymer chains are densely distributed, spatial repulsion causes the polymer to elongate and form a dense polymer brush layer of a certain thickness on the surface of the substrate [39]. In aqueous environments, the polymer brushes have a high penetration pressure and thus exhibit excellent lubricating properties with high load carrying.

Based on the inherent lubricating properties of natural polymers in living organisms, natural polymers were first modified, and their tribological properties were investigated. By functionalizing the natural macromolecules present in synovial joints, such as hyaluronic acid [40–42], polysaccharide [43–45], and phospholipid [46–48], researchers well modeled the tribological properties of mammalian joint lubrication systems. Based on the promoted understanding of the hydration lubrication mechanism of articular cartilage surfaces, the researchers further synthesized various types of cartilage-mimicking surface lubrication materials by various methods, such as surface-grafted polymer brush layers, surface-adsorbed polymer brush layers, and gel matrices with intrinsic surface lubrication [49].

In addition to the use of polymer brushes to achieve efficient lubrication, by adjusting the external conditions to apply stimuli to the lubrication layer, such as solvent [50], light [51], temperature [52], pH [53], electric field [54], and shear stress [55], the conformation of some polymers can be changed accordingly to achieve further modulation of the interface lubricating properties. For example, based on the mimicry of the lubrication performance of fish skin, Wu et al. [56] further introduced the pH-sensitive monomers sodium methacrylate (NaMA) and 2-(dimethylamino)ethyl methacrylate into the temperature-sensitive graphene-pNIPAM gel system, obtaining a hydrogel with the dual responsiveness of the pH and temperature (Figure 6a–d). The hydrogel has an ultra-low COF ($\approx$0.05), which can be gradually varied from 0.05 to 1.2 by sequentially adjusting the pH and temperature of the solution reversibly, without structural damage to the gel. Wang et al. [57] prepared semi-transformable hydrogels with reversible photo-responsive supramolecular lubrication properties by integrating a responsive supramolecular system of α-cyclodextrin/poly(ethylene glycol) (α-CD/PEG) and a competing guest, 1-[p-(Phenylazo)benzyl]pyridinium bromide (AzoPB), into the frameworks of poly(vinyl alcohol) (PVA) and PAAm. Upon irradiation using UV and visible light, respectively, the competitive host–guest interactions between the α-CD/PEG supramolecular network and AzoPB led to the repeated formation and disappearance of sol–gel layers on the surface of the hydrogels, whereas the PVA and PAAm were unaffected and maintained their backbone properties, thus providing a reversible photo-responsive lubrication capability with variable toughness (Figure 6e,f). Inspired by the mechanism of transition from lubrication to astringency in the oral environment, Deng et al. [58] simulated this transition from ultra-low friction to a high friction state by combining mucin with PVA and achieved a large span of lubrication state switching (μ~0.009 to μ~0.47) by the interactions between mucin and tannic acid (Figure 6g–i).

Inspired by the lubrication mechanism in living organisms, the modulation of friction through the chemical properties of surfaces, as exemplified by polymer brushes, can fundamentally regulate the lubricating properties by controlling the degree of hydration to change the molecular state of the surface and achieve a significant reduction or reversible modulation of the COF in aqueous environments, which has brought great convenience and manipulability. However, most of the strategies for modulating interfacial interactions through interfacial chemistry find it difficult to take into account the surface roughness, hardness, deformability, and other factors that may result in a non-ideal contact state under real conditions, which may lead to a significant reduction in the lubricating performance of the material under real conditions. In addition, for friction modulation systems with stimulus-response capability, the surface's molecular state or the response layer's size limits the magnitude of the lubrication regulation. In addition, the actual application environment is far less stable than in the laboratory, and the required conditions imposed in the response process may be difficult to realize precisely in real use.

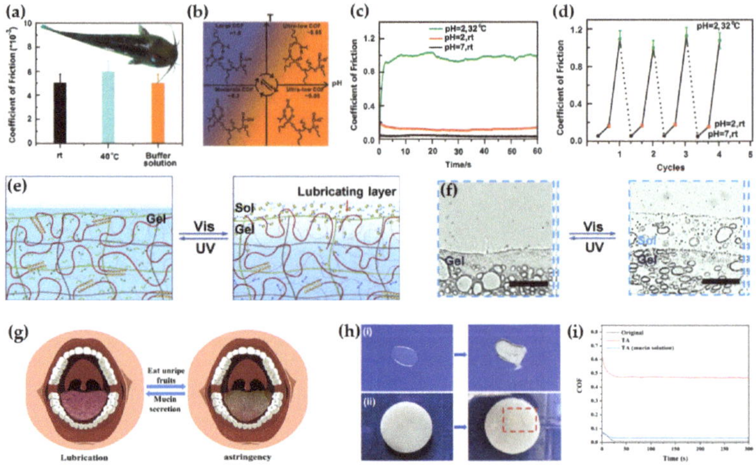

Figure 6. (**a**) Images of catfish skin under different conditions and the corresponding COF, (**b**) pH and temperature changes induce a reversible swelling–collapse cycle, (**c**) the COF curves of the pNIPAM$_{11}$–NaMA$_3$ gel at pH = 7 and 2, rt, and pH = 2 at 32 °C, and (**d**) the switchable COF of the pNIPAM$_{11}$–NaMA$_3$ gel with the stimuli of the pH and temperature [56]. Copyright Permission from Springer Nature, 2014. (**e**) Schematic of the possible photo-responsive lubrication mechanism of PSCHs, (**f**) images of the sol–gel transition in UV and visible light [57]. Copyright Permission from Elsevier, 2021. (**g**) Schematic representation of the switchable lubrication behavior in the oral cavity, (**h**) interaction of mucin solution and PVA/mucin hydrogel with tannins, (**i**) COFs of pristine hydrogel (black line), TA-treated hydrogel (red line), and TA-treated and incubated hydrogel in mucin solution (blue line) [58]. Copyright Permission from Elsevier, 2023.

3. Surface Structure-Dominated Friction

The successful application of the surface structure in tribological performance optimization dates back to the 1940s, and surface geometry has also been extensively studied as an essential influence in tribology, in addition to interfacial chemistry [59,60]. Researchers have long noted that many organisms in nature have evolved various types and scales of surface structures to significantly change the tribological properties to adapt to complex living environments. We can find many examples in nature, such as lotus leaves [61], gecko toe pads [3], shark skin [62], and snake skin [63], where the structures of different surfaces confer different tribological properties (Figure 7a–f). Accordingly, researchers have designed a variety of surface-structured arrays to modulate the contact condition at the interface, thus obtaining tribological properties similar to those of biological surfaces.

For example, the lotus, one of the most famous organisms with superhydrophobic surfaces in nature, has attracted the attention of biologists and materials scientists since the last century and has been extensively studied in the field of drag reduction at solid–liquid interfaces [64]. The surface of the lotus leaf is rough and randomly distributed with many microcapillaries with branching nano-stratified structures at the top of the papillae. Thus, an air-lubricated membrane layer can be formed between the solid phase surface and the liquid phase due to the combined effect of the micropapillary structure and epidermal waxes [65]. The lubrication reduces the frictional resistance at the air–liquid interface, which allows the water droplets to roll easily on the surface of the leaf [66]. Bidkar et al. [67] further demonstrated the drag reduction capability of this type of hydrophobic surface by preparing randomly textured surfaces on flat plates and performing turbulence experiments. The skin-friction resistance was reduced by 20~30% in the experiments. Inspired by the surface structure of the lotus leaf, researchers have also prepared various surfaces with micro- and nano-graded structures, which have been widely used in waterproofing [68], ice-proofing [69], and self-cleaning [70,71].

Shark skin is also a rough surface capable of providing less frictional resistance, with oriented ribs of ordered size and space covering the shark's dermis [72]. The rib-like teeth of the skin are present as grooves along the direction of the water flow, which reduces the friction between the water and the surface of the shark's skin by decreasing the intensity of the turbulence [4]. At the same time, the interstices between these grooved structures also reduce the adhesion of the surface, making it difficult for tiny aquatic organisms to adhere to the shark's body. Inspired by the shark skin structure, researchers have conducted a series of studies on this type of drag-reducing surface. Berchert et al. theoretically investigated the effect of several types of rib geometries on drag reduction, providing theoretical guidance for subsequent designs [73]. Shark skin-inspired rib structures have been demonstrated to reduce drag by up to 9.9% [74]. Xing et al. [75] prepared bionic shark skin textures with DLC coatings on Si_3N_4 ceramic. The sample exhibited a COF of 0.21 at 300 °C, which was 37.26% lower than that of the blank ceramic. Qin et al. [76] investigated the friction behavior of soft materials by preparing a bionic shark texture on polydimethylsiloxane (PDMS). Based on the synergistic effect of the bionic aligned texture and plasma treatment, the friction on the PDMS surface was effectively reduced. In addition, this type of structure has also been used in various applications, such as fluid drag reduction and antifouling [77].

As a limbless reptile with an elongated body covered with scales, snakes rely on friction between their body and the ground for locomotion [78]. This type of locomotion requires that the scales on their body surfaces generate sufficient friction to support the forward movement of the body but also provide a low coefficient of friction when the body is sliding. Researchers have studied the tribological properties of snakes' body surfaces in different locomotion states and found that snakes exhibit significant anisotropy when moving in other directions. The COF was higher when the snake moved in the other direction and 1/4 to 1/2 of the other direction when moving forward [79,80]. The snake's scales have a multiscale surface structure, with fibrous structures constituting micrometer-scale fiber waves with asymmetric tips. During changes in a snake's state of locomotion, the interface between the fibers and the ground constantly changes between the tips and the lateral, causing the contact area to change, resulting in the snake's body surface displaying different friction coefficients in different motion directions. The regulation mechanism of the snake's skin originates from the variations brought about by the multilayers of the surface structure and asymmetries in the contact interfaces. The researchers have already achieved drag reduction and the lubrication effect by mimicking and optimizing this microstructure in the wet and dry state and on various organic and inorganic surfaces [81–84].

Figure 7. *Cont.*

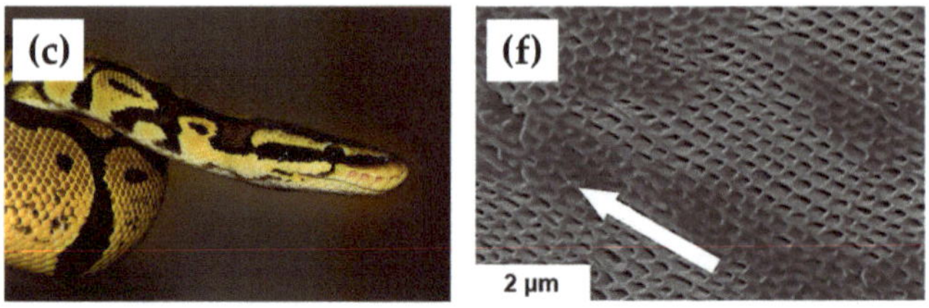

Figure 7. Images of a (**a**) lotus leaf, (**b**) shark, and (**c**) snake and the corresponding surface structures. (**d**) Papillae on the surface of the lotus leaf [85]. Copyright Permission from ACS, 2004. (**e**) Grooves on the scales of the shark [86]. Copyright Permission from Wiley, 2013. (**f**) Periodic structures on the body surface of the snake [83]. Copyright Permission from Springer Nature, 2023.

With the wide variety of plants and animals in nature, researchers have developed a variety of biomimetic lubrication materials with unique tribological properties inspired by various surface structures, either through alternative approaches or by focusing on their strengths. For example, based on the microstructure of the head of the dung beetle, You et al. [87] developed a structured surface that reduces friction by decreasing the contact area and trapping abrasive particles, and its resistance to cutting decreased by 30.41% compared with conventional materials. The gill covers of water snails continuously rub against their hard shells without significant wear. Xu et al. [88] revealed the fluid lubrication mechanism by observing and numerically analyzing the microgroove structure on the gill cover surface, which provides a COF as low as 0.012 in a liquid environment. Gregory et al. [89] studied the low resistance structures of rice blades and butterfly wings and coated nanostructures with the lotus effect onto polyurethane products with shark skin structures. The composite surfaces successfully mimicked the functions of rice blades and butterfly wings, advancing the understanding of surface design elements of biomimetic structures.

Currently, the means of obtaining the surface structures of materials include 3D printing technology and photolithography, which are greatly affected by the manufacturing cost, process precision, and time required. A large part of the material is still in the stage of laboratory preparation in small quantities, making it difficult to realize large-scale industrial preparation. For rigid substrates, the excessive load in the loading, friction, and unloading process is prone to cause severe damage to the structure, which leads to a decline in tribological performance and to lubrication failure; for soft substrates, the deformation after loading will also have an impact on the actual state of the surface structure during the friction process. Especially for systems that require the adaptive adjustment of the tribological performance, it is often difficult to achieve satisfying and continuous lubrication in complex working conditions by relying on only the surface structure to reduce friction.

4. Mechanics-Dominated Friction

Many organisms in nature have evolved surfaces with unique tribological properties and, at the same time, functional organs with specific or adjustable mechanical properties to maintain the adaptive working status under extreme conditions or to switch working states rapidly [90,91]. For example, sea cucumbers can escape danger by hardening the dermis to achieve sudden changes in surface stiffness, and many fish can escape from their captors by contraction hardening and deformation of the muscles [92,93] (Figure 8a,b). The mechanical properties of material surfaces greatly influence the contact state of the surface interface and directly affect the total friction force [94]. Researchers have long been concerned with deformation due to differences in the surface mechanical properties when studying elastomers such as rubber and the significant effect of hysteresis and loss on the total friction. In studies on the tribological behavior of human skin, friction brings about a large amount of lateral deformation, and the contribution of deformation friction to the total friction can be close to 50% at high speeds [95,96] (Figure 8c). For the mammalian joint system, the orderly hierarchical fibrous structure of nano/micro-collagen fibers endows the articular cartilage with excellent mechanical properties, which allows the shear forces in joint motion to be well carried and dispersed, thus cooperating with the synovial fluid and hydrophilic polymer layer on the cartilage to provide long-lasting adaptive lubrication [97,98] (Figure 8d).

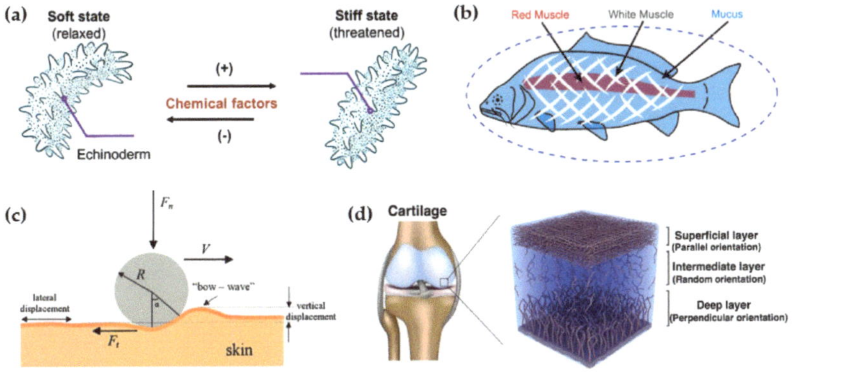

Figure 8. Schematic representation of (**a**) a sea cucumber [92] and (**b**) a fish [93]. The sea cucumber reversibly transforms its dermal hardness when threatened, while the fish escapes by struggling with epidermal mucus. Copyright Permission from Wiley, 2022; Copyright Permission from Springer Nature, 2022. (**c**) Schematic representation of the deformation of the skin surface during friction [95]. Copyright Permission from Elsevier, 2009. (**d**) Layered and organized structure of articular cartilage [98]. Copyright Permission from Wiley, 2017.

In the friction lubrication system described above, the surface mechanics of the material greatly determine the contact state between the biological surface and the target substrate. If we can further modulate the mechanical properties based on considering the chemical properties and structure of the surface, further optimization of specific tribological properties or inducible switching of lubrication states can be achieved. Materials that change surface/subsurface stiffness in response to stimuli have been used in soft actuators and soft robotics research [99–101]. Modulating friction and lubrication performance via changes in the surface mechanical properties is easier for engineering applications than materials that modulate friction through interfacial chemistry. However, obtaining good friction and lubrication properties is difficult when relying on only a single change in mechanical properties without structuring or chemically treating the material's surface.

5. Multiple Strategies Coupling-Dominated Friction

Many lubricating materials and devices based on a single biomimetic design strategy have been reported. However, obvious functional limitations still make it difficult to fully meet people's production and life needs. Some of the materials remain at the stage of conceptual design and laboratory validation, and it is not easy to advance to the level of actual technological transformation. Therefore, it has become a hotspot and a challenge to study and understand the friction control mechanism of biological organs in nature from multiple perspectives and to develop high-performance or intelligent materials by coupling the design strategies of interfacial chemistry, surface structure, and surface mechanics.

5.1. Surface Chemistry Coupling Structure

The strategy of coupling the interfacial chemistry and surface structure enables friction reduction by reducing the contact area through the surface structure and further enhancing the lubrication effect through chemicals on the surface. For example, plants such as the Nepenthes pitcher plant use the structure on the surface to lock in the mucus it secretes [102]. Through an excellent match of solid and liquid surface energies, coupled with the roughness due to the microstructure, the surface can form a stable and effective liquid film that allows errant insects to slide down [103,104]. Wong et al. [105] were the first to introduce the concept of a slippery liquid-infused porous surface (SLIPS) and prepared surfaces with excellent stability, liquid repellency, and adhesion resistance using inexpensive materials such as Teflon. Ma et al. [106] used a simple nanosecond laser treatment method to prepare SLIPSs on carbon steel substrates. In addition to excellent hydrophobicity and corrosion resistance, the tribological properties of the smooth surfaces were improved, with the COF decreasing from about 0.52 to about 0.13 for the base steel. Tong et al. [107] further prepared a smart SLIPS coating inspired by the mucus-secreting behavior of the blind eel. Based on the responsive supramolecular interactions between azobenzene and α-cyclodextrin, the surface could achieve self-replenishment of the lubricant on the surface by contraction of the polymer chains under visible light or thermal stimulation.

The epidermal friction reduction of earthworms is also based on the synergistic effect of mucus secreted by their epidermal glands and annular grooves on the body surface [108,109]. The mucus secreted by the glands forms a lubricating layer on the earthworm's body, while the grooves store the mucus and keep the lubricating layer stable while forming a gap between the body and the soil. Zhao et al. [110] mimicked the lubrication mechanism of earthworms and introduced textured structures onto the liquid-releasing polymer coatings, and the lubricants were stored as discrete droplets in a supramolecular matrix prepared from urea and polydimethylsiloxane copolymers. When the rough surface is subjected to localized pressure, the lubricant is released from the matrix and covers the corresponding area, achieving self-replenishing lubrication. Ruan et al. [111] combined the advantages of porous polyimide and phase change materials by impregnating paraffin wax into the porous material. They constructed smart lubrication materials with the ability to self-repair the lubrication layer. The material can release the internal lubricant under thermal stimulation and form a new paraffin lubrication layer on the surface quickly after the original layer is worn out. This type of coupling strategy can optimize the contact condition of the interface to some extent and improve the interfacial interaction, as well as optimize the stability and continuity of the lubrication layer. However, for solid lubrication, most interfacial chemical interactions are complicated to regulate and require specific means to immobilize the corresponding molecules onto the structured surface, making it difficult to achieve stable and rapid preparation in practical applications.

5.2. Surface Chemistry Coupling Mechanics

Combining the surface chemistry with the surface mechanics can lead to materials with outstanding performance through specific surface modification and substrate stiffness selection, as well as realize large-span lubricating state switching through the change in mechanical properties. The superior lubrication performance of mammalian articular cartilage is attributed to the dense and stable hydrophilic macromolecular layer on its surface and the well-organized layered structure with excellent adaptive load-bearing capacity. Inspired by the lubrication mechanism of articular cartilage, researchers have designed and synthesized a variety of high-performance propriety polymer lubrication materials [112–114] and surface-modified polymer lubrication materials [115–118], aiming to realize the effective combination of surface lubrication and propriety load-bearing of real cartilage. In addition, researchers have also achieved substantial tuning of the lubricating properties by hydrating the lubrication layer with a responsive substrate. Liu et al. [119] reported a temperature-responsive layered material prepared by brush-grafting the poly(potassium salt of 3-sulfopropyl methacrylate) onto the sub-surface of an initiator-embedded, high-strength hydrogel [poly(N-isopropylacrylamide-co-acrylic acid-co-initiator/Fe^{3+})] [P(NIPAAm-AA-iBr/Fe^{3+})]. The soft hydrogel/brush on the top layer provides hydration lubrication, and the temperature-sensitive hydrogel layer at the bottom provides adaptive load-bearing capacity, exhibiting tunable mechanical properties in response to temperatures above or below the lower critical solubilization temperature (LCST) (Figure 9a,b). Fish exhibit unique locomotion and lubrication mechanisms based on a highly hydrated body surface with modulus-adaptive muscle enhancement. Zhang et al. [93] proposed a modulation strategy for interfacial lubrication control based on modulus changes. The modulus-adaptive lubrication hydrogel (MALH) consists of a hydrophilic lubrication layer at the top and a thermally hardened phase-separated layer at the bottom, in which the bottom hydrogel can change from a soft state (20 °C, modulus of elasticity ~0.3 MPa) to a rigid state (80 °C, modulus of elasticity ~120 MPa), which enables the material to achieve switchable lubrication states in water when heated (COF from ~0.37 to ~0.027) (Figure 9c–g). The researchers further designed the Modulus Adaptive Switching Lubrication Device (MASLD) and demonstrated the promising application of this regulatory strategy in flexible devices and smart lubrication systems. The above strategy optimizes the lubrication performance through further knowledge and understanding of the lubrication mechanism of biological organs. The coupling of the two possible means of responsive modulation makes substantial tribological performance tuning possible. However, most of the materials studied so far are limited to single-component externally stimulated modulation, and it is not easy to realize synergistic modulation between the lubrication layer and the substrate material, which makes it difficult to realize a wide range of applications.

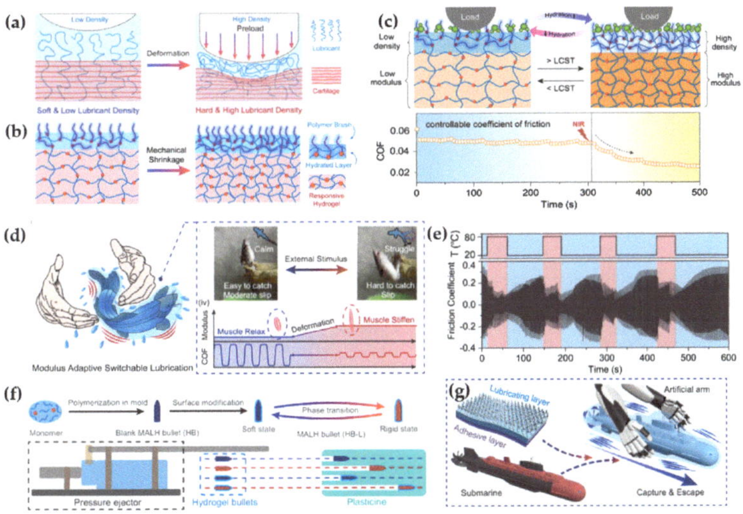

Figure 9. (**a**) Schematic representation of adaptive mechanically controlled lubrication in the joint during unloading and loading; (**b**) schematic representation of the change in density of the top hydrogel/brush composite layer and the change in the mechanical strength of the underlying hydrogel layer as the material network shrinks during heating; (**c**) improvement of the interfacial lubrication of this hydrogel during the heating process, which exhibits dynamic adaptation [119]. Copyright Permission from ACS, 2020. (**d**) Struggling behavior of a fish during capture and its skin muscle modulus versus COF; (**e**) evolution of the COF during in situ heating and cooling of the MALH; (**f**) demonstration of the MALH as a smart bullet; (**g**) schematic diagram of the underwater in situ capture device of the MASLD [93]. Copyright Permission from Springer Nature, 2022.

5.3. Simultaneous Coupling of Three Strategies

Considering the advantages and disadvantages of the above two design strategies, we can combine the three previous single strategies to develop novel biomimetic lubricating materials. Interfacial chemistry provides specific interaction force properties and regulatory mechanisms, surface structure provides optimized contact states, and surface mechanics provide the desired load-bearing capacity and dynamically tunable response states. Zhang et al. [120] proposed a method to synthesize a large-span viscous-slip switchable hydrogel by combining dynamic multiscale contact and coordinate regulation, which can achieve temperature-responsive viscous-slip switching. The responsive process mainly consists of molecular-scale chemical modulation that mimics the adhesion mechanism of mussels and mesoscale modulation based on surface roughness and modulus changes (Figure 10a–c). This smart hydrogel (DMCS-hydrogel) with dynamic multiscale contact synergistic modulation can be applied to various substrate surfaces and exhibits fast switching capability. Considering the coupled design of the three factors, Liu et al. [121] created a biomimetic high-strength anisotropic layered lubrication hydrogel (ALLH) with an ultra-low COF by coupling a hydrophilic polyelectrolyte brush, an anisotropic surface microstructure inspired by scallion leaf, and a high-mechanical-strength substrate mimicking human cartilage (Figure 10d–g). The artificial scallion leaves exhibit low friction (COF < 0.01) in different sliding directions under a wide range of contact stresses (≈0.2 to 2.4 MPa, corresponding to loads of ≈5 to 60 N) and ultra-low friction (COF ≈ 0.006) along the microstrip structure. For high contact pressure and long-term durability tests, the material achieves almost zero surface wear, which mimics human cartilage's physiological function.

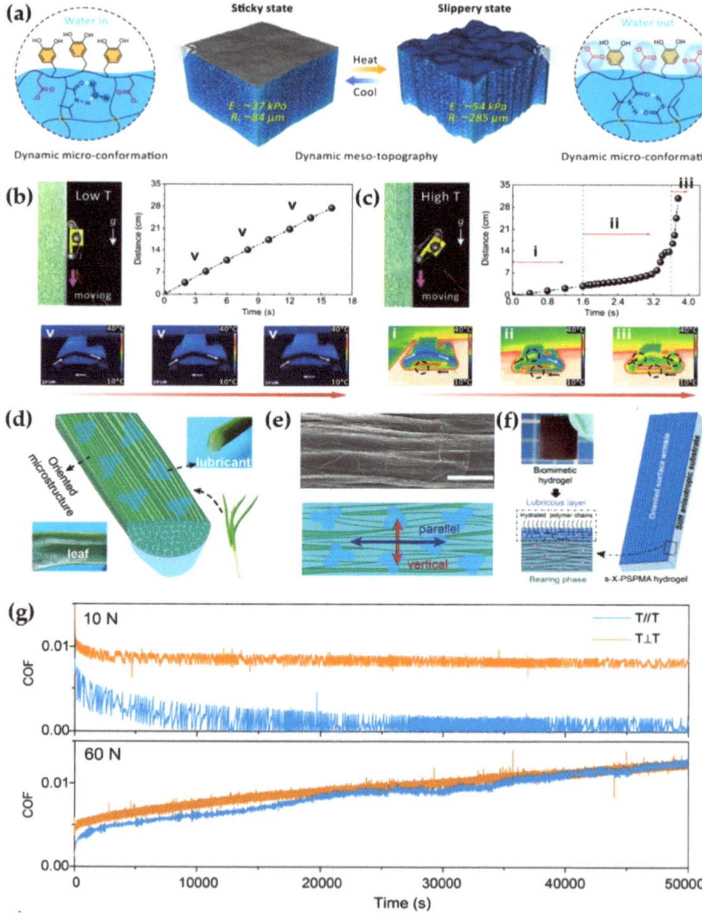

Figure 10. (**a**) DMCS-hydrogel based on dynamic multiscale contact synergy originating from the dynamic meso-topography and micro-conformation to realize the stick-slip switching; mobile device

at low (**b**) and high (**c**) temperatures in the images and time–distance curves of crawling on a vertical metal plate, and infrared images showing the heat transfer process from the substrate to the DMSC-hydrogel [120]. Copyright Permission from Springer Nature, 2022. (**d**) Oriented distribution of substrate fibers and well-hydrated mucilage on the surface; (**e**) SEM images and schematic diagrams of the surface structure of scallion leaves; (**f**) schematic diagrams of ALLH samples; (**g**) the COF curves of the ALLH sample in the entire 50,000 sliding cycles under the normal loads of 10 N (contact pressure = 0.4 MPa) and 60 N (contact pressure = 2.4 MPa) in two perpendicular directions with water as the lubricant [121]. Copyright Permission from Wiley, 2024.

By coupling the three strategies, the problems of the single chemical property of the material itself and the unsatisfactory interfacial contact state are improved. However, the more factors that are combined, the more complicated the process links that need to be considered and the more parameters that to be regulated during the material preparation. Considering the laboratory operation limitations, most current materials and devices are multi-material composites, and the performance differences between different materials and the weak interfacial bonding remain to be solved. In the future, integrating the advantages of various materials and developing propriety functional materials with excellent performance or easy modification to realize the on-demand design and manufacturing of bionic lubrication materials will remain a significant challenge.

6. Summary and Perspective

With people's deepening understanding of the mechanism of biological lubrication, a variety of biomimetic lubricating materials with better design strategies have been reported one after another. The single biomimetic lubrication strategy has been widely used in developing practical and functional lubricating materials. Polymer brush systems inspired by articular cartilage have made a big splash in water lubrication systems and bio-lubrication, while structuring processes based on animal and plant surface structures have been widely used in self-cleaning, fluid drag reduction and antifouling. The comprehensive influence of surface mechanics on the friction or lubrication performance of materials has also been gradually emphasized by researchers.

However, most of the development of bionic lubrication materials and devices is still limited to the laboratory stage. It is difficult to meet the harsh conditions of use, which puts forward new practical requirements. Given the inadequacy of a single biomimetic strategy, researchers have begun to develop biomimetic materials by coupling multiple factors. Surface modifications such as hydrophilicity/hydrophobicity can further optimize the properties of structured surfaces produced by conventional processes. In contrast, the surface structure, in turn, improves the contact state of the host material or optimizes the durability and stability of tribological properties. In addition, by introducing the factor of the surface/subsurface mechanical properties, the lubrication state is expected to be further optimized and drastically regulated.

Coupling strategies can compensate for the shortcomings of a single strategy to a certain extent while highlighting its advantages and maximizing the utility of each mechanism. However, multiple regulatory factors often bring about more complex design strategies and manufacturing processes, and researchers often need to integrate and regulate the performance of multiple functionalized systems. The differences in properties of various materials can easily bring about insufficient bonding power and difficulties in regulation. In this case, how to reasonably couple the advantages to obtain a responsive propriety functionalized material that is easy to regulate or develop a composite material with better performance is still a great challenge. From the engineering point of view, to ensure the efficient, continuous and reliable lubrication performance of materials or devices under complex and harsh conditions, realizing adaptive lubrication performance under real and variable working conditions is also a major focus and difficulty. With the deepening research and understanding of the interfacial lubrication mechanism in the biological movement process and the continuous innovation of the material synthesis process, the development of new biomimetic friction lubrication materials with the ability to adapt to the working conditions or environment will become one of the critical development directions in the field in the future, with a focus on the adjustable interfacial contact state, through the combination of polymer design and synthesis, multiscale surface structuring, surface mechanical property regulation and mechanical deformation, and so on. In the future, these materials are expected to shine in biomedicine, intelligent electronic sensor devices, soft robots, and precision manufacturing.

Author Contributions: Concept, supervision, revision and editing of manuscript by S.M. and F.Z.; figures and writing of most of the manuscript by Y.K. All authors have read and agreed to the published version of the manuscript.

Funding: This research was funded by the Strategic Priority Research Program of the Chinese Academy of Sciences (XDB 0470201), National Natural Science Foundation of China (52075522, 52322506).

Data Availability Statement: There were no new data created.

Conflicts of Interest: The authors declare no conflicts of interest.

References

1. Li, X.; Guo, Z.; Huang, Q.; Yuan, C. Application of Bionic Tribology in Water-Lubricated Bearing: A Review. *J. Bionic Eng.* **2022**, *19*, 902–934. [CrossRef]
2. Coles, J.M.; Chang, D.P.; Zauscher, S. Molecular Mechanisms of Aqueous Boundary Lubrication by Mucinous Glycoproteins. *Curr. Opin. Colloid Interface Sci.* **2010**, *15*, 406–416. [CrossRef]
3. Autumn, K.; Liang, Y.A.; Hsieh, S.T.; Zesch, W.; Chan, W.P.; Kenny, T.W.; Fearing, R.; Full, R.J. Adhesive Force of a Single Gecko Foot-Hair. *Nature* **2000**, *405*, 681–685. [CrossRef] [PubMed]
4. Zhang, D.; Luo, Y.; Li, X.; Chen, H. Numerical Simulation and Experimental Study of Drag-Reducing Surface of a Real Shark Skin. *J. Hydrodyn.* **2011**, *23*, 204–211. [CrossRef]
5. Wilkie, I.C. Is Muscle Involved in the Mechanical Adaptability of Echinoderm Mutable Collagenous Tissue? *J. Exp. Biol.* **2002**, *205 Pt 2*, 159–165. [CrossRef] [PubMed]
6. Jusufi, A.; Vogt, D.M.; Wood, R.J.; Lauder, G.V. Undulatory Swimming Performance and Body Stiffness Modulation in a Soft Robotic Fish-Inspired Physical Model. *Soft Robot.* **2017**, *4*, 202–210. [CrossRef] [PubMed]
7. Klein, J. Hydration Lubrication. *Friction* **2013**, *1*, 1–23. [CrossRef]
8. Ma, L.; Gaisinskaya-Kipnis, A.; Kampf, N.; Klein, J. Origins of Hydration Lubrication. *Nat. Commun.* **2015**, *6*, 6060. [CrossRef] [PubMed]
9. Raviv, U.; Laurat, P.; Klein, J. Fluidity of Water Confined to Subnanometre Films. *Nature* **2001**, *413*, 51–54. [CrossRef]
10. Pashley, R.M. Hydration Forces between Mica Surfaces in Aqueous Electrolyte Solutions. *J. Colloid Interface Sci.* **1981**, *80*, 153–162. [CrossRef]
11. Lane, J.M.D.; Chandross, M.; Stevens, M.J.; Grest, G.S. Water in Nanoconfinement between Hydrophilic Self-Assembled Monolayers. *Langmuir* **2008**, *24*, 5209–5212. [CrossRef] [PubMed]
12. Xu, J.; Luo, J.B.; Liu, S.H.; Xie, G.X.; Ma, L. Tribological Characteristics of Aloe Mucilage. *Tribol. –Mater. Surf. Interfaces* **2008**, *2*, 72–76. [CrossRef]
13. Hakala, T.J.; Saikko, V.; Arola, S.; Ahlroos, T.; Helle, A.; Kuosmanen, P.; Holmberg, K.; Linder, M.B.; Laaksonen, P. Structural Characterization and Tribological Evaluation of Quince Seed Mucilage. *Tribol. Int.* **2014**, *77*, 24–31. [CrossRef]
14. Arad, S.; Rapoport, L.; Moshkovich, A.; Van Moppes, D.; Karpasas, M.; Golan, R.; Golan, Y. Superior Biolubricant from a Species of Red Microalga. *Langmuir* **2006**, *22*, 7313–7317. [CrossRef] [PubMed]
15. Li, J.; Liu, Y.; Luo, J.; Liu, P.; Zhang, C. Excellent Lubricating Behavior of Brasenia Schreberi Mucilage. *Langmuir* **2012**, *28*, 7797–7802. [CrossRef] [PubMed]
16. Pult, H.; Tosatti, S.G.P.; Spencer, N.D.; Asfour, J.-M.; Ebenhoch, M.; Murphy, P.J. Spontaneous Blinking from a Tribological Viewpoint. *Ocul. Surf.* **2015**, *13*, 236–249. [CrossRef] [PubMed]
17. Lin, C.X.; Li, W.; Deng, H.Y.; Li, K.; Zhou, Z.R. Friction Behavior of Esophageal Mucosa Under Axial and Circumferential Extension. *Tribol. Lett.* **2019**, *67*, 9. [CrossRef]
18. Veeregowda, D.H.; Busscher, H.J.; Vissink, A.; Jager, D.-J.; Sharma, P.K.; Van Der Mei, H.C. Role of Structure and Glycosylation of Adsorbed Protein Films in Biolubrication. *PLoS ONE* **2012**, *7*, e42600. [CrossRef] [PubMed]
19. Cleather, D.J.; Goodwin, J.E.; Bull, A.M.J. Hip and Knee Joint Loading during Vertical Jumping and Push Jerking. *Clin. Biomech.* **2013**, *28*, 98–103. [CrossRef]
20. Gong, J.P. Friction and Lubrication of Hydrogels—Its Richness and Complexity. *Soft Matter* **2006**, *2*, 544–552. [CrossRef]
21. Klein, J. Repair or Replacement--A Joint Perspective. *Science* **2009**, *323*, 47–48. [CrossRef] [PubMed]
22. Xue, F.; Zhang, H.; Hu, J.; Liu, Y. Hyaluronic Acid Nanofibers Crosslinked with a Nontoxic Reagent. *Carbohydr. Polym.* **2021**, *259*, 117757. [CrossRef] [PubMed]

23. Bełdowski, P.; Yuvan, S.; Dėdinaitė, A.; Claesson, P.M.; Pöschel, T. Interactions of a Short Hyaluronan Chain with a Phospholipid Membrane. *Colloids Surf. B Biointerfaces* **2019**, *184*, 110539. [CrossRef] [PubMed]
24. Klein, J.; Kumacheva, E.; Mahalu, D.; Perahia, D.; Fetters, L.J. Reduction of Frictional Forces between Solid Surfaces Bearing Polymer Brushes. *Nature* **1994**, *370*, 634–636. [CrossRef]
25. Tadmor, R.; Janik, J.; Klein, J.; Fetters, L.J. Sliding Friction with Polymer Brushes. *Phys. Rev. Lett.* **2003**, *91*, 115503. [CrossRef] [PubMed]
26. Kobayashi, M.; Takahara, A. Tribological Properties of Hydrophilic Polymer Brushes under Wet Conditions. *Chem. Rec.* **2010**, *10*, 208–216. [CrossRef]
27. Jahn, S.; Klein, J. Hydration Lubrication: The Macromolecular Domain. *Macromolecules* **2015**, *48*, 5059–5075. [CrossRef]
28. De Beer, S.; Kenmoé, G.D.; Müser, M.H. On the Friction and Adhesion Hysteresis between Polymer Brushes Attached to Curved Surfaces: Rate and Solvation Effects. *Friction* **2015**, *3*, 148–160. [CrossRef]
29. Vandeweerd, J.-M.; Innocenti, B.; Rocasalbas, G.; Gautier, S.E.; Douette, P.; Hermitte, L.; Hontoir, F.; Chausson, M. Non-Clinical Assessment of Lubrication and Free Radical Scavenging of an Innovative Non-Animal Carboxymethyl Chitosan Biomaterial for Viscosupplementation: An in-Vitro and Ex-Vivo Study. *PLoS ONE* **2021**, *16*, e0256770. [CrossRef]
30. Bauerova, K.; Ponist, S.; Kuncirova, V.; Mihalova, D.; Paulovicova, E.; Volpi, N. Chondroitin Sulfate Effect on Induced Arthritis in Rats. *Osteoarthr. Cartil.* **2011**, *19*, 1373–1379. [CrossRef]
31. Sawitzke, A.D.; Shi, H.; Finco, M.F.; Dunlop, D.D.; Bingham III, C.O.; Harris, C.L.; Singer, N.G.; Bradley, J.D.; Silver, D.; Jackson, C.G.; et al. The Effect of Glucosamine and/or Chondroitin Sulfate on the Progression of Knee Osteoarthritis: A Report from the Glucosamine/Chondroitin Arthritis Intervention Trial. *Arthritis Rheum.* **2008**, *58*, 3183–3191. [CrossRef] [PubMed]
32. Wathier, M.; Lakin, B.A.; Bansal, P.N.; Stoddart, S.S.; Snyder, B.D.; Grinstaff, M.W. A Large-Molecular-Weight Polyanion, Synthesized via Ring-Opening Metathesis Polymerization, as a Lubricant for Human Articular Cartilage. *J. Am. Chem. Soc.* **2013**, *135*, 4930–4933. [CrossRef] [PubMed]
33. Hartung, W.; Drobek, T.; Lee, S.; Zürcher, S.; Spencer, N.D. The Influence of Anchoring-Group Structure on the Lubricating Properties of Brush-Forming Graft Copolymers in an Aqueous Medium. *Tribol. Lett.* **2008**, *31*, 119–128. [CrossRef]
34. Perry, S.S.; Yan, X.; Limpoco, F.T.; Lee, S.; Müller, M.; Spencer, N.D. Tribological Properties of Poly(L-Lysine)-*Graft*-Poly(Ethylene Glycol) Films: Influence of Polymer Architecture and Adsorbed Conformation. *ACS Appl. Mater. Interfaces* **2009**, *1*, 1224–1230. [CrossRef] [PubMed]
35. Hartung, W.; Rossi, A.; Lee, S.; Spencer, N.D. Aqueous Lubrication of SiC and Si3N4 Ceramics Aided by a Brush-like Copolymer Additive, Poly(l-Lysine)-Graft-Poly(Ethylene Glycol). *Tribol. Lett.* **2009**, *34*, 201–210. [CrossRef]
36. Pettersson, T.; Naderi, A.; Makuška, R.; Claesson, P.M. Lubrication Properties of Bottle-Brush Polyelectrolytes: An AFM Study on the Effect of Side Chain and Charge Density. *Langmuir* **2008**, *24*, 3336–3347. [CrossRef] [PubMed]
37. Crockett, R. Boundary Lubrication in Natural Articular Joints. *Tribol. Lett.* **2009**, *35*, 77–84. [CrossRef]
38. Raviv, U.; Giasson, S.; Kampf, N.; Gohy, J.-F.; Jérôme, R.; Klein, J. Lubrication by Charged Polymers. *Nature* **2003**, *425*, 163–165. [CrossRef] [PubMed]
39. Zhao, B.; Brittain, W.J. Polymer Brushes: Surface-Immobilized Macromolecules. *Prog. Polym. Sci.* **2000**, *25*, 677–710. [CrossRef]
40. Cai, Z.; Zhang, H.; Wei, Y.; Wu, M.; Fu, A. Shear-Thinning Hyaluronan-Based Fluid Hydrogels to Modulate Viscoelastic Properties of Osteoarthritis Synovial Fluids. *Biomater. Sci.* **2019**, *7*, 3143–3157. [CrossRef]
41. Lin, W.; Mashiah, R.; Seror, J.; Kadar, A.; Dolkart, O.; Pritsch, T.; Goldberg, R.; Klein, J. Lipid-Hyaluronan Synergy Strongly Reduces Intrasynovial Tissue Boundary Friction. *Acta Biomater.* **2019**, *83*, 314–321. [CrossRef] [PubMed]
42. Momose, T.; Amadio, P.C.; Sun, Y.-L.; Zhao, C.; Zobitz, M.E.; Harrington, J.R.; An, K.-N. Surface Modification of Extrasynovial Tendon by Chemically Modified Hyaluronic Acid Coating. *J. Biomed. Mater. Res.* **2002**, *59*, 219–224. [CrossRef] [PubMed]
43. Xu, R.; Ma, S.; Wu, Y.; Lee, H.; Zhou, F.; Liu, W. Adaptive Control in Lubrication, Adhesion, and Hemostasis by Chitosan–Catechol–pNIPAM. *Biomater. Sci.* **2019**, *7*, 3599–3608. [CrossRef] [PubMed]
44. Valle-Delgado, J.J.; Johansson, L.-S.; Österberg, M. Bioinspired Lubricating Films of Cellulose Nanofibrils and Hyaluronic Acid. *Colloids Surf. B Biointerfaces* **2016**, *138*, 86–93. [CrossRef] [PubMed]
45. Røn, T.; Javakhishvili, I.; Hvilsted, S.; Jankova, K.; Lee, S. Ultralow Friction with Hydrophilic Polymer Brushes in Water as Segregated from Silicone Matrix. *Adv. Mater. Interfaces* **2016**, *3*, 1500472. [CrossRef]
46. Gaisinskaya-Kipnis, A.; Jahn, S.; Goldberg, R.; Klein, J. Effect of Glucosamine Sulfate on Surface Interactions and Lubrication by Hydrogenated Soy Phosphatidylcholine (HSPC) Liposomes. *Biomacromolecules* **2014**, *15*, 4178–4186. [CrossRef] [PubMed]
47. Wang, M.; Zander, T.; Liu, X.; Liu, C.; Raj, A.; Florian Wieland, D.C.; Garamus, V.M.; Willumeit-Römer, R.; Claesson, P.M.; Dėdinaitė, A. The Effect of Temperature on Supported Dipalmitoylphosphatidylcholine (DPPC) Bilayers: Structure and Lubrication Performance. *J. Colloid Interface Sci.* **2015**, *445*, 84–92. [CrossRef] [PubMed]
48. Duan, Y.; Liu, Y.; Zhang, C.; Chen, Z.; Wen, S. Insight into the Tribological Behavior of Liposomes in Artificial Joints. *Langmuir* **2016**, *32*, 10957–10966. [CrossRef]
49. Adibnia, V.; Mirbagheri, M.; Faivre, J.; Robert, J.; Lee, J.; Matyjaszewski, K.; Lee, D.W.; Banquy, X. Bioinspired Polymers for Lubrication and Wear Resistance. *Prog. Polym. Sci.* **2020**, *110*, 101298. [CrossRef]
50. Kim, H.S.; Crosby, A.J. Solvent-Responsive Surface via Wrinkling Instability. *Adv. Mater.* **2011**, *23*, 4188–4192. [CrossRef]
51. Wang, D.; Liu, Y.; Liu, X.; Zhou, F.; Liu, W.; Xue, Q. Towards a Tunable and Switchable Water Adhesion on a TiO_2nanotube Film with Patterned Wettability. *Chem. Commun.* **2009**, *45*, 7018–7020. [CrossRef]

52. Moon, H.J.; Ko, D.Y.; Park, M.H.; Joo, M.K.; Jeong, B. Temperature-Responsive Compounds as in Situ Gelling Biomedical Materials. *Chem. Soc. Rev.* **2012**, *41*, 4860–4883. [CrossRef] [PubMed]
53. Yu, X.; Wang, Z.; Jiang, Y.; Shi, F.; Zhang, X. Reversible pH-Responsive Surface: From Superhydrophobicity to Superhydrophilicity. *Adv. Mater.* **2005**, *17*, 1289–1293. [CrossRef]
54. Lahann, J.; Mitragotri, S.; Tran, T.-N.; Kaido, H.; Sundaram, J.; Choi, I.S.; Hoffer, S.; Somorjai, G.A.; Langer, R. A Reversibly Switching Surface. *Science* **2003**, *299*, 371–374. [CrossRef] [PubMed]
55. Zhang, X.; Shi, K.; Yang, X.; Wang, J.; Liu, H.; Song, W.; Wang, S. Supramolecular Semi-Convertible Hydrogel Enabled Self-Healing Responsive Lubrication under Dynamic Shearing. *CCS Chem.* **2023**, *5*, 2482–2496. [CrossRef]
56. Wu, Y.; Pei, X.; Wang, X.; Liang, Y.; Liu, W.; Zhou, F. Biomimicking Lubrication Superior to Fish Skin Using Responsive Hydrogels. *NPG Asia Mater* **2014**, *6*, e136. [CrossRef]
57. Wang, J.; Zhang, X.; Zhang, S.; Kang, J.; Guo, Z.; Feng, B.; Zhao, H.; Luo, Z.; Yu, J.; Song, W.; et al. Semi-Convertible Hydrogel Enabled Photoresponsive Lubrication. *Matter* **2021**, *4*, 675–687. [CrossRef]
58. Deng, S.; Wu, L.; Lai, J.; Wang, L.; Yan, G.; Zhao, C.; Xiang, D.; Li, H.; Wang, B.; Li, Z.; et al. Bioinspired Ultra-Lubrication Hydrogels with Controllably Switchable Lubrication. *Colloids Surf. A Physicochem. Eng. Asp.* **2023**, *675*, 132033. [CrossRef]
59. Willis, E. Surface Finish in Relation to Cylinder Liners. *Wear* **1986**, *109*, 351–366. [CrossRef]
60. Etsion, I. State of the Art in Laser Surface Texturing. *J. Tribol.* **2005**, *127*, 248–253. [CrossRef]
61. Lu, Y. Superior Lubrication Properties of Biomimetic Surfaces with Hierarchical Structure. *Tribol. Int.* **2018**, *119*, 131–142. [CrossRef]
62. Li, X.; Deng, J.; Yue, H.; Ge, D.; Zou, X. Wear Performance of Electrohydrodynamically Atomized WS2 Coatings Deposited on Biomimetic Shark-Skin Textured Surfaces. *Tribol. Int.* **2019**, *134*, 240–251. [CrossRef]
63. Hazel, J.; Stone, M.; Grace, M.S.; Tsukruk, V.V. Nanoscale Design of Snake Skin for Reptation Locomotions via Friction Anisotropy. *J. Biomech.* **1999**, *32*, 477–484. [CrossRef] [PubMed]
64. Su, Y.; Ji, B.; Zhang, K.; Gao, H.; Huang, Y.; Hwang, K. Nano to Micro Structural Hierarchy Is Crucial for Stable Superhydrophobic and Water-Repellent Surfaces. *Langmuir* **2010**, *26*, 4984–4989. [CrossRef] [PubMed]
65. Gogolides, E.; Ellinas, K.; Tserepi, A. Hierarchical Micro and Nano Structured, Hydrophilic, Superhydrophobic and Superoleophobic Surfaces Incorporated in Microfluidics, Microarrays and Lab on Chip Microsystems. *Microelectron. Eng.* **2015**, *132*, 135–155. [CrossRef]
66. Burton, Z.; Bhushan, B. Hydrophobicity, Adhesion, and Friction Properties of Nanopatterned Polymers and Scale Dependence for Micro- and Nanoelectromechanical Systems. *Nano Lett.* **2005**, *5*, 1607–1613. [CrossRef] [PubMed]
67. Bidkar, R.A.; Leblanc, L.; Kulkarni, A.J.; Bahadur, V.; Ceccio, S.L.; Perlin, M. Skin-Friction Drag Reduction in the Turbulent Regime Using Random-Textured Hydrophobic Surfaces. *Phys. Fluids* **2014**, *26*, 085108. [CrossRef]
68. Gao, N.; Yan, Y.Y.; Chen, X.Y.; Mee, D.J. Superhydrophobic Surfaces with Hierarchical Structure. *Mater. Lett.* **2011**, *65*, 2902–2905. [CrossRef]
69. Qin, L.; Wu, Y.; Ma, Z.; Li, H.; Lu, S.; Wang, Z.; Liu, J.; Dong, M.; Zhang, Y. Fluorine-Free and Highly Durable Superhydrophobic Textiles for Antifouling and Anti-Icing Applications. *ACS Appl. Polym. Mater.* **2023**, *5*, 6278–6287. [CrossRef]
70. Zhang, X.; Guo, Y.; Zhang, Z.; Zhang, P. Self-Cleaning Superhydrophobic Surface Based on Titanium Dioxide Nanowires Combined with Polydimethylsiloxane. *Appl. Surf. Sci.* **2013**, *284*, 319–323. [CrossRef]
71. Dalawai, S.P.; Saad Aly, M.A.; Latthe, S.S.; Xing, R.; Sutar, R.S.; Nagappan, S.; Ha, C.-S.; Kumar Sadasivuni, K.; Liu, S. Recent Advances in Durability of Superhydrophobic Self-Cleaning Technology: A Critical Review. *Prog. Org. Coat.* **2020**, *138*, 105381. [CrossRef]
72. Bechert, D.W.; Bruse, M.; Hage, W. Experiments with Three-Dimensional Riblets as an Idealized Model of Shark Skin. *Exp. Fluids* **2000**, *28*, 403–412. [CrossRef]
73. Bechert, D.W.; Bartenwerfer, M. The Viscous Flow on Surfaces with Longitudinal Ribs. *J. Fluid Mech.* **1989**, *206*, 105–129. [CrossRef]
74. Bechert, D.W.; Bruse, M.; Hage, W.; Hoeven, J.G.T.V.D.; Hoppe, G. Experiments on Drag-Reducing Surfaces and Their Optimization with an Adjustable Geometry. *J. Fluid Mech.* **1997**, *338*, 59–87. [CrossRef]
75. Xing, Y.; Zhu, M.; Wu, Z.; Li, Z.; Bai, S.; Zhang, K.; Liu, L. High-Temperature Tribological Properties of Si3N4/TiC Ceramic with Bionic Surface Textures and DLC Coatings. *Tribol. Int.* **2023**, *186*, 108648. [CrossRef]
76. Qin, L.; Huang, X.; Sun, Z.; Ma, Z.; Mawignon, F.J.; Lv, B.; Shan, L.; Dong, G. Synergistic Effect of Sharkskin-Inspired Morphologies and Surface Chemistry on Regulating Stick-Slip Friction. *Tribol. Int.* **2023**, *187*, 108765. [CrossRef]
77. Ghimire, A.; Dahl, R.B.; Shen, S.; Chen, P. Shark Skin Denticles: From Morphological Diversity to Multi-functional Adaptations and Applications. *Adv. Funct. Mater.* **2024**, 2307121. [CrossRef]
78. Abdel-Aal, H.A. On Surface Structure and Friction Regulation in Reptilian Limbless Locomotion. *J. Mech. Behav. Biomed. Mater.* **2013**, *22*, 115–135. [CrossRef] [PubMed]
79. Benz, M.J.; Kovalev, A.E.; Gorb, S.N. Anisotropic Frictional Properties in Snakes. In *Bioinspiration, Biomimetics, and Bioreplication 2012*; SPIE: Bellingham, WA, USA, 2012; Volume 8339, pp. 256–261. [CrossRef]
80. Baum, M.J.; Kovalev, A.E.; Michels, J.; Gorb, S.N. Anisotropic Friction of the Ventral Scales in the Snake Lampropeltis Getula Californiae. *Tribol. Lett.* **2014**, *54*, 139–150. [CrossRef]
81. Saito, Y.; Yabu, H. Bio-Inspired Low Frictional Surfaces Having Micro-Dimple Arrays Prepared with Honeycomb Patterned Porous Films as Wet Etching Masks. *Langmuir* **2015**, *31*, 959–963. [CrossRef]

82. Ballesteros, L.M.; Zuluaga, E.; Cuervo, P.; Rudas, J.S.; Toro, A. Tribological Behavior of Polymeric 3D-Printed Surfaces with Deterministic Patterns Inspired in Snake Skin Morphology. *Surf. Topogr.-Metrol. Prop.* **2021**, *9*, 014002. [CrossRef]
83. Zhao, J.; Ji, K.; Chen, Q.; Khan, M.N.; Tu, C.; Ma, Z.; Wu, J.; Chen, J.; Dai, Z. Resistance Reduction of Patterned Surface Inspired by Cuticle Structure of Achalinus Spinalis. *Friction* **2023**, *11*, 1359–1370. [CrossRef]
84. Zhao, Y.; Mei, H.; Chang, P.; Yang, Y.; Cheng, L.; Zhang, L. High-Strength Printed Ceramic Structures for Higher Temperature Lubrication. *Compos. Part B Eng.* **2021**, *221*, 109013. [CrossRef]
85. Patankar, N.A. Mimicking the Lotus Effect: Influence of Double Roughness Structures and Slender Pillars. *Langmuir* **2004**, *20*, 8209–8213. [CrossRef] [PubMed]
86. Jung, Y.C.; Bhushan, B. Biomimetic Structures for Fluid Drag Reduction in Laminar and Turbulent Flows. *J. Phys.-Condens. Matter* **2010**, *22*, 035104. [CrossRef] [PubMed]
87. You, C.; Zhao, G.; Chu, X.; Zhou, W.; Long, Y.; Lian, Y. Design, Preparation and Cutting Performance of Bionic Cutting Tools Based on Head Microstructures of Dung Beetle. *J. Manuf. Process.* **2020**, *58*, 129–135. [CrossRef]
88. Xu, X.; Wu, J.; Yang, Y.; Zhu, R.; Yan, S. Operculum of a Water Snail Is a Hydrodynamic Lubrication Sheet. *J. Bionic Eng.* **2018**, *15*, 471–480. [CrossRef]
89. Bixler, G.D.; Bhushan, B. Bioinspired Rice Leaf and Butterfly Wing Surface Structures Combining Shark Skin and Lotus Effects. *Soft Matter* **2012**, *8*, 11271. [CrossRef]
90. Radhakrishnan, V. Locomotion: Dealing with Friction. *Proc. Natl. Acad. Sci. USA* **1998**, *95*, 5448–5455. [CrossRef]
91. Dickinson, M.H.; Farley, C.T.; Full, R.J.; Koehl, M.A.R.; Kram, R.; Lehman, S. How Animals Move: An Integrative View. *Science* **2000**, *288*, 100–106. [CrossRef]
92. Ming, X.; Yao, L.; Zhu, H.; Zhang, Q.; Zhu, S. Dramatic and Reversible Water-Induced Stiffening Driven by Phase Separation within Polymer Gels. *Adv. Funct. Mater.* **2022**, *32*, 2109850. [CrossRef]
93. Zhang, Y.; Zhao, W.; Ma, S.; Liu, H.; Wang, X.; Zhao, X.; Yu, B.; Cai, M.; Zhou, F. Modulus Adaptive Lubricating Prototype Inspired by Instant Muscle Hardening Mechanism of Catfish Skin. *Nat. Commun.* **2022**, *13*, 377. [CrossRef] [PubMed]
94. Popov, V.L. *Contact Mechanics and Friction*; Springer: Berlin/Heidelberg, Germany, 2010. [CrossRef]
95. Kwiatkowska, M.; Franklin, S.E.; Hendriks, C.P.; Kwiatkowski, K. Friction and Deformation Behaviour of Human Skin. *Wear* **2009**, *267*, 1264–1273. [CrossRef]
96. Mahdi, D.; Riches, A.; Gester, M.; Yeomans, J.; Smith, P. Rolling and Sliding: Separation of Adhesion and Deformation Friction and Their Relative Contribution to Total Friction. *Tribol. Int.* **2015**, *89*, 128–134. [CrossRef]
97. Ge, Z.; Li, C.; Heng, B.C.; Cao, G.; Yang, Z. Functional Biomaterials for Cartilage Regeneration. *J. Biomed. Mater. Res. Part A* **2012**, *100A*, 2526–2536. [CrossRef] [PubMed]
98. Zhao, Z.; Fang, R.; Rong, Q.; Liu, M. Bioinspired Nanocomposite Hydrogels with Highly Ordered Structures. *Adv. Mater.* **2017**, *29*, 1703045. [CrossRef] [PubMed]
99. Buckner, T.L.; Yuen, M.C.; Kim, S.Y.; Kramer-Bottiglio, R. Enhanced Variable Stiffness and Variable Stretchability Enabled by Phase-Changing Particulate Additives. *Adv. Funct. Mater.* **2019**, *29*, 1903368. [CrossRef]
100. Sharifi, S.; Rux, C.; Sparling, N.; Wan, G.; Mohammadi Nasab, A.; Siddaiah, A.; Menezes, P.; Zhang, T.; Shan, W. Dynamically Tunable Friction via Subsurface Stiffness Modulation. *Front. Robot. AI* **2021**, *8*, 691789. [CrossRef]
101. Liu, Y.; Wang, Y.; Yang, X.; Huang, W.; Zhang, Y.; Zhang, X.; Wang, X. Stiffness Variable Polymer for Soft Actuators with Sharp Stiffness Switch and Fast Response. *ACS Appl. Mater. Interfaces* **2023**, *15*, 26016–26027. [CrossRef]
102. Bohn, H.F.; Federle, W. Insect Aquaplaning: Nepenthes Pitcher Plants Capture Prey with the Peristome, a Fully Wettable Water-Lubricated Anisotropic Surface. *Proc. Natl. Acad. Sci. USA* **2004**, *101*, 14138–14143. [CrossRef]
103. Bauer, U.; Federle, W. The Insect-Trapping Rim of Nepenthes Pitchers: Surface Structure and Function. *Plant Signal. Behav.* **2009**, *4*, 1019–1023. [CrossRef] [PubMed]
104. Federle, W.; Riehle, M.; Curtis, A.S.G.; Full, R.J. An Integrative Study of Insect Adhesion: Mechanics and Wet Adhesion of Pretarsal Pads in Ants1. *Integr. Comp. Biol.* **2002**, *42*, 1100–1106. [CrossRef] [PubMed]
105. Wong, T.-S.; Kang, S.H.; Tang, S.K.Y.; Smythe, E.J.; Hatton, B.D.; Grinthal, A.; Aizenberg, J. Bioinspired Self-Repairing Slippery Surfaces with Pressure-Stable Omniphobicity. *Nature* **2011**, *477*, 443–447. [CrossRef] [PubMed]
106. Ma, Q.; Wang, W.; Dong, G. Facile Fabrication of Biomimetic Liquid-Infused Slippery Surface on Carbon Steel and Its Self-Cleaning, Anti-Corrosion, Anti-Frosting and Tribological Properties. *Colloids Surf. A Physicochem. Eng. Asp.* **2019**, *577*, 17–26. [CrossRef]
107. Tong, Z.; Song, L.; Chen, S.; Hu, J.; Hou, Y.; Liu, Q.; Ren, Y.; Zhan, X.; Zhang, Q. Hagfish-Inspired Smart SLIPS Marine Antifouling Coating Based on Supramolecular: Lubrication Modes Responsively Switching and Self-Healing Properties. *Adv. Funct. Mater.* **2022**, *32*, 2201290. [CrossRef]
108. Zhang, D.; Chen, Y.; Ma, Y.; Guo, L.; Sun, J.; Tong, J. Earthworm Epidermal Mucus: Rheological Behavior Reveals Drag-Reducing Characteristics in Soil. *Soil Tillage Res.* **2016**, *158*, 57–66. [CrossRef]
109. Huang, C.; Guo, Z. Fabrications and Applications of Slippery Liquid-Infused Porous Surfaces Inspired from Nature: A Review. *J. Bionic Eng.* **2019**, *16*, 769–793. [CrossRef]
110. Zhao, H.; Sun, Q.; Deng, X.; Cui, J. Earthworm-Inspired Rough Polymer Coatings with Self-Replenishing Lubrication for Adaptive Friction-Reduction and Antifouling Surfaces. *Adv. Mater.* **2018**, *30*, 1802141. [CrossRef] [PubMed]

111. Ruan, H.; Zhang, Y.; Wang, Q.; Wang, C.; Wang, T. A Novel Earthworm-Inspired Smart Lubrication Material with Self-Healing Function. *Tribol. Int.* **2022**, *165*, 107303. [CrossRef]
112. Yang, F.; Zhao, J.; Koshut, W.J.; Watt, J.; Riboh, J.C.; Gall, K.; Wiley, B.J. A Synthetic Hydrogel Composite with the Mechanical Behavior and Durability of Cartilage. *Adv. Funct. Mater.* **2020**, *30*, 2003451. [CrossRef]
113. Sakakibara, K.; Maeda, K.; Yoshikawa, C.; Tsujii, Y. Water Lubricating and Biocompatible Films of Bacterial Cellulose Nanofibers Surface-Modified with Densely Grafted, Concentrated Polymer Brushes. *ACS Appl. Nano Mater.* **2021**, *4*, 1503–1511. [CrossRef]
114. Wei, Z.; He, J.; Liang, T.; Oh, H.; Athas, J.; Tong, Z.; Wang, C.; Nie, Z. Autonomous Self-Healing of Poly(Acrylic Acid) Hydrogels Induced by the Migration of Ferric Ions. *Polym. Chem.* **2013**, *4*, 4601–4605. [CrossRef]
115. Gan, D.; Wang, Z.; Xie, C.; Wang, X.; Xing, W.; Ge, X.; Yuan, H.; Wang, K.; Tan, H.; Lu, X. Mussel-Inspired Tough Hydrogel with In Situ Nanohydroxyapatite Mineralization for Osteochondral Defect Repair. *Adv. Healthc. Mater.* **2019**, *8*, 1901103. [CrossRef] [PubMed]
116. Rong, M.; Liu, H.; Scaraggi, M.; Bai, Y.; Bao, L.; Ma, S.; Ma, Z.; Cai, M.; Dini, D.; Zhou, F. High Lubricity Meets Load Capacity: Cartilage Mimicking Bilayer Structure by Brushing Up Stiff Hydrogels from Subsurface. *Adv. Funct. Mater.* **2020**, *30*, 2004062. [CrossRef]
117. Qu, M.; Liu, H.; Yan, C.; Ma, S.; Cai, M.; Ma, Z.; Zhou, F. Layered Hydrogel with Controllable Surface Dissociation for Durable Lubrication. *Chem. Mater.* **2020**, *32*, 7805–7813. [CrossRef]
118. Zhao, W.; Zhang, Y.; Zhao, X.; Ji, Z.; Ma, Z.; Gao, X.; Ma, S.; Wang, X.; Zhou, F. Bioinspired Design of a Cartilage-like Lubricated Composite with Mechanical Robustness. *ACS Appl. Mater. Interfaces* **2022**, *14*, 9899–9908. [CrossRef] [PubMed]
119. Liu, H.; Zhao, X.; Zhang, Y.; Ma, S.; Ma, Z.; Pei, X.; Cai, M.; Zhou, F. Cartilage Mimics Adaptive Lubrication. *ACS Appl. Mater. Interfaces* **2020**, *12*, 51114–51121. [CrossRef] [PubMed]
120. Zhang, Z.; Qin, C.; Feng, H.; Xiang, Y.; Yu, B.; Pei, X.; Ma, Y.; Zhou, F. Design of Large-Span Stick-Slip Freely Switchable Hydrogels via Dynamic Multiscale Contact Synergy. *Nat. Commun.* **2022**, *13*, 6964. [CrossRef]
121. Liu, H.; Zhao, W.; Zhang, Y.; Zhao, X.; Ma, S.; Scaraggi, M.; Zhou, F. Robust Super-Lubricity for Novel Cartilage Prototype Inspired by Scallion Leaf Architecture. *Adv. Funct. Mater.* **2024**, 2310271. [CrossRef]

biomimetics

Article

Indentation and Detachment in Adhesive Contacts between Soft Elastomer and Rigid Indenter at Simultaneous Motion in Normal and Tangential Direction: Experiments and Simulations

Iakov A. Lyashenko [1,2,*], Valentin L. Popov [1] and Vadym Borysiuk [1,3]

1 Department of System Dynamics and Friction Physics, Institute of Mechanics, Technische Universität Berlin, 10623 Berlin, Germany; v.popov@tu-berlin.de (V.L.P.); v.borysiuk@campus.tu-berlin.de (V.B.)
2 Department of Applied Mathematics and Complex Systems Modeling, Faculty of Electronics and Information Technology, Sumy State University, 40007 Sumy, Ukraine
3 Department of Nanoelectronics and Surface Modification, Faculty of Electronics and Information Technology, Sumy State University, 40007 Sumy, Ukraine
* Correspondence: i.liashenko@tu-berlin.de; Tel.: +49-(0)30-314-75917

Abstract: In reported experiments, a steel indenter was pressed into a soft elastomer layer under varying inclination angles and subsequently was detached under various inclination angles too. The processes of indentation and detachment were recorded with a video camera, and the time dependences of the normal and tangential components of the contact force and the contact area, as well as the average contact pressure and average tangential stresses, were measured as functions of the inclination angle. Based on experimental results, a simple theoretical model of the indentation process is proposed, in which tangential and normal contacts are considered independently. Both experimental and theoretical results show that at small indentation angles (when the direction of motion is close to tangential), a mode with elastomer slippage relative to the indenter is observed, which leads to complex dynamic processes—the rearrangement of the contact boundary and the propagation of elastic waves (similar to Schallamach waves). If the angle is close to the normal angle, there is no slipping in the contact plane during the entire indentation (detachment) phase.

Keywords: adhesion; simultaneous normal and tangential contact; elastomer; friction; adhesive strength; work of adhesion; experiment; simulation

Citation: Lyashenko, I.A.; Popov, V.L.; Borysiuk, V. Indentation and Detachment in Adhesive Contacts between Soft Elastomer and Rigid Indenter at Simultaneous Motion in Normal and Tangential Direction: Experiments and Simulations. *Biomimetics* **2023**, *8*, 477. https://doi.org/10.3390/biomimetics8060477

Academic Editor: Thies Büscher

Received: 18 August 2023
Revised: 14 September 2023
Accepted: 15 September 2023
Published: 7 October 2023

1. Introduction

Mechanical contacts involving adhesion are very common in biological systems. Adhesive interactions occur both at the microscopic level (the surface adhesion of living cells, viruses and bacteria) and macro level (the ability of certain animals to move on vertical surfaces and ceilings) [1–5]. Different animal species may use different mechanisms of adhesion for their motion abilities. For instance, certain species exhibit attachment to the surface by means of capillarity or by direct adhesive contact or a combination of both mechanisms [6].

Among the most popular examples that illustrate adhesive contacts in nature is the attachment and detachment of the feet of gecko lizards to surfaces that can be smooth or rough. The extraordinary ability of geckos to walk on ceilings is ensured by many keratinous hairs, called setae, on their pads, where each seta is 2–10 μm wide and about 100 μm long. At the submicron level, setae consist of numerous protruding structures known as spatula (≈200 nm wide and long) [6,7]. Thus, in general, the attachment of the gecko foot to the surface can be considered as a large number of the much smaller discrete contacts. Complex structures that consist of setae are also found in other animals. For example, abalones have setae with an average diameter of around 1 μm [8], which is similar to the size of a gecko's setae.

The strength of the mechanical contact, which must be strong enough to hold the weight of the animal, and their ability to quickly switch between attachment and detachment are also of big interest. It is known that this feature is ensured by different angles at which the animal places and detaches its foot to and from the surface [7,9].

Directly studying the adhesive contacts involving biological objects is a difficult challenge even in laboratory conditions as it typically involves experiments with animals [9]. As was mentioned above, one of the distinct features of the adhesive contacts between the feet of certain animals and a surface is the hierarchical structure of the pads, which firmly covers the surface roughness, providing a larger contact area. Therefore, one possible way to mimic these contacts in laboratory conditions without using animals is to study the indentation of the elastomer substrate with a rigid indenter. The soft surface of the elastomer can also fill the small asperities at the rigid indenter because a soft substrate can be significantly deformed and can fill gaps between asperities, especially in adhesive contacts.

It is worth noting that the problem of the tangential contact and detachment of the soft surface involving adhesion arises not only in the biological environment; it is also an important topic among many scientific groups working in the fields of tribology, contact mechanics, engineering and even modern robotics [10]. Classical adhesion theories such as JKR, DMT and Maugis theory [11–13] cannot provide exact general solutions even in the easiest case of pure normal contact, and these theories cannot describe significant differences between the indentation and detachment phases, which are detected in real contacts [14,15]. Much more complicated tangential adhesive contacts are often described within various computational approaches. During recent decades, several techniques for studying the tangential contacts of elastomers involving adhesion that focus on numerical simulations and modeling of the contact phenomena were developed. For example, in [16], the authors use finite element analysis (FEA) to study the adhesion strength of the sliding contacts, taking into account the surface roughness. Moreover, FEA was successfully applied to study adhesion in living cells in [17], where the authors also used atomic force microscopy in their experiments. In [18], the authors developed a model for mixed lubrication that includes adhesion, plastic deformation and surface topography by using the finite element method. Besides FEA, a boundary element method (BEM) is another powerful computing tool that is frequently used to study mechanical contacts. Recently, it was applied to study the effect of adhesion and surface roughness on friction hysteresis [19]. Computational studies of friction and adhesion between polymers at the nanoscale are typically performed within the framework of molecular dynamics (MD) simulations where special approaches for tangential contacts were also developed (see, for example, [20,21]). As another illustrative example of the modeling of tangential adhesive contact, we can refer to the model based on traction–separation laws developed in [22], the n-point asperity model proposed in [23], the fracture mechanics model [24] and many others (see, for example, [25,26]).

Besides theoretical studies, tangential contacts are also investigated experimentally. For this purpose, special laboratory facilities were developed. A famous example of experimental techniques for simultaneous measurements of both the normal and tangential forces between soft surfaces is the special version of the classical surface force apparatus [27]. In a more recent study, a rotary shear apparatus was used to study the tangential adhesion strength between clay and steel in [28]. Important examples of laboratory equipment developed to measure tangential adhesion are special tribometers (see, for instance, [29,30]). There are many studies in which the authors use relatively simple self-developed experimental setups. As a rule, that is enough for the investigation of adhesive contacts in biological macroscopic organisms (animals) [8]. But, despite the large number of experimental and theoretical works, adhesive tangential contact is still the object of hot discussions in various scientific groups. This is due to complex processes during contact area restructuring at tangential motion (attachment of new contact areas, peculiarities of detachment, adhesive hysteresis at changing of motion direction, partially sliding, elastic waves propagation, pores formation during motion, etc.) and importance for practical applications.

Here, we present a series of experiments and mathematical simulations concerning both indentation of the rigid indenter into soft elastomer and also withdrawal of the indenter from elastomer simultaneously in normal and tangential directions. Also, we propose a simple model for normal and tangential contact within the method of dimensionality reduction (MDR) for adhesive contact [31]. We believe that the performed study can bring new insights into understanding of formation and detachment of the biomechanical contacts.

2. Experimental Set-Up

To perform described below experiments a special type of laboratory equipment, that is shown in Figure 1 was designed and assembled.

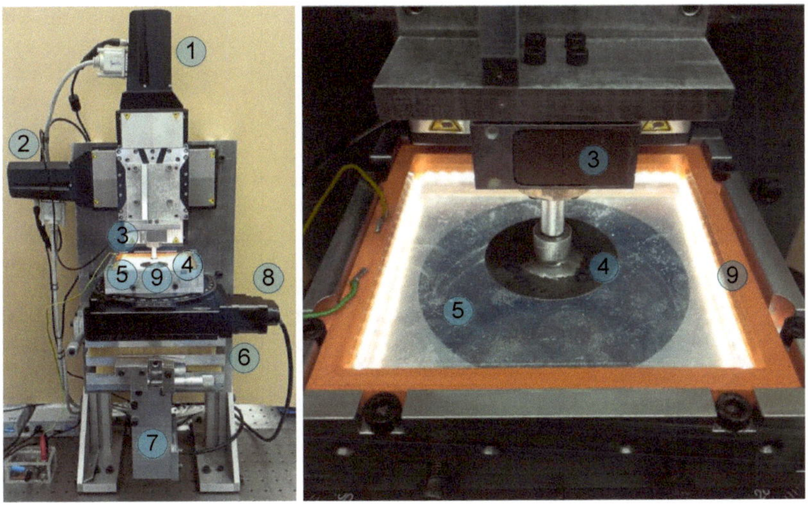

Figure 1. (**left panel**) Photo of the whole experimental facility; (**right panel**) enlarged photo of the contact area between a spherical indenter 4 and elastomer 5 with all-sides LED lighting 9.

The left panel of the figure shows a general view of the designed facility, whereas the right panel demonstrates only the area of the contact. Both photos of the facility depict the main parts of the experimental device. The spherical indenter made of steel (denoted as (4) in the figure), mounted onto three-axial force sensor ME K3D40 (denoted as (3)). During the experiment, the force sensor measures all three components of normal force. Electric signal from the sensor (3) is amplified by 4-channel amplifier GSV-1A4 SubD37/2 with three out of four active channels for each component of the force. Outputs of the amplifier paired with desktop PC through the 16-bit analogue to digital converter NI USB-6211. Indenter is capable of moving in both normal and tangential directions as it is driven by two high-precise motors PI L-511.24AD00 (1) and (2), respectively, that are governed by USB controllers PI C-863. For more precise positioning, the mentioned motors are equipped with a feedback mechanism that automatically measures and corrects (if needed) the coordinate after movement. With the presence of the feedback, the accuracy of positioning may reach values up to ± 0.2 µm. Another important option of the drives that are used in the facility is a lack of hysteresis of coordinate after changing the direction of motion. Such a type of hysteresis almost always exists in devices without feedback due to the inevitable backlash in the mechanical converters of rotary motion into translational motion, even in devices with more accurate and expensive ball-screw gears.

Sheet of transparent rubber TARNAC CRG N3005 with thickness $h = 5$ mm was used as an elastomer substrate (5), and almost perfect transparence of this rubber ensures the possibility of direct observation of the contact area. This type of rubber is characterized by elastic modulus $E \approx 0.324$ MPa and Poisson's ratio $\nu \approx 0.48$, which were estimated in [32] by generalizing a large amount of experimental data. Elastic modulus of the indenter material (steel) is 5 orders of magnitude higher and equals $E \approx 2 \times 10^5$ MPa, therefore in experiments the indenter can be considered as absolutely rigid and deformations only occur inside the rubber substrate.

Observation of the contact area in experiments is performed from the bottom part of the device through the rubber sheet by digital camera Ximea 2.2MP MQ022CG-CM with FUJINON HF16SA-1, 2/3″ lens. In Figure 1 the position of the camera is denoted by (7), whereas the camera itself is closed from the observer by an aluminum plate. Tilt mechanism (6) can be used to manually change the position of the elastomer in the horizontal plane, which is critically important for experiments with tangential motion. Motorized rotation stage 8MR190-90-4247-MEn1 (8), governed by 8SMC5-USB-B8-1 USB-controller, is typically used for the rotation of elastomer; however, in the described experiments, it is idle. The facility shown in Figure 1 was described in detail in our recent work [33], and the provided therein "Supplementary Video" shows the performance of the device in real time.

Here, we describe a series of experiments performed according to scenarios (A) and (B), which are shown in Figure 2. In both cases, in the first phase of experiment during the movement of indenter with radius R in normal direction, the point of the first contact between indenter and rubber was detected, and a related normal coordinate was considered as zero indentation depth $d_0 = 0$ mm. Then, in scenario (A), indenter was immersed into the depth $d_{\max} = 0.3$ mm at an angle α to the elastomer surface (see Figure 2). Movement of the indenter at a certain angle was performed by its simultaneous displacement, driven by the motors (1) and (2) (see Figure 1) with different velocities v_z and v_x. Velocities v_z and v_x were chosen in such a way that the absolute value of the resulting velocity in each case was the same $v = \sqrt{v_z^2 + v_x^2} = 1$ µm/s (see Figure 2). Thus, at a given indentation angle α absolute values of velocities can be determined as $v_x = v \cos\alpha$ and $v_z = v \sin\alpha$. Note that at such a small velocity of indenter motion, the contact can be considered as a quasi-static (viscoelastic effects do not appear or are very small) as it was experimentally shown in our previous work [34]. It means that the obtained experimental data can be described by classical theories of adhesion of elastic bodies as JKR-type theories.

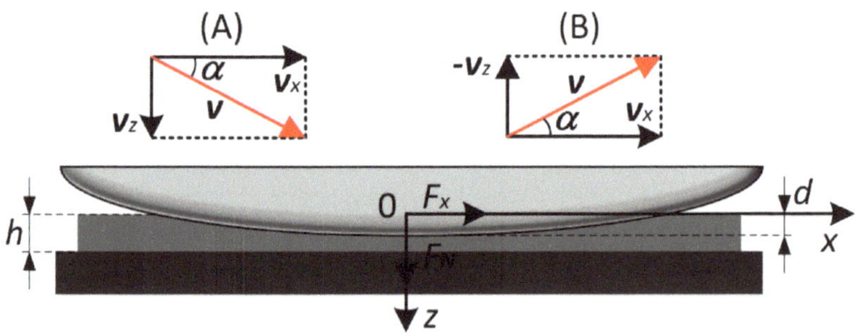

Figure 2. Schematics of two experiments: (**A**) immersion of the indenter into elastomer, and its withdrawal (**B**). In both experiments indenter moves with velocity v that has normal v_z and tangential v_x components. Figure shows the configuration with thickness of an elastomer h, indentation depth d, normal and tangential components of contact force F_N and F_x, and indentation angle α.

In scenario (B), the indenter was initially immersed into the elastomer to the depth $d_{\max} = 0.3$ mm at motion in a normal direction, and then it was withdrawn from the substrate at an angle α to its surface up to full detachment of the contact. Vectors of velocities for both scenarios (A) and (B) are shown in the Figure 2. Several series of experiments were performed with two different indenters of radius $R = 30$ mm and $R = 100$ mm and at angles $\alpha = 10°, 20°, 30°, 40°, 50°, 60°, 70°, 80°, 90°$ (angle $\alpha = 90°$ related to normal contact). Separately, experiments with pure tangential movement of an indenter were performed (at $\alpha = 0°$), where prior to tangential shift indenter was immersed into elastomer to the depth d_{max} = 0.3 mm. In this case, the conditions of the experiments differ from both scenarios (A) and (B).

3. Experimental Results

Movement of the indenter at an angle α to the elastomer surface includes both normal and tangential shift. Therefore, prior to analyses of the data obtained from indentation at an angle, it is necessary to have an insight of what is happening in simpler cases of pure normal and tangential contacts. In several of our previous works, various aspects concerning normal [32–36] and tangential [33,37,38] contact were considered. However, due to the complexity of the considered phenomena, here we will not only cite our previous works, but additionally will conduct experiments on both normal indentation and tangential shear (the descriptions of the experiments are given in the next two subsections). It is worth noting that these experiments were performed with the same elastomer and indenter, as the experiments with the indentation at an angle α. This will help to obtain additional information about the studying system.

3.1. Normal Contact

Figure 3 shows results of experiments where indenter was moved at an angle α = 90° to elastomer plane (see schematics in Figure 2). As it was mentioned above, this situation relates to the indentation in pure normal direction. The figure shows five full indentation/detachment cycles. Overlapping of the curves, corresponding to different cycles, indicates good reproducibility of the experiments.

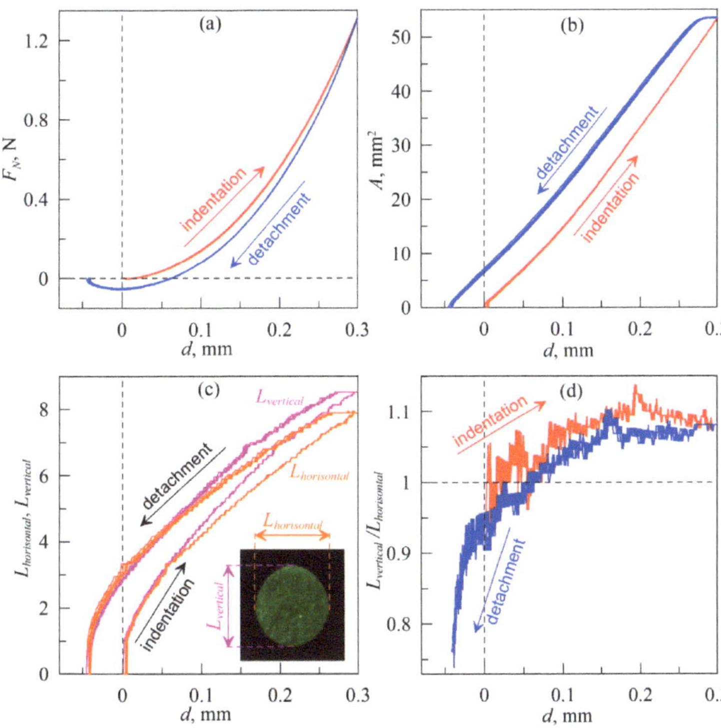

Figure 3. Dependencies of normal force F_N (**a**), contact area A (**b**), size of the contact in vertical $L_{vertical}$ and horizontal $L_{horisontal}$ directions (**c**) and ratio $L_{vertical}/L_{horisontal}$ (**d**) on indentation depth d. Radius of an indenter R = 30 mm, elastomer (CRG N3005) thickness h = 5 mm. Supplementary Video S1 is also available.

As it was observed in various experiments on different scales [32–40], in the indentation phase, adhesion is neglectable and contact can be considered as non-adhesive. However, in the detachment phase, adhesion significantly affects the properties of the contact, which results in an additional force ($F_N < 0$ N at $d < 0$ mm) that must be applied for complete detachment of the indenter. Mentioned peculiarities lead to adhesive hysteresis of the second art in the dependencies of the force $F_N(d)$ and the contact area $A(d)$, which is clearly visible in the Figure 3a,b. Reasons which lead to hysteresis are not completely established yet, and in the literature it is explained by the influence of humidity, viscoelasticity or roughness of the contacting surfaces [39]. The presence of the hysteresis causes additional dissipation of the mechanical energy in oscillating adhesive contacts [35]. In a situation where contacting surfaces have a non-uniform distribution of surface energy (which determines the specific work of adhesion), such hysteresis arises naturally within the framework of classical adhesion theories, like JKR [35]. If two hard contacting surfaces have surface energy densities γ_1 and γ_2, and γ_{12} is an interfacial energy density for interface between these surfaces in the contact, the specific work of adhesion can be expressed as [41]

$$\Delta\gamma = \gamma_1 + \gamma_2 - \gamma_{12}. \quad (1)$$

In the case if $\Delta\gamma > 0$, the contacted surfaces adhere to each other. A simple phenomenological approach to the description of adhesive hysteresis of the second art can be to introduce two effective values of the specific work of adhesion at the stages of indentation $\Delta\gamma_{eff,0}$ and detachment $\Delta\gamma_{eff,1}$, where $\Delta\gamma_{eff,1} > \Delta\gamma_{eff,0}$. Such a simple assumption makes it possible to simulate the normal adhesive contact with a sufficient accuracy of reproducibility of experimental results [34,36,40]; however, it does not explain the cause of the hysteresis.

It is worth noting that in a real experiment the shape of the indenter is not absolutely spherical and the rubber surface is also not ideally smooth. Therefore, the contact area will deviate from the expected round shape. In this case, one parameter (contact radius) used for description of the contact geometry becomes insufficient. In this regard, in all experiments described below, additional dependencies of the "width" in horizontal $L_{horisontal}$ and vertical $L_{vertical}$ directions are introduced (see Figure 3c). Here and below mutually perpendicular "horizontal" $L_{horisontal}$ and "vertical" $L_{vertical}$ directions are chosen from the location of the camera in the device shown in Figure 1. For experiments with indentation in the normal direction, these are arbitrary directions that have no geometric meaning. However, if there is tangential movement, the direction $L_{horisontal}$ coincides with the direction of the tangential shift of the indenter. At the same time, $L_{vertical}$ will automatically coincide with the direction perpendicular to the direction of the tangential shift. In the case of a circular contact patch, its "width" is the same in both directions and reduced to the diameter of the contact.

3.2. Tangential Contact

Figure 4 shows results of an experiment in which the indenter was first immersed in the normal direction into the rubber to a depth d_{max} = 0.3 mm, and then shifted in a tangential direction at a fixed depth d_{max}. After shifting to a distance Δx = 3 mm, the indenter was withdrawn from the elastomer until the moment of complete detachment of the contact. The figure shows time dependencies of the normal force F_N, tangential force F_x, contact area A, average contact pressure $\langle p \rangle = F_N/A$, averaged tangential stresses $\langle \tau \rangle = F_x/A$, and ratio $L_{vertical}/L_{horisontal}$. Two vertical dashed lines 1 and 2 are present in all panels of the figure, where up to line 1, indentation in the normal direction is performed (in this case, the tangential force F_x and corresponding tangential stresses are equal to zero) Between lines 1 and 2, the indenter is shifted in tangential direction at fixed indentation depth d_{max}, and after line 2, the indenter is pulled off the elastomer until the detachment.

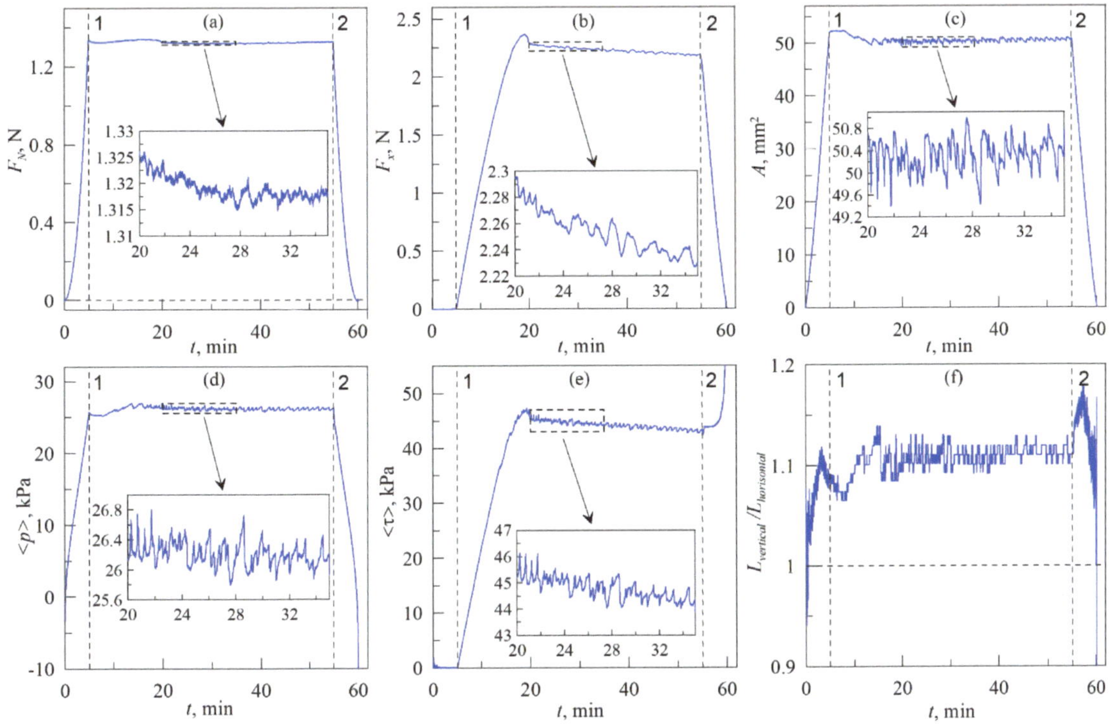

Figure 4. Time dependencies of the normal F_N (**a**) and tangential F_x (**b**) forces, contact area A (**c**), average contact pressure $<p> = F_N/A$ (**d**), averaged tangential stresses $<\tau> = F_x/A$ (**e**) and ratio $L_{vertical}/L_{horisontal}$ (**f**). Radius of the indenter R = 30 mm, elastomer thickness (CRG N3005) h = 5 mm, indentation depth during tangential shift $d_{\max}$ = 0.3 mm, velocity of the indenter motion v = 1 µm/s. Supplementary Video S2 is also available.

During the phase of experiment until line 1, the dependencies in Figure 4 relate to normal indentation, therefore they have the same features as the data in Figure 3. Curves plotted in Figure 3 relate to the dependencies on the indentation depth d, whereas Figure 4 shows time dependencies of the main measured quantities in the experiment. Nevertheless, indentation is performed with constant velocity, therefore both figures have the same peculiarities, such as a linear increase in the size of contact area A in Figures 3 and 4 with the growth of indentation depth d and time t.

A tangential shift of the indenter begins over the line 1 in panels of Figure 4, and after that, the tangential force F_x first increases to a certain maximum value, then decreases. After this decrease in the F_x, the stationary sliding mode begins. In all panels of the figure, except panel (f), immediately after the establishment of the stationary sliding mode, enlarged parts of the curves (that are bounded by dashed rectangles in the main figures) are shown for better clearance. In the enlarged regions, it is easier to trace the relationships between the quantities that are measured in the experiment. For example, in the insets to the panels of the figure, it is clearly visible that all measured quantities change nonmonotonically. This is due to the fact that during the sliding, complex processes of contact propagation occur in the system. These processes are associated with the abrupt appearance of new contact areas between the rubber and the indenter at the leading edge of the contact, with more monotonous-like separation of the rubber at the trailing edge, and also with the propagation of elastic waves in the contact area. These waves are known as "Schallamach waves" [42–46] and it is one of the least understood effects in adhesive contacts. Each subsequent propagation of an elastic wave leads to a decrease in the friction force and tangential stresses, which increase again with further shear. In our previous works [33,37,38], these processes are described in detail, especially in the recent work [38], where we discuss the inhomogeneous nature of slip phase during tangential shear of the indenter. Moreover, in Supplementary Video S2, features of contact propagation during tangential motion are clearly visible.

One particular feature, which can be observed in Figure 4, needs additional explanation. As it follows from the figure, in the mode of stationary sliding, normal force F_N and contact area A fluctuate around certain average magnitudes, without any radical changes (either decrease or growth) in time. This means that during tangential movement, the indentation depth d = 0.3 mm remains constant, i.e., there are no explicit irregularities on the surface of this particular rubber substrate in the direction of indenter movement. However, despite constant values of the force F_N and area A, tangential force F_x decreases in time (see Figure 4b), together with tangential stresses $<\tau> = F_x/A$. In the considered case of adhesive contact in the established stationary sliding mode, the friction force, regardless of the indentation depth d_{max}, is determined by expression (without considering friction at the contact boundary [37]):

$$F_x \approx \tau_0 A, \tag{2}$$

where stationary tangential stresses τ_0 depend on adhesive forces between contacting surfaces. In this case, the coefficient of friction μ loses its physical meaning as it becomes dependent on indentation depth d_{max} [38]. According to (2), friction force F_x is proportional to the contact area A, which has almost constant value $<A> \approx 50$ mm^2 in stationary mode (see Figure 4c). Therefore, the observed decrease in the friction force is possible only due to the reduction of the stationary stresses τ_0, which can be observed in the Figure 4e. The decreasing of stresses occurs due to the degradation of the adhesive properties of the contact, caused by contamination and oxidation of the indenter surface during the tangential shift. In this case, the specific work of adhesion decreases, which leads to a decrease in the adhesive strength in normal contact (by adhesive strength here we assume the normal force that must be applied to completely detach the indenter from the rubber) [36]. In the case of tangential contact, as it can be seen, the degradation of adhesive properties leads to a decrease in the stationary value of tangential stresses τ_0, and, as a result, to a decreasing of friction force F_x.

Decrease in the adhesive strength of the contact makes comparative analysis of experimental data more difficult, since each individual experiment will have its own value of the specific work of adhesion, on which all other characteristics of the contact such as its radius, contact forces, etc., depend. In further analysis of the experimental data concerning indentation at different angles α, we will neglect the changes in the adhesive strength of the contact. In the case considered below, such constraint is relevant, since a change in the angle α at which the indentation is performed has a much stronger effect on the contact properties than the observed change in the specific work of adhesion due to degradation of the contact properties.

3.3. Immersion of the Indenter at an Angle to the Surface, Scenario (A)

Figure 5 shows results of experiments on indentation of the metallic spherical indenter with radius R = 30 mm into elastomer sheet TARNAC CRG N3005 with thickness h = 5 mm, that was performed according to scenario A (see Figure 2 and related description). The figure shows time dependencies of the normal F_N (panel a) and tangential F_x (panel b) components of the contact force, contact area A (panel c), average normal pressures $<p> = F_N/A$ (panel d), average tangential stresses $<\tau> = F_x/A$ (panel e) and ratios $L_{vertical}/L_{horisontal}$ (panel f). The group of curves, corresponding to experiments on indentation at angles α from 10° to 90° with 10° increments, are plotted. Dependencies corresponding to the same experiment (the same angle α), plotted in different panels of the figure in the same color. Moreover, each panel contains nine curves, except for dependencies of the tangential force $F_x(t)$ and tangential stresses $<\tau>(t)$ (panels (b) and (e)), where the results of the experiments with α = 90° are not shown, because under pure normal load the tangential force and related stresses τ do not occur.

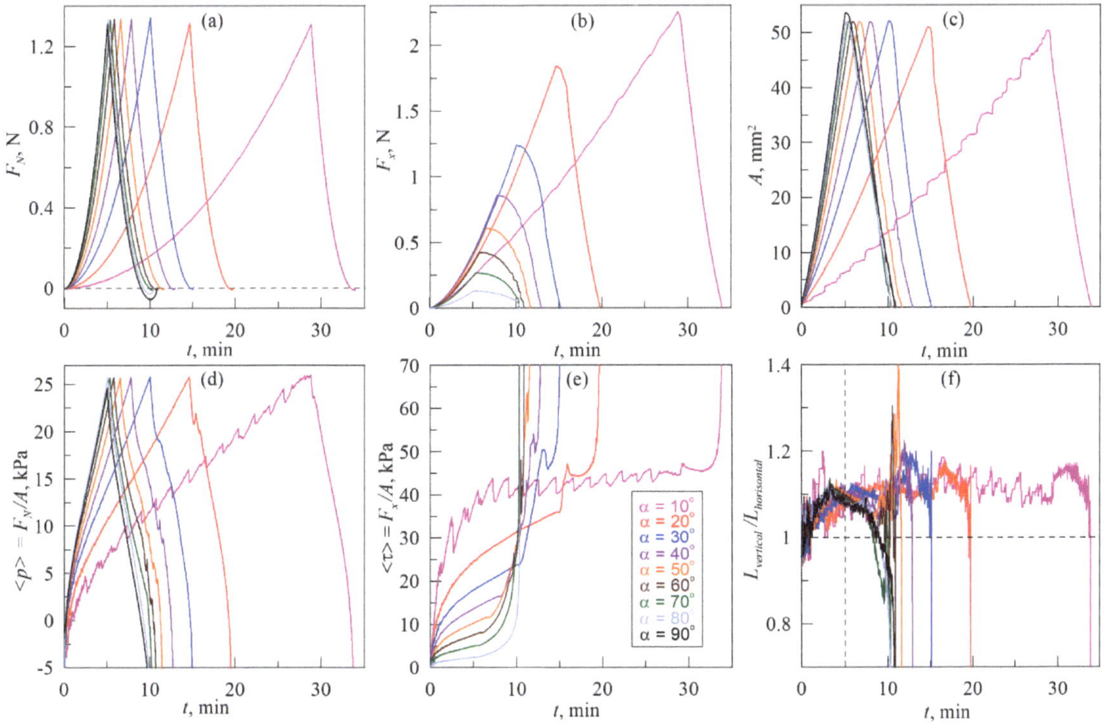

Figure 5. Time dependencies of the normal F_N (**a**) and tangential F_x (**b**) forces, contact area A (**c**), average contact pressure <p> (**d**), tangential stresses <τ> (**e**) and ratio $L_{vertical}/L_{horisontal}$ (**f**). Radius of the indenter R = 30 mm, elastomer thickness (CRG N3005) h = 5 mm, maximal indentation depth $d_{\max}$ = 0.3 mm, in the experiment according to scenario (A) (see Figure 2). Dependencies corresponding to the experiment with the same angle α (from 10° to 90°), plotted in different panels of the figure in the same color. Supplementary Video S3 is available.

In all experiments, presented in Figure 5, after the appearance of the first contact point between the indenter and the rubber, the indenter was pressed into the elastomer at an angle α to its surface until reaching the indentation depth $d_{\max}$ = 0.3 mm. As soon as the depth $d_{\max}$ was reached, the indenter was withdrawn from the elastomer in a normal direction until the complete detachment of the contact. As it follows from the figure, during the indentation phase, magnitudes of normal F_N and tangential F_x forces as well as contact area A and contact pressure <p> are increasing. As indentation always continues up to the depth $d = d_{\max}$ = 0.3 mm, F_N, A and <p> grow to approximately the same maximum values regardless the magnitude of an angle α. This happens because the quantities F_N, A and <p> characterize a normal contact, which is mainly determined by the indentation depth $d_{\max}$. The main difference between curves $F_N(t)$, $A(t)$ and <p>(t), corresponding to different angles α, consists of different rates of growth, since with an increasing of the angle α indenter reaches the maximal depth $d_{\max}$ faster.

After reaching the maximum values, F_N, A and <p> begin to decrease, as the phase of indentation at an angle is replaced by the phase of the detachment, where indenter moves in the pure normal direction (detachment in all cases occurs at α = 90°).

Let us note some peculiarities of contact propagation during indentation at different angles. As it was observed in previous works, stationary sliding mode is established in the studied system during pure tangential motion. In stationary mode, regardless of the value of the indentation depth d, tangential stresses $\tau_0 = F_x/A$ occur over all contact area of magnitude from 30 to 50 kPa in different experiments (see Figure 4e, and also experimental data reported in [37,38]). Magnitudes of τ_0 vary in different experiments as τ_0 strongly depends on the current state of the contacting surfaces, namely, on the presence of chemical and physical contaminations, roughness, etc. As it follows from the Figure 5, only in the case of indentation at a minimum angle $\alpha = 10°$the indenter is shifted in the tangential direction on the distance enough for tangential stresses τ to reach the critical value $\tau_0 \approx 40$ kPa, above which tangential slip begins. After the stresses reach the value τ_0, it does not increase with further tangential shift, although the friction force F_x and contact area A continue to increase. These parameters grow due to the fact that when indenting at an angle α to the surface, shift in tangential and normal directions is performed simultaneously, and, due to the increasing of the indentation depth d, the contact area A growth together with friction force $F_x \approx \tau_0 A$ (2).

In the stationary sliding mode, contact propagation has certain features that can be traced on the dependencies $A(t)$, $\langle p\rangle(t)$ and $\langle\tau\rangle(t)$ in Figure 5. To explain these features, let us return to the case of solely tangential shear, which is shown in Figure 4. Here, during the contact propagation, smooth detachment of the rubber from the indenter occurs on the back front, so the boundary of a contact on the back front practically does not change its shape. At the leading age of the contact, new areas of rubber are involved in the contact with the indenter's surface in an abrupt manner (see Supplementary Video S2), therefore contact area $A(t)$ also increases abruptly. Since the new parts of the rubber that have just come into contact are not loaded in the tangential direction, they do not contribute to the friction force F_x immediately after contact. Therefore, such an increase in the contact area A without corresponding growth of the friction force F_x leads to a sharp decrease in the average tangential stresses $\langle\tau\rangle = F_x/A$. With further shifting, the tangential stresses $\langle\tau\rangle$ increase again due to the loading of the "new" areas of the rubber, at the same time the contact area A monotonically decreases. The described contact propagation process is repeated cyclically, which can be seen in the dependencies plotted in Figure 4 and in Supplementary Video S2 (for more details, please see related description in [38]).

In experiments on indentation at an angle α to the surface of the elastomer, a mode similar to the tangential shift is established in the system at small angles α. In this mode, there is also a jump-like mechanism of contact propagation, so dependence $A(t)$ in Figure 5c at $\alpha = 10°$ exhibits regions with sharp growth of the area A. However, in contrast to a pure tangential shift, new contact areas appear in an abrupt manner not only at the leading edge of the contact, but along the entire contact boundary, which is clearly visible in Supplementary Video S3. The reason for such behavior is the fact that along with the tangential movement, there is a constant increase in the indentation depth d. As in the case of a pure tangential shift, immediately after the appearance of "fresh" areas of contact, they appear to be unloaded in the tangential direction, which is verified by the absence of abrupt-like changes in the tangential force F_x after the attachment of new areas of rubber (see Figure 5b at $\alpha = 10°$). As it follows from Figure 5, abrupt growth of the contact area A does not lead to sharp changes in the normal force F_N also. Therefore, after the following act of contact propagation, both calculated values of the averaged tangential stresses $\langle\tau\rangle = F_x/A$ and contact pressure $\langle p\rangle = F_N/A$ decrease. For an adequate description of the evolution of these parameters, it is necessary to take into account the spatial distribution of stresses, as it was discussed in our previous work [38].

Above we described the case of indentation at an angle $\alpha = 10°$, which is close to tangential sliding. In this case, the tangential stresses reach a critical value $\tau_0 \approx 40$ kPa during the tangential shift, after which the elastomer slips over the surface of the indenter. In all other experiments with $\alpha > 10°$, as it follows from the Figure 5e, tangential stresses $\langle\tau\rangle$ did not reach a critical value throughout the entire indentation up to the critical indentation depth $d = d_{max}$. Therefore, in these experiments, global slippage as well as abrupt-like expansion of the contact area were not observed, during indentation. As it follows from the data plotted in Figure 5 and from the Supplementary Video S3, at $\alpha > 10°$ all main parameters, such as both components of contact forces F_x and F_N, contact area A, as well as contact pressure <p> and tangential stresses <τ> increased smoothly and monotonously during the whole indentation phase.

3.4. Pull-Off of the Indenter at an Angle to the Surface, Scenario (B)

In the experiments, the results of which are shown in Figure 6, the indenter was first immersed into the elastomer to a depth d_{max} = 0.3 mm (moved in the normal direction) and then was withdrawn from the elastomer at an angle α to its surface until the complete detachment.

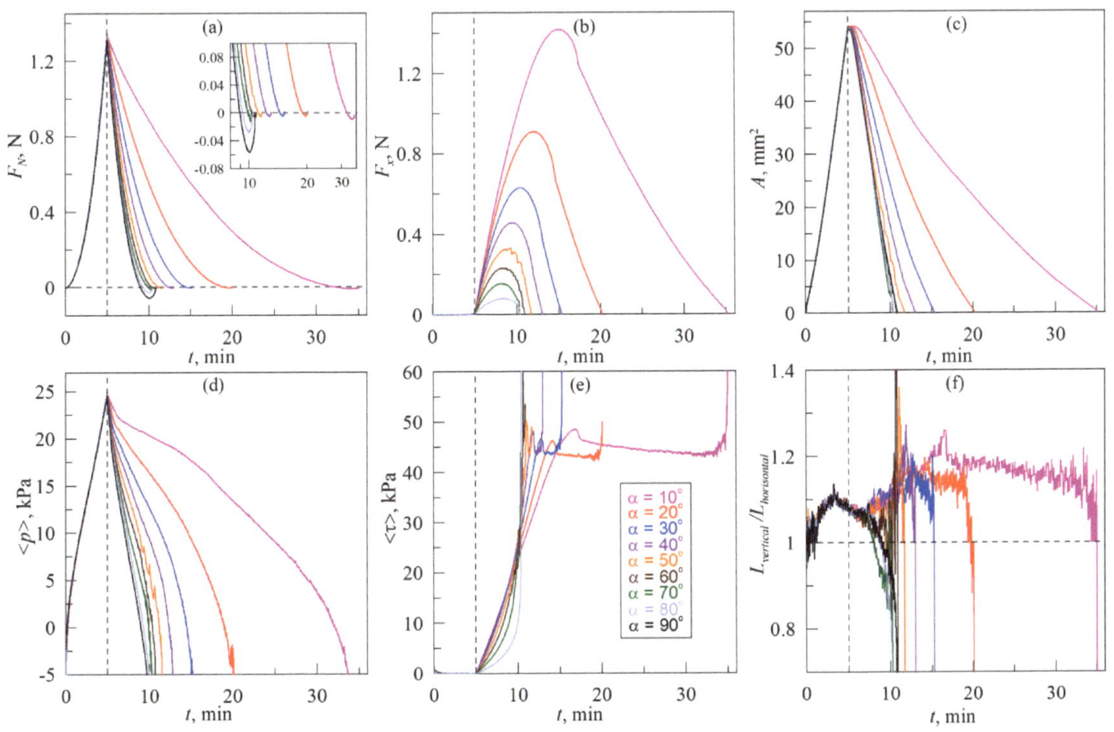

Figure 6. Time dependencies of the normal F_N (**a**) and tangential F_x (**b**) forces, contact area A (**c**), average contact pressure <p> (**d**), tangential stresses <τ> (**e**) and ratio $L_{vertical}/L_{horisontal}$ (**f**). Radius of the indenter R = 30 mm, elastomer thickness (CRG N3005) h = 5 mm, maximal indentation depth d_{max} = 0.3 mm, in the experiment according to scenario (B) (see Figure 2). Dependencies corresponding to the experiment with the same angle α (from 10° to 90°), plotted in different panels of the figure in the same color. Supplementary Video S4 is also available.

Here, the case with $\alpha = 90°$, as in previous Figure 5, relates to only normal indentation. Therefore, analogously to Figure 5, in the panels (b) and (e) of Figure 6 there are no dependencies related to the case with $\alpha = 90°$.

Vertical dashed lines in all panels of Figure 6 show the point of time when indentation in the normal direction ends and the indenter is pulled off of the elastomer when it moves at an angle α to its surface until the complete detachment of the contact. It is worth noting that all dependencies $F_x(t)$, plotted in Figure 6b, have maximums. Two main factors affect the friction force F_x during the pull-off of the indenter. First, friction force F_x growth due to the increase in the tangential stresses τ, as the tangential movement is performed. The second factor that causes the reduction of the tangential force is the decreasing of the contact area A due to the detachment of the indenter in normal direction. Moreover, stresses τ, and contact area A equally affect the friction force, as it is defined by the expression $F_x \approx <\tau>A$ (2). Based on the fact that in all cases the force F_x first increases and then decreases, we can conclude that at the beginning stresses τ grow faster than area A decreases. However, with further movement of the indenter, at certain moment the stresses reach a critical value $\tau_0 \approx 45$ kPa, after which slip occurs in the system, and the shear stresses no longer increase (see Figure 6e). However, in the sliding mode, the contact area A continues to decrease due to the pull-off of the indenter in the normal direction, which leads to the reduction of tangential force F_x.

The behavior of the system in the case shown in Figure 6, has a significant difference comparing to the series of experiments described above and presented in Figure 5. In experiment, shown in Figure 6, movement in the tangential direction begins only after the indenter reaches its maximum indentation depth. The contact area in this case is maximal. Therefore, the subsequent tangential movement loads the entire contact at once, while simultaneously moving the indenter in the normal direction. In this case, there is no contact propagation when attaching "new" regions that are non-loaded in the tangential direction, which were observed during indentation according to the scenario (A) (see Figure 5). Dependencies shown in Figure 6, as well as in Supplementary Video S4, show a smooth decrease in the contact area regardless the angle α, at which the indenter is pulled off. Therefore, in this case, we can assume that during the entire process of withdrawing of the indenter, new contact regions between the rubber and the indenter are not formed, and only the destruction of the existing contact at its boundary occurs. Such an assumption significantly simplifies the understanding of the processes occurring in the system. If there is only destruction of the contact without its propagation and corresponding rearrangement of the contact boundary, a simpler numerical model can be used for the description, such as the method of dimensionality reduction (MDR) or method of the boundary elements (BEM). Note that these methods give the exact solution of the contact problem for normal contact with adhesive interaction of a JKR type [31,32].

Let us note one important feature. The assumption that during the withdrawing of the indenter there is no contact propagation, but only its destruction, will fail in two cases. The first case is when indenter is pulled off at a very small angle α, as at small α, contact approaches a completely tangential motion mode, in which the rubber slips and contact inevitably propagates during movement (see Figure 4). The second case is weak adhesive interaction between contacting bodies. Such contacts are characterized by small magnitudes of critical stresses τ_0, at which slips that are always associated with the spread of the contact in the tangential direction, occur. Therefore, at weak adhesion, when moving at an angle, after a relatively small shift of the indenter, the sliding mode will occur, and the accurate description of this mode requires a serious modification of existing models (see, for example, [38]).

4. Simulation of the Indentation/Detachment Process

4.1. Formalism of the Model

In this section, we describe the simulation of the indentation at an angle in an adhesive contact that is developed from experimental data provided in the previous sections of the article. Simulation setup is based on the method of dimensionality reduction (MDR [31]), the schematics of which for normal adhesive contact are presented in Figure 7.

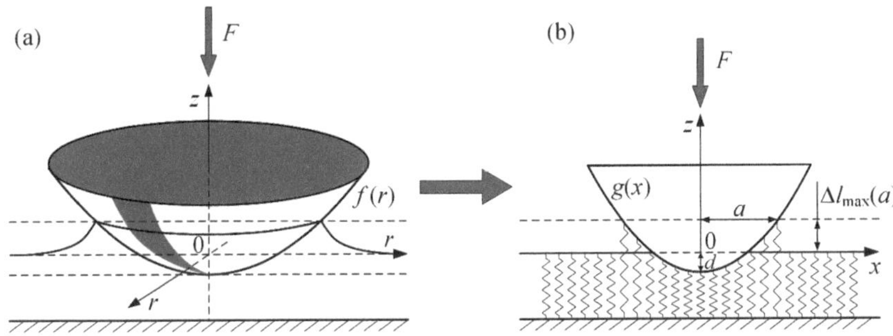

Figure 7. Schematics of the method of dimensionality reduction (MDR) for normal adhesive contact—initial three-dimensional profile (**a**) and corresponding one-dimensional configuration (**b**).

Within MDR for an axially symmetrical indenter with three-dimensional profile $f(r)$ an equivalent one-dimensional profile $g(x)$ must be found according to Abel transform:

$$g(x) = |x| \int_0^{|x|} \frac{f'(r)}{\sqrt{x^2 - r^2}} \mathrm{d}r. \tag{3}$$

In our experiment, we use a spherical indenter, which, at small indentation depths d, can be replaced by a paraboloid $f(r) = r^2/(2R)$ with sufficient accuracy, thus, according to the procedure (3) we obtain:

$$g(x) = \frac{x^2}{R}. \tag{4}$$

Then, the elastic half-space is replaced by an array of non-interacting springs (see Figure 7b), each of them has both normal Δk_z and tangential Δk_x stiffness:

$$\Delta k_z = E^* \Delta x, \ \Delta k_x = G^* \Delta x, \tag{5}$$

where Δx determines the discretization of the space (the numerical solution does not depend on Δx). In the considered case of contact of a rigid indenter with a soft elastic material, effective elastic E^* and shear G^* modules are defined by the expressions

$$E^* = \frac{E}{1 - \nu^2}, \ G^* = \frac{2E^*(1 - \nu)}{2 - \nu}. \tag{6}$$

where E and ν—elastic modulus and Poisson ratio of the material being indented. When profile $g(x)$ is immersed into an array of non-interacting springs to a depth d, compression of an individual spring with coordinate x in the normal direction is

$$u_z(x) = d - g(x) = d - \frac{x^2}{R}. \tag{7}$$

If there is an adhesion in the system, the outer springs are "pulled" to the indenter. As it can be seen from the Figure 7b, these "adhesive" springs reduce the value of the normal force and increase the contact radius a, which is calculated according to the Hess rule [31]:

$$\Delta l_{\max} = -\sqrt{\frac{2\pi a \Delta\gamma}{E^*}}, \tag{8}$$

where $\Delta l_{\max}$ is a magnitude of the enlargement of border springs, and $\Delta\gamma$ is an adhesion specific work. As it follows from (7) and (8) with account of $\Delta l_{\max} \equiv u(a)$

$$d(a) = \frac{a^2}{R} - \sqrt{\frac{2\pi a \Delta\gamma}{E^*}}. \tag{9}$$

Normal contact force F_N is defined as a sum of the forces from individual springs in contact (both stretched and compressed):

$$F_N = E^* \int_{-a}^{a} u_z(x)\mathrm{d}x = 2E^* \int_{0}^{a} \left(d - \frac{x^2}{R}\right)\mathrm{d}x = \frac{4E^*a^3}{3R} - \sqrt{8\pi a^3 E^* \Delta\gamma}. \tag{10}$$

The results (9), (10) are exactly the same as the relations from JKR theory [11], concerning the normal contact force F_N, indentation depth d and contact radius a in the case of adhesive contact. A distinct feature of the experiment, schematically shown in Figure 2, is the indentation at an angle α to the elastomer surface where a tangential shift of the indenter is also present together with normal motion. Therefore, after the contact, the springs shown in the Figure 7b undergo a tangential shift. The experiment shows that complex processes of contact boundary restructuring, propagation of elastic waves, etc., occur during the tangential shift. Moreover, the contact becomes asymmetrical (see Videos S2–S4). As we have already noted above, various descriptions of these processes exist in the literature; however, there is still no complete understanding of what happens in adhesive contact during tangential shift. Here, we propose a simple model that neglects the effects of elastic wave propagation and the jump-like reconstruction of the contact boundary, but at the same time allows us to describe the evolution of such quantities as both components of the contact force and the contact area during the process of indentation at an angle precisely enough.

As it follows from numerous experiments [37,38], during tangential shear in the stationary sliding mode, shear stresses preserve their values at almost constant level $\tau_0 = F_x/A$ and do not depend on indentation depth d. The model will be developed from this assumption. The following assumptions should be considered not as a rigorous MDR model but as an estimation of the sliding conditions. Let us consider the stationary sliding mode, with corresponding experimental data shown in Figure 4. Estimation for the average tangential stress can be made using the parameters of the MDR model as

$$\tau = \frac{F_x}{A} = \frac{\Delta k_x \sum\limits_{cont} u_x(x_i)}{\pi a^2}, \tag{11}$$

where tangential shift of all springs $u_x(x_i)$ in contact is accumulated in sum. Here, a is a radius of a contact, which within a discrete model is defined as $a = (n/2)\Delta x$, where n is the total number of springs in contact, and Δx is the space discretization step defined above, which can also be interpreted as a width of the spring or the distance between them. In the case of stationary sliding, when all springs in contact have the same shift $u_x(x) \equiv u_x^{crit}$, with respect to (5) and (6), we have

$$\tau_0 = \frac{G^* \Delta x \cdot n \cdot u_x^{crit}}{\pi n^2 \Delta x^2/4} = \frac{2G^* u_x^{crit}}{\pi a}. \tag{12}$$

Expression (12) sets the maximum tension of the springs during tangential shear in the form

$$u_x^{crit}(a) = \frac{\pi a \tau_0}{2G^*}. \tag{13}$$

When exceeding this value, springs begin to slide. At this point, we would like to stress again that this is a condition of an "adhesive-like" sliding criterion, which is not valid for the considered system, but will provide a correct estimation of the critical tangential stress. It is worth noting that u_x^{crit} depends on the contact radius a. Hence, with an increase in the indentation depth, the maximum tangential tension of the springs also increases.

Let us consider the simulation procedure using an example according to scenario (A) that is shown in Figure 2. At a given angle α the indenter moves in the normal and tangential directions with velocities $v_z = v\sin\alpha$ and $v_x = v\cos\alpha$, which specify its normal and tangential displacements as functions of time t:

$$d = vt\sin\alpha, \; x = vt\cos\alpha. \tag{14}$$

During the indenter motion, profile $g(x)$ is immersed at depth d. At the same time, at every iteration, according to Hess rule (8) the springs that are in contact are determined, which defines the contact radius $a = (n/2)\Delta x$. According to (7), the strain of the individual spring in normal direction can be found. Shifts of the springs in tangential direction $u_x(x_i)$ are the same as the shift of the indenter after the moments of the contact of these springs with an indenter, while $u_x(x_i) < u_x^{crit}(a)$.

For those springs where tension $u_x(x_i)$ exceeds critical value $u_x^{crit}(a)$, the tension magnitude is set to be equal $u_x(x_i) = u_x^{crit}(a)$. Normal F_N and tangential F_x forces are calculated as a sum of all forces in contact:

$$F_N = E^*\Delta x \sum_{cont} u_z(x_i), \; F_x = G^*\Delta x \sum_{cont} u_x(x_i). \tag{15}$$

After the indenter reaches the maximum indentation depth $d_{\max}$, it is pulled out of the elastomer in the vertical direction with the velocity components $v_z = -v$, $v_x = 0$. In this case, all the above described actions are performed. Another feature of the simulation is that when new springs come into contact (in the indentation phase), the value of the specific work of adhesion $\Delta\gamma_0$ is assigned to them, while during the detachment (pull-off phase), a different value $\Delta\gamma_1$ is used. When the condition $\Delta\gamma_1 > \Delta\gamma_0$ is true, the developed model reproduces experimentally observed secondary adhesive hysteresis [34,39].

Since in the pull-off phase the value $\Delta\gamma_1$ is applied only to those springs that have already come into contact, when changing the direction of movement of the indenter, the size of a contact area A is preserved for some time, which is clearly observable in experiments on normal indentation (see Figure 3).

The model reported in this subsection, describes the contact between a rigid indenter and an elastic half-space, where the tangential and normal contacts are considered independently of each other. In fact, MDR makes it possible to take into account the thickness of the elastomer being indented [47], as well as the relation between normal and tangential contact, as it was carried out, for example, in [48]. A simpler model was deliberately chosen by us in order to show the main patterns of indentation at an angle, which are described in the next subsection of the work. Description of more subtle effects, such as the effect of tangential shear on contact strength in the normal direction, propagation of elastic waves, friction at the contact boundary, loss of contact area symmetry, etc., requires the construction of more complex models and is not the aim of this work. Some of the abovementioned effects were described in the framework of a dynamic model, proposed by us in [38].

4.2. Results of the Simulation

Figure 8 shows the dependencies of the main parameters, obtained from the simulation of indentation according to scenario (A), the schematics of which are shown in Figure 2. The dependencies, shown in Figure 8, relate to the conditions of a real-life experiment with the experimental data presented in Figure 5.

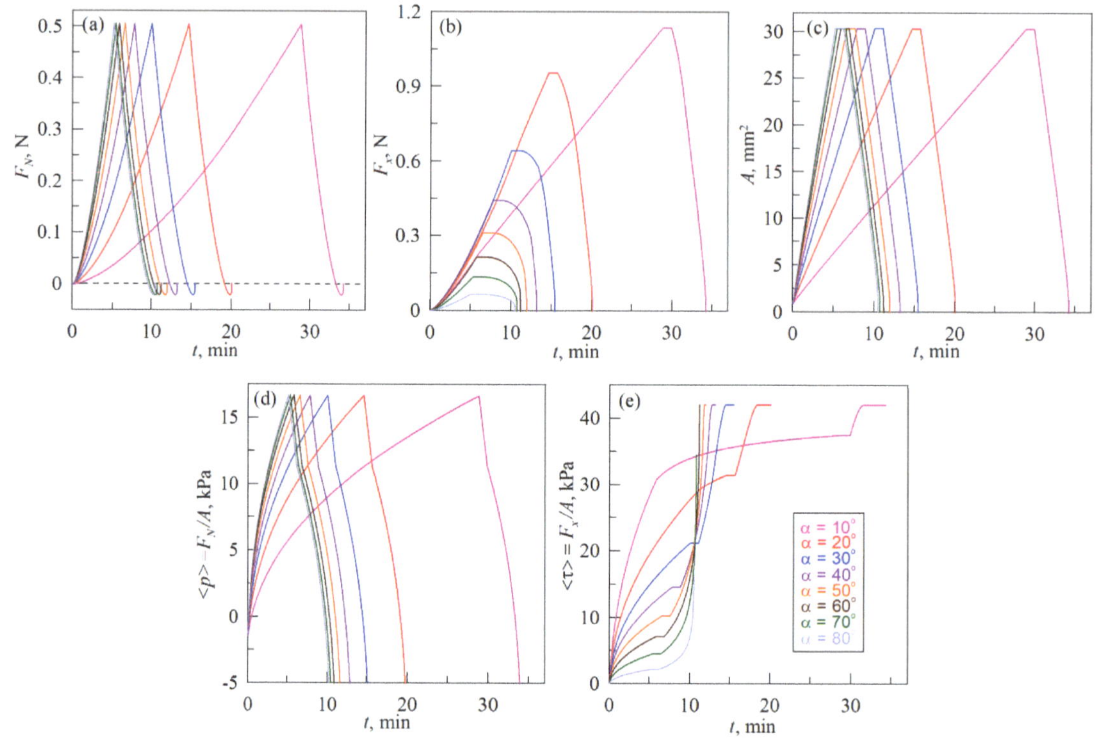

Figure 8. Time dependencies of the normal F_N (**a**) and tangential F_x (**b**) forces, contact areas A (**c**), average normal pressure <p> (**d**) and average tangential stresses <τ> (**e**), obtained in MDR simulations for indentation according to scenario (A), that is explained in Figure 2. Dependencies corresponding to the experiment with the same angle α (from 10° to 80°), plotted in different panels of the figure in the same color. Parameters of simulations: elastic modulus of the half-space E = 0.324 MPa, its Poisson ratio ν = 0.48, specific adhesion work during indentation $\Delta\gamma_0$ = 0.01 J/m^2, and during pull-off $\Delta\gamma_1$ = 0.15 J/m^2, stationary tangential stresses in the sliding mode τ_0 = 42 kPa, space discretization parameter $\Delta x = 10^{-6}$ m. All other parameters such as indenter radius R, its velocity of movement v, maximal indentation depth $d_{\max}$, as well as indentation angle α are exactly the same as in the experiment.

Figure 8 does not show the case with pure normal indentation at α = 90°. Additionally, the panel (f) that shows relation $L_{vertical}/L_{horisontal}$ in experimental Figure 5 is also absent in the Figure 8 due to the fact that in MDR simulations the contact area is considered as circular and ratio $L_{vertical}/L_{horisontal}$ is always equal to one.

As it follows from the Figure 8, the simulation results reproduce performed experiment qualitatively correct (see Figure 5). The most important differences between the experiment and simulation, as well as possible explanations for such differences and suggestions for further improvement of the model, are described below.

(1) As Figure 5a shows, the adhesive strength of the contact in the normal direction decreases with decreasing of the angle value α. Here, by adhesive strength we mean the absolute value of the normal force $|F_{N,\min}|$ during pull-off phase at $F_N < 0$ N, i.e., the force caused by adhesion, because the same external force must be applied to completely destroy the contact. The maximum adhesive strength is observed in a purely normal contact (see dependence $F_N(d)$ in Figure 3a for pull-off phase). At the presence of tangential shift, the strength of the contact in the normal direction decreases. In the simulation, we considered tangential and normal contact independently, therefore, for all angles α magnitudes of

$|F_{N,min}|$ are the same (see Figure 8a). Moreover, independent consideration of normal and tangential contact leads to the situation where dependencies $A(t)$ in Figure 8c at the beginning of the pull-off phase exhibit intervals with the constant size of the contact area A. Tangential force F_x also does not change within these time intervals of constant contact area A, due to the absence of the tangential movement during the indenter pull-off in the pure normal direction. This constrain can be avoided by introducing the coupling between normal and tangential contact as it was carried out in [48] for instance. However, application of the criterium, proposed in [48], in our simulation did not lead to the expected results and therefore additional studies on this matter are needed.

(2) Maximum values of the normal F_N and tangential F_x forces, as well as the size of a contact area A and average contact pressure $<p>$ obtained in experiments exceed those from simulations. The reason for this is that in the experiment, indentation was performed in an elastomer layer with a limited thickness h = 5 mm, while simulation was carried out for a half-space, for which $h \to \infty$. In the case of an elastic layer, the stiffness of the contact is significantly higher (especially for elastomers) [32], which leads to increased values compared to the half-space case. To take into account the limited thickness of the layer, it is possible to use the generalized MDR proposed in a recent work [47], which, however, contains a description of the modeling procedure only for normal contact.

(3) In the pull-off phase of the detachment of the indenter from the elastomer in the normal direction the tangential stresses in all cases increase to the maximum stationary value τ_0 = 42 kPa (see Figure 8e) in the simulation, while in the experiment stresses $<\tau>$ in the pull-off phase are characterized by a rapid growth (see Figure 5e). In simulations, the limit for τ_0 values caused by the use of the springs sliding criteria (12), (13)—during the pull-off phase contact radius a decreases, therefore, according to (13), there comes a moment when all the springs in contact begin to slide due to a decrease in the value of their critical tension u_x^{crit}, and thus providing constant tangential stresses τ_0 over all contact area. The rapid growth of stresses τ_0, observed in experiments during pull-off, may be related to contact strengthening in time, and also to the viscoelasticity of the elastomer. Strengthening of the contact leads to the fact that the value τ_0 increases as well as adhesive strength in normal direction $|F_{N,min}|$. At the same time, viscoelasticity leads to a decrease in the velocity of rubber slip over the indenter, and, as a result, to the growth of τ_0 during the indenter motion. In the used estimation based on MDR, it is possible to take into account both contact strengthening (by increasing the value of the specific work of adhesion $\Delta\gamma$ with time) and viscoelasticity (by using Kelvin–Voigt elements instead of springs shown in Figure 7b). However, before such a modification of the model, it is necessary to find out the true causes of the described behavior first, which requires additional experiments that are beyond the scope of the proposed work.

(4) The experiment shows the moments of abrupt increase in the contact area A when new regions of rubber are attached to the indenter, and the rearrangement of the contact boundary develops in different ways on the front and back sides of the contact. It is clearly observable in the experimental dependencies $A(t)$, $<p>(t)$ and $\tau(t)$ presented in the Figure 5 for an angle $\alpha = 10°$ (see also Supplementary Video S3), at conditions close to tangential shear, as well as during pure tangential contact (see Figure 4 and Supplementary Video S2). In the simulation, the contact area grows smoothly, since its increase occurs only due to indentation in the normal direction (adhesive JKR contact). This simplification of the model is related to the paragraph (2) above, when the normal and tangential contact are considered independently.

Note that mentioned simplifications of the model lead to visible differences between experiment and simulation only when the angle at which the indentation is performed deviates significantly from the value $\alpha = 90°$, that is corresponding to normal contact. In the case of indentation at angles close to 90° (for example, 80°, 70° and 60°), the experiment and theory give almost qualitatively identical results, since at such angles, in the indentation phase newly attached rubber regions move together with the indenter without slipping.

Taking into account the spring slip criterion (13), and assuming that indenter trajectory is known (14), the conditions when slippage of the springs begins can be easily determined. Now, we consider indentation according to scenario (A) in the phase of indentation, when the specific work of adhesion $\Delta\gamma_0$ is small, therefore, with sufficient accuracy, the contact can be considered as non-adhesive in the normal direction. At this, the contact radius is defined as [49]

$$a = \sqrt{Rd}. \tag{16}$$

According to (13) and (14), sliding of springs in contact in the phase of immersion of the indenter at an angle α occurs when the general displacement of the indenter $x = vt\cos\alpha$ becomes equal to critical displacement $u_x^{crit}(a)$ (13). If the contact radius is defined according to (16), it can be determined that slippage will start if the indentation depth d exceeds a critical value:

$$d_{\text{crit}} = \frac{\pi^2\tau_0^2 R}{4(G^*)^2}\tan^2\alpha, \tag{17}$$

which depends on the indentation angle α. According to (17), at indentation angles $\alpha < 90°$ slipping will occur anyways (with the growth of α slipping appears at larger indentation depth d). Dependence $d_{\text{crit}}(\alpha)$ is shown in Figure 9a in solid line.

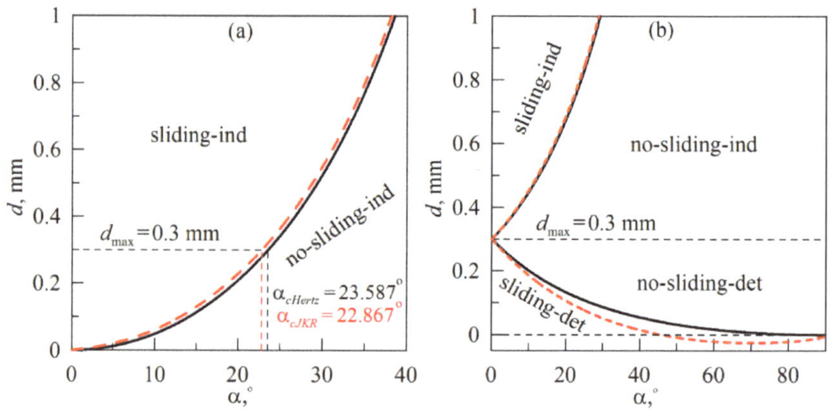

Figure 9. Dependences of the critical indentation depth d_{crit}, at which slip begins, on the indentation angle α. Panel (**a**) shows dependencies for indentation scenario (A), panel (**b**) for scenario (B), both indentation scenarios are shown schematically in Figure 2.

Figure 9a represents a diagram with two regions, the "sliding-ind" (here "ind" denotes indentation) region, characterized by sliding of the springs, and the "no-sliding-ind" region, where during the entire indentation phase, new springs move in the tangential direction together with the indenter without slipping after attachment. The solid curve in the figure shows the dependence $d_{\text{crit}}(\alpha)$, defined by the expression (17). According to this dependence, at maximal indentation depth $d_{\text{max}} = 0.3$ mm sliding in the indentation phase takes place if $\alpha < \alpha_{cHertz} \approx 23.59°$. However, main disadvantage of estimation (17) is the definition of the contact radius a (16) obtained for non-adhesive contact. At small magnitudes of the specific work of adhesion $\Delta\gamma$, such an approach is precise enough. However, if $\Delta\gamma$ is not necessarily small, for example in the pull-off phase, adhesion plays a crucial role. When the simple definition (16) is not adopted, dependence $d_{\text{crit}}(\alpha)$ is defined by the solution of the system of equations:

$$\begin{cases} d_{\text{crit}} = \dfrac{\pi a \tau_0 \tan\alpha}{2G^*}, \\ d_{\text{crit}} = \dfrac{a^2}{R} - \sqrt{\dfrac{2\pi a\Delta\gamma}{E^*}}, \end{cases} \tag{18}$$

where first equation follows from (13) and (14) at $x = u_x^{crit}(a)$, while second equation is a relation (9), needed for contact radius a determination. Obtained from (18) dependence $d_{\rm crit}(\alpha)$, calculated at $\Delta\gamma = 0.01\ {\rm J/m^2}$ for the indentation phase (see parameters in the caption of Figure 8), is shown in Figure 9a in dashed line and located slightly above the "non-adhesive" dependence (17). Thus, taking into account adhesion widens the diagram region related to the absence of sliding ("no-sliding-ind"), at the same time at the maximum indentation depth $d_{\rm max}$ = 0.3 mm slips occur at smaller, comparing to non-adhesive case, angle $\alpha_{cJKR} \approx 22.87°$.

Dependencies shown in Figure 8, automatically relate to the above analysis of the slippage criterion, since they are obtained from the simulations. However, experimental data presented in Figure 5 also show the absence of the slipping mode at large angles α, that confirms Equations (17) and (18). Note that the above-mentioned complex dynamic processes take place within the slip mode, associated with a jump-like increase in the contact area, the propagation of elastic waves, etc. Qualitatively, these processes were described in our recent paper [38], where the influence of the indentation depth $d_{\rm max}$ on the tangential contact in the presence of adhesion was studied. In [38], an experiment was also conducted and a numerical model was proposed.

Figure 10 shows the results of the simulation according to scenario (B), shown in Figure 2. Thus, results presented in Figure 10, should be compared to the experimental data shown in Figure 6. As it follows from the comparison, the simulation results and experimental data differ similarly to the scenario (A), therefore we will not discuss them in detail again. Let us discuss, however, the sliding criterion, similar to the above described expressions (17) and (18), that were obtained for scenario (A). In scenario (B) after the indentation to the maximal depth $d_{\rm max}$ = 0.3 mm, the indenter is shifted along the trajectory that is defined by equations:

$$d = d_{\rm max} - v\tilde{t}\sin\alpha,\ x = v\tilde{t}\cos\alpha, \tag{19}$$

where $\tilde{t}$ is time, measured from the moment when the indenter stopped to immerse, i.e., from the beginning of the tangential shift. Expressions (13), (16) and (19) together lead to the quadratic equation for critical value of the indentation depth $d_{\rm crit}$, at which springs start to slip:

$$d_{\rm crit}^2 - \left[2d_{\rm max} + R\left(\frac{\pi\tau_0\tan\alpha}{2G^*}\right)^2\right]d_{\rm crit} + d_{\rm max}^2 = 0 \tag{20}$$

with two positive roots.

The smaller root $d_{\rm crit} \leq d_{\rm max}$ defines the slip condition when pulling-off the indenter according to the scenario (B). In Figure 9b, this root relates to the lowest solid line $d_{\rm crit}(\alpha)$, that divides the diagram region $d < d_{\rm max}$ = 0.3 mm into two parts: "sliding-det" with sliding (here "det" denotes detachment) and "no-sliding-det" without sliding. The diagram region at $d < d_{\rm max}$ relates to the performed experiment (see Figure 6) and simulation (see Figure 10). As it follows from Figure 9b, during the indenter pull-off from the initial indentation depth $d_{\rm max}$ at an angle α, if $\alpha < 90°$ springs begin to slip in a certain moment of time.

Moreover, the smaller the angle, the earlier the slip begins, since the depth of indentation d in the experiment decreases, starting from the value of $d_{\rm max}$ (see the first equation in (19)). Note that with a purely tangential movement ($\alpha = 0°$), slippage on the diagram occurs already at $d = d_{\rm max}$, since the indentation depth d in the experiment is equal to the maximum value $d_{\rm max}$ and does not change (case with the absence of normal movement of the indenter, is shown in Figure 4). When the indenter is pulled-off in the normal direction ($\alpha = 90°$), there is no slippage (case with the absence of tangential movement, is shown in Figure 3). The above conclusions are confirmed by the experiment, in which a region τ_0 = const is observed within whole range of angles α (see Figure 6e). The length of this region, however, significantly decreases with the growth of an angle α. Similar behavior is also observed in simulations (see Figure 10e).

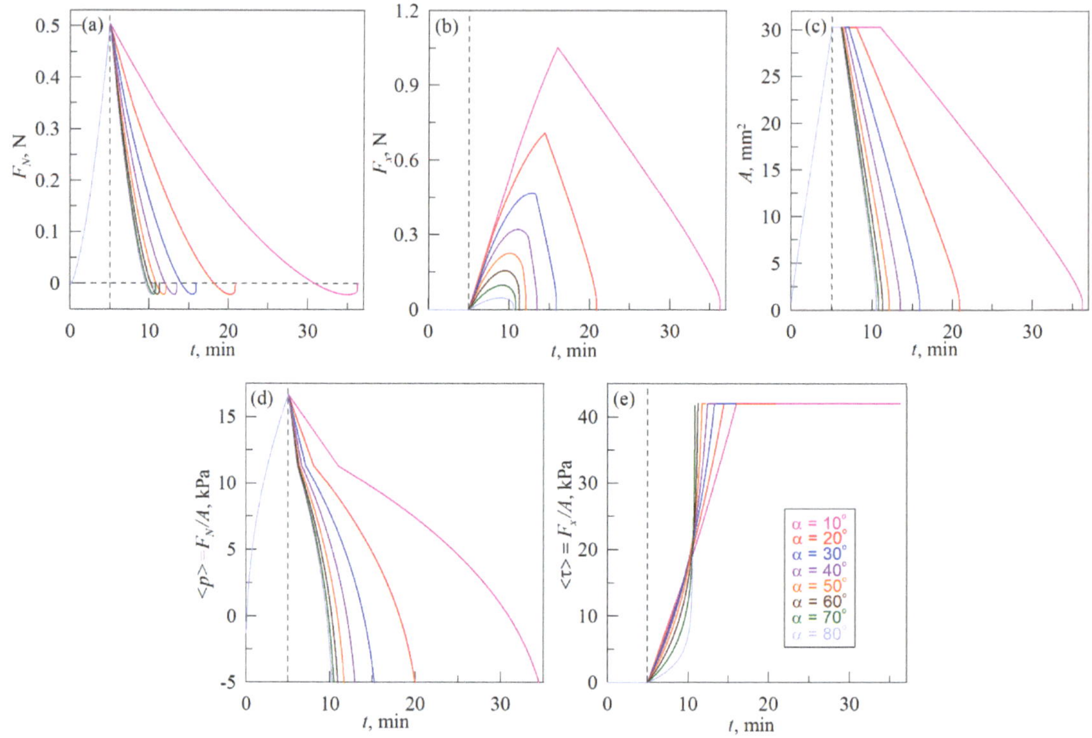

Figure 10. Time dependencies of the normal F_N (**a**) tangential F_x (**b**) forces, contact areas A (**c**), average normal pressure $<p>$ (**d**) and average tangential stresses $<\tau>$ (**e**), obtained from the MDR simulations for the scenario (B), schematics of which are shown in Figure 2. Dependencies corresponding to the experiment with the same angle α (from 10° to 80°), plotted in different panels of the figure in the same color.

Dependence $d_{\text{crit}}(\alpha)$, which is located above the value $d = d_{\text{max}} = 0.3$ mm in the Figure 9b, relates to the situation when the indenter is not withdrawing from the elastomer after reaching the critical indentation depth, but continues to immerse into elastomer, moving at an angle α instead of indentation in the normal direction. This curve also divides the area of the diagram at $d > d_{\text{max}} = 0.3$ mm into two parts with ("sliding-ind") and without ("no-sliding-ind") sliding. Plotted together by solid lines in the Figure 9b, both dependencies $d_{\text{crit}}(\alpha)$ define a complete diagram of indentation modes, for the cases of indentation and pull-off of the indenter at an angle α from the initial value $d = d_{\text{max}} = 0.3$ mm.

The refined dependence $d_{\text{crit}}(\alpha)$ with taking into account the adhesion in the normal direction is given by the solution of the system of equations (compare with (18))

$$\begin{cases} d_{\text{crit}} = d_{\text{max}} \pm \dfrac{\pi a \tau_0 \tan\alpha}{2G^*}, \\ d_{\text{crit}} = \dfrac{a^2}{R} - \sqrt{\dfrac{2\pi a \Delta\gamma}{E^*}}, \end{cases} \tag{21}$$

where in first equation "–" sign relates to the scenario (B), when the trajectory of the indenter is defined by Equation (19). The choice of the "+" sign in this equation describes a situation in which the indenter after reaching the maximum indentation depth d_{max} during the indentation in the normal direction continues immersing into the elastomer at an angle α. Dependencies, defined by Equation (21), are shown in the Figure 9b by dashed lines

The upper dashed curve was obtained at a smaller value of the specific work of adhesion $\Delta\gamma = \Delta\gamma_0 = 0.01\ \mathrm{J/m^2}$, because it relates to the indentation phase. Therefore, the difference between the "adhesion-free" curve (solid line) and the curve with adhesion in the normal direction (dashed line) is not significant here. However, for the pull-off phase (curves below d_{max} value), the difference is more significant, since the pull-off phase is characterized by a significantly larger value $\Delta\gamma = \Delta\gamma_1 = 0.15\ \mathrm{J/m^2}$. Moreover, the dashed curve, obtained by taking into account the adhesion in the normal direction, in a certain range of angles α located below the $d = 0$ mm axis, since due to adhesion contact it also exists at $d < 0$ mm.

Besides the experiments described above performed with an indenter with a radius of $R = 30$ mm, we conducted a similar series of experiments with an indenter of a larger radius $R = 100$ mm. In order to not overload the article, we do not describe these experiments here, but they are available as video files in the Supplementary Materials (Videos S5–S8), which show the evolution of the contact area and main parameters. The difference between the corresponding experiments with indenters of different radii is only the value of the radius R, all other conditions of the experiment were the same. In addition to the video files, the Supplementary Materials contain a file named "Figures S1–S4" with the dependences of the main parameters of the system obtained for the indenter with the radius $R = 100$ mm. Presented dependencies are similar to the data described above for the case $R = 30$ mm.

5. Conclusions

We performed two series of experiments, in which a steel indenter is indented into a soft elastomer at an angle α to its surface. In the first series of experiments, the indentation is carried out at an angle, i.e., the indenter immerses into the elastomer, and the contact area increases. The second series relate to the pull-off phase, when the indenter that is immersed to a certain depth is pulled out of the elastomer when moving at an angle the contact area thereby decreases. With the aim to obtain the full picture, experiments were separately conducted for cases with normal ($\alpha = 90°$) and tangential ($\alpha = 0°$) contact. Experiments in which the indenter is immersed at an angle are of particular interest, since in this case, during the entire indentation, new areas of the substrate that are not loaded in the tangential direction are involving in the contact, which leads to an inhomogeneous distribution of tangential stresses in the contact zone. The case where the indenter is pulled-off at an angle is a simpler situation, since the movement occurs from a fixed depth of the elastomer, while the entire contact is loaded during tangential movement.

For both cases, a simple numerical model based on the method of dimensionality reduction (MDR) was developed. The model makes it possible to obtain the dependences of the main parameters that are also were measured in the experiment (both components of the contact force and the contact area) on the angle α at which indentation was performed and on the displacement of the indenter in the tangential and normal directions. Although the model is simple, it produces correct dependencies that are also in good agreement with the experimental results. The simulation allowed us to build an indentation regime diagram, which contains areas with the presence and absence of slippage. In the presence of slip, complex dynamic effects such as the propagation of elastic waves and the constant rearrangement of the contact boundary are observed. The description of such effects is impossible within the framework of the proposed quasi-static contact model. Due to the complexity of these processes, additional research is needed, which is, however, beyond the scope of the presented study. In the no-slip (stick) mode, the simulation shows a much better agreement with the experiment, since the dynamic effects mentioned above are absent.

The performed study may help to better understand what happens when a contact occurs and when it is destroyed if the contacting objects move at an angle to one another. Such a situation is typical for biological organisms that use adhesion to move along inclined surfaces. An important feature of our work is that we separately consider the cases of contact occurrence (when indenting into an elastomer) and its destruction (pulling an indenter out of an elastomer), as these two phases are necessary for animals to complete their movement. Even though animals were not used in the presented study, the considered system of contact of a rigid object (indenter) with a much softer elastomer is very close to the contact of the adhesive surfaces of animals (for example, the paws of some frog species) and hard surfaces along which they successfully move (stone, wood, etc.). In this case, the soft material, due to its good ability to deform, is able to fill up existing roughness in the solid body during contact even with the application of small forces. Due to such filling, there is a significant increase in the real contact area and, as a consequence, the adhesive strength at detachment. These features allow animals to form a strong adhesive contact that can hold their weight on inclined and even vertical surfaces.

Supplementary Materials: The following supporting information can be downloaded at: https://www.mdpi.com/article/10.3390/biomimetics8060477/s1, Video S1: An experiment concerning indentation in the normal direction of a steel indenter with a radius R = 30 mm into a layer of TARNAC CRG N3005 rubber with a thickness h = 5 mm. The video shows the evolution of the contact zone, as well as the dependence of the normal force F_N and the contact area A on the indentation depth d, where the regions of the dependences corresponding to the indentation and the pull-off phase are shown in red and blue, respectively. The velocity of the indenter movement in the indentation and pull-off phases v = 1 μm/s. The upper left corner shows the real time t from the beginning of the experiment in seconds (from the moment of the first contact between the elastomer and the indenter). Video relates to Figure 3 in the article. Video S2: An experiment: concerning tangential shift of the indenter. Initially steel indenter with R = 30 mm immersed into a layer of TARNAC CRG N3005 rubber with a thickness h = 5 mm to the depth d_{max} = 0.3 mm. Then, the indenter is shifted to a distance x = 3 mm in the tangential direction at a fixed indentation depth d_{max}, after which it is pulled out until the complete detachment of the contact in the normal direction. Indenter's velocity in all phases of the experiment v = 1 μm/s. The video shows the evolution of the contact zone, as well as the time dependencies of the normal force F_N, tangential force F_x and contact area A. The top left corner shows the current indentation depth d and tangential shift x. Video relates to Figure 4 in the article. Video S3: An experiment performed according to the scenario (A), schematic of which is shown in Figure 2 in the article. Indenter with the radius R = 30 mm was immersed to depth d_{max} = 0.3 mm into a layer of TARNAC CRG N3005 rubber with a thickness h = 5 mm. In the indentation phase, the indenter was shifted at an angle α to the elastomer surface with the velocities v_z in normal and v_x in tangential direction, where $v_x = v\cos\alpha$, $v_z = v\sin\alpha$, and $v = \sqrt{v_z^2 + v_x^2} = 1$ μm/s The video shows the series of experiments with angles α = 80°, 70°, 60°, 50°, 40°, 30°, 20° and 10°. After the indentation to the depth d_{max} indenter was pulled off the elastomer in the normal direction with the velocity v = 1 μm/s. Separate panels in video show time dependencies of the normal F_N and tangential F_x forces, contact area A, average contact pressure $\langle p \rangle = F_N/A$ and average tangential stresses $\langle \tau \rangle = F_x/A$. In addition, the lower left panel shows the evolution of the contact zone, it also shows the current values of the indentation depth d, the tangential shift of the indenter x, and the time t that has passed since the beginning of indentation (since the first contact between the rubber and the indenter). Video relates to Figure 5 in the article. Video S4: An experiment performed according to the scenario (B), schematics of which is shown in Figure 2 in the article First, indenter with the radius R = 30 mm was immersed to the depth d_{max} = 0.3 mm into a layer of TARNAC CRG N3005 rubber with a thickness h = 5 mm at motion in normal direction with the velocity v = 1 μm/s. After this indenter was pulled off from the elastomer at an angle α to its surface with the velocities v_z in normal and v_x in tangential directions, where $v_x = v\cos\alpha$, $v_z = -v\sin\alpha$, and $v = \sqrt{v_z^2 + v_x^2} = 1$ μm/s. The video shows the series of experiments with angles α = 80°, 70°, 60° 50°, 40°, 30°, 20° and 10°. Separate panels in the video show time dependencies of the normal F_N and tangential F_x forces, contact area A, average contact pressure $\langle p \rangle = F_N/A$ and average tangential stresses $\langle \tau \rangle = F_x/A$. In addition, the lower left panel shows the evolution of the contact area, and it also shows the current values of the indentation depth d, the tangential shift of the indenter x, and the time t that has passed since the beginning of indentation (since the first contact between the rubber and the indenter). Video relates to Figure 6 in the article. Video S5: Same as the Video S1, but for the indenter with radius R = 100 mm. Video relates to Figure S1 in Supplementary Materials. Video S6 Same as the Video S2, but for the indenter with radius R = 100 mm. Video relates to Figure S2 in Supplementary Materials. Video S7: Same as the Video S3, but for the indenter with radius R = 100 mm. Video relates to Figure S3 in Supplementary Materials. Video S8: Same as the Video S4 but for the indenter with radius $R = 100$ mm. Video relates to Figure S4 in Supplementary Materials. Figures S1–S4: Experimental dependencies of the main parameters obtained for the indenter with the radius $R = 100$ mm. Figures S1–S4 similar to Figures 3–6 presented in the article (In the article these dependencies are obtained for the indenter with the radius of $R = 30$ mm).

Author Contributions: Conceptualization, scientific supervision, project administration, writing—review and editing, V.L.P.; methodology, hardware, software, validation, experiments and experimental data analysis, theoretical model construction, simulations and theoretical data analysis, analytical analysis, visualization, Supplementary Videos, writing—original draft preparation, I.A.L.; analysis of literary sources for Section 1, writing—original draft preparation, V.B. All authors have read and agreed to the published version of the manuscript.

Funding: This research was funded by Deutsche Forschungsgemeinschaft (Project DFG PO 810-55-3). V.B. is grateful to Technische Universität Berlin and German Academic Exchange Service (DAAD) for support.

Institutional Review Board Statement: Not applicable.

Informed Consent Statement: Not applicable.

Data Availability Statement: The datasets generated for this study are available upon request from the corresponding authors.

Conflicts of Interest: The authors declare no conflict of interest.

References

1. Springer, T.A. Adhesion receptors of the immune system. *Nature* **1990**, *346*, 425–434. [CrossRef] [PubMed]
2. Hermansson, M. The DLVO theory in microbial adhesion. *Colloids Surf. B Biointerfaces* **1999**, *14*, 105–119. [CrossRef]
3. Berne, C.; Ellison, C.K.; Ducret, A.; Brun, Y.V. Bacterial adhesion at the single-cell level. *Nat. Rev. Microbiol.* **2018**, *16*, 616–627. [CrossRef]
4. Gao, H.; Wang, X.; Yao, H.; Gorb, S.; Arzt, E. Mechanics of hierarchical adhesion structures of geckos. *Mech. Mater.* **2005**, *37*, 275–285. [CrossRef]
5. Büscher, T.H.; Buckley, T.R.; Grohmann, C.; Gorb, S.N.; Bradler, S. The Evolution of Tarsal Adhesive Microstructures in Stick and Leaf Insects (*Phasmatodea*). *Front. Ecol. Evol.* **2018**, *6*, 69. [CrossRef]
6. Huber, G.; Mantz, H.; Spolenak, R.; Mecke, K.; Jacobs, K.; Gorb, S.N.; Arzt, E. Evidence for capillarity contributions to gecko adhesion from single spatula nanomechanical measurements. *Proc. Natl. Acad. Sci. USA* **2005**, *102*, 16293. [CrossRef] [PubMed]
7. Higham, T.E.; Russell, A.P.; Niewiarowski, P.H.; Wright, A.; Speck, T. The Ecomechanics of Gecko Adhesion: Natural Surface Topography, Evolution, and Biomimetics. *Commun. Integr. Biol.* **2019**, *59*, 148–167. [CrossRef]
8. Li, J.; Zhang, Y.; Liu, S.; Liu, J. Insights into adhesion of abalone: A mechanical approach. *J. Mech. Behav.* **2018**, *77*, 331–336. [CrossRef]
9. van den Boogaart, L.M.; Langowski, J.K.A.; Amador, G.J. Studying Stickiness: Methods, Trade-Offs, and Perspectives in Measuring Reversible Biological Adhesion and Friction. *Biomimetics* **2022**, *7*, 134. [CrossRef]
10. Tang, T.; Hou, X.; Xiao, Y.; Su, Y.; Shi, Y.; Rao, X. Research on motion characteristics of space truss-crawling robot. *Int. J. Adv. Robot. Syst.* **2019**, *16*, 1. [CrossRef]
11. Johnson, K.L.; Kendall, K.; Roberts, A.D. Surface energy and the contact of elastic solids. *Proc. R. Soc. Lond. A* **1971**, *324*, 301–313. [CrossRef]
12. Derjaguin, B.V.; Muller, V.M.; Toporov, Y.P. Effect of contact deformations on the adhesion of particles. *JCIS* **1975**, *53*, 314–326. [CrossRef]
13. Maugis, D. Adhesion of spheres: The JKR-DMT transition using a dugdale model. *JCIS* **1992**, *150*, 243–269. [CrossRef]
14. Waters, J.F.; Guduru, P.R. Mode-mixity-dependent adhesive contact of a sphere on a plane surface. *Proc. R. Soc. A* **2010**, *466*, 1303–1325. [CrossRef]
15. Attard, P. Interaction and Deformation of Elastic Bodies: Origin of Adhesion Hysteresis. *J. Phys. Chem. B* **2000**, *104*, 10635–10641. [CrossRef]
16. Diao, D.; Kandori, A. Finite element analysis of the effect of interfacial roughness and adhesion strength on the local delamination of hard coating under sliding contact. *Tribol. Int.* **2006**, *39*, 849–855. [CrossRef]
17. Nguyen, T.; Gu, Y. Investigation of Cell-Substrate Adhesion Properties of Living Chondrocyte by Measuring Adhesive Shear Force and Detachment Using AFM and Inverse FEA. *Sci. Rep.* **2016**, *6*, 38059. [CrossRef]
18. Lorentz, B.; Albers, A. A numerical model for mixed lubrication taking into account surface topography, tangential adhesion effects and plastic deformations. *Tribol. Int.* **2013**, *59*, 259–266. [CrossRef]

19. Bazrafshan, M.; de Rooij, M.B.; Schipper, D.J. The effect of adhesion and roughness on friction hysteresis loops. *Int. J. Mech. Sci.* **2019**, *155*, 9–18. [CrossRef]
20. de Beer, S.; Kenmoé, G.D.; Müser, M.H. On the friction and adhesion hysteresis between polymer brushes attached to curved surfaces: Rate and solvation effects. *Friction* **2015**, *3*, 148–160. [CrossRef]
21. Khajeh Salehani, M.; Irani, N.; Müser, M.H.; Nicola, L. Modelling coupled normal and tangential tractions in adhesive contacts. *Tribol. Int.* **2018**, *124*, 93–101. [CrossRef]
22. Pérez-Ràfols, F.; Nicola, L. Incipient sliding of adhesive contacts. *Friction* **2022**, *10*, 963–976. [CrossRef]
23. Waghmare, A.K.; Sahoo, P. Adhesive friction at the contact between rough surfaces using n-point asperity model. *Eng. Sci. Technol. Int. J.* **2015**, *18*, 463–474. [CrossRef]
24. Papangelo, A.; Scheibert, J.; Sahli, R.; Pallares, G.; Ciavarella, M. Shear-induced contact area anisotropy explained by a fracture mechanics model. *Phys. Rev. E* **2019**, *99*, 053005. [CrossRef] [PubMed]
25. Immel, K.; Nguyen, V.-H.; Haïat, G.; Sauer, R.A. Modeling the debonding process of osseointegrated implants due to coupled adhesion and friction. *Biomech. Model Mechanobiol.* **2023**, *22*, 133–158. [CrossRef]
26. Nadimi, S.; Otsubo, M.; Fonseca, J.; O'Sullivan, C. Numerical modelling of rough particle contacts subject to normal and tangential loading. *Granul. Matter* **2019**, *21*, 108. [CrossRef]
27. Homola, A.M.; Israelachvili, J.N.; McGuiggan, P.M.; Gee, M.L. Fundamental experimental studies in tribology: The transition from "interfacial" friction of undamaged molecularly smooth surfaces to "normal" friction with wear. *Wear* **1990**, *136*, 65–83. [CrossRef]
28. Pengfei, L.; Shuying, W.; Yufeng, S.; Junsheng, Y.; Jinyang, F.; Feng, Y. Tangential Adhesion Strength between Clay and Steel for Various Soil Softnesses. *J. Mater. Civ. Eng.* **2019**, *31*, 04019048. [CrossRef]
29. Yao, H.; Della Rocca, G.; Guduru, P.R.; Gao, H. Adhesion and sliding response of a biologically inspired fibrillar surface: Experimental observations. *J. R. Soc. Interface* **2008**, *5*, 723–733. [CrossRef]
30. Sahli, R.; Pallares, G.; Ducottet, C.; Ben Ali, I.E.; Al Akhrass, S.; Guibert, M.; Scheibert, J. Evolution of real contact area under shear and the value of static friction of soft materials. *Proc. Natl. Acad. Sci. USA* **2018**, *115*, 471–476. [CrossRef]
31. Popov, V.L.; Heß, M.; Willert, E. *Handbook of Contact Mechanics: Exact Solutions of Axisymmetric Contact Problems*; Springer: Berlin, Germany, 2019. [CrossRef]
32. Lyashenko, I.A.; Popov, V.L.; Borysiuk, V. Experimental verification of the boundary element method for adhesive contacts of a coated elastic half-space. *Lubricants* **2023**, *11*, 84. [CrossRef]
33. Lyashenko, I.A.; Popov, V.L.; Pohrt, R.; Borysiuk, V. High-Precision Tribometer for Studies of Adhesive Contacts. *Sensors* **2023**, *23*, 456. [CrossRef]
34. Lyashenko, I.A.; Popov, V.L. Hysteresis in an adhesive contact upon a change in the indenter direction of motion: An experiment and phenomenological model. *Tech. Phys.* **2021**, *66*, 672–690. [CrossRef]
35. Lyashenko, I.A.; Popov, V.L. Dissipation of mechanical energy in an oscillating adhesive contact between a hard indenter and an elastomer. *Tech. Phys. Lett.* **2020**, *46*, 1092–1095. [CrossRef]
36. Lyashenko, I.A.; Li, Q.; Popov, V.L. Influence of chemical heterogeneity and third body on adhesive strength: Experiment and simulation. *Front. Mech. Eng.* **2021**, *7*, 658858. [CrossRef]
37. Popov, V.L.; Li, Q.; Lyashenko, I.A.; Pohrt, R. Adhesion and friction in hard and soft contacts: Theory and experiment. *Friction* **2021**, *9*, 1688–1706. [CrossRef]
38. Lyashenko, I.A.; Filippov, A.E.; Popov, V.L. Friction in Adhesive Contacts: Experiment and Simulation. *Machines* **2023**, *11*, 583 [CrossRef]
39. Greenwood, J.A. Reflections on and Extensions of the Fuller and Tabor Theory of Rough Surface Adhesion. *Tribol. Lett.* **2017**, *65*, 159. [CrossRef]
40. Deng, W.; Kesari, H. Depth-dependent hysteresis in adhesive elastic contacts at large surface roughness. *Sci. Rep.* **2019**, *9*, 1639 [CrossRef]
41. Liu, J.-L.; Xia, R. A unified analysis of a micro-beam, droplet and CNT ring adhered on a substrate: Calculation of variation with movable boundaries. *Acta Mech. Sin.* **2013**, *29*, 62–72. [CrossRef]
42. Schallamach, A. How does rubber slide? *Wear* **1971**, *17*, 301–312. [CrossRef]
43. Barquins, M. Sliding Friction of Rubber and Schallamach Waves—A Review. *Mater. Sci. Eng.* **1985**, *73*, 45–63. [CrossRef]
44. Rand, C.J.; Crosby, A.J. Insight into the periodicity of Schallamach waves in soft material friction. *Appl. Phys. Lett.* **2006**, *89*, 261907. [CrossRef]
45. Viswanathan, K.; Chandrasekar, S. Fifty years of Schallamach waves: From rubber friction to nanoscale fracture. *Philos. Trans. R Soc. A* **2022**, *380*, 20210339. [CrossRef] [PubMed]
46. Zhibo, C.; Zhaoqian, S.; Dandan, H.; Genzong, L.; Jian, W.; Benlong, S.; Yuyan, L.; Youshan, W. From small wrinkles to Schallamach waves during rubber friction: In situation experiment and 3D simulation. *Polym. Test.* **2021**, *96*, 107084. [CrossRef]
47. Forsbach, F. A Simple Semi-Analytical Method for Solving Axisymmetric Contact Problems Involving Bonded and Unbonded Layers of Arbitrary Thickness. *Machines* **2023**, *11*, 474. [CrossRef]

48. Popov, V.L.; Lyashenko, I.A.; Filippov, A.E. Influence of tangential displacement on the adhesion strength of a contact between a parabolic profile and an elastic half-space. *R. Soc. Open Sci.* **2017**, *4*, 161010. [CrossRef]
49. Hertz, H. Ueber die Berührung fester elastischer Körper. *J. Reine Angew. Math.* **1881**, *92*, 156–171. [CrossRef]

biomimetics

MDPI

Article

Characterization of Morphologically Distinct Components in the Tarsal Secretion of *Medauroidea extradentata* (Phasmatodea) Using Cryo-Scanning Electron Microscopy

Julian Thomas *, Stanislav N. Gorb and Thies H. Büscher

Functional Morphology and Biomechanics, Institute of Zoology, Kiel University, Am Botanischen Garten 9, 24118 Kiel, Germany; sgorb@zoologie.uni-kiel.de (S.N.G.); tbuescher@zoologie.uni-kiel.de (T.H.B.)
* Correspondence: jthomas@zoologie.uni-kiel.de

Abstract: Attachment to the substrate is an important phenomenon that determines the survival of many organisms. Most insects utilize wet adhesion to support attachment, which is characterized by fluids that are secreted into the interface between the tarsus and the substrates. Previous research has investigated the composition and function of tarsal secretions of different insect groups, showing that the secretions are likely viscous emulsions that contribute to attachment by generating capillary and viscous adhesion, leveling surface roughness and providing self-cleaning of the adhesive systems. Details of the structural organization of these secretions are, however, largely unknown. Here, we analyzed footprints originating from the arolium and euplantulae of the stick insect *Medauroidea extradentata* using cryo-scanning electron microscopy (cryo-SEM) and white light interferometry (WLI). The secretion was investigated with cryo-SEM, revealing four morphologically distinguishable components. The 3D WLI measurements of the droplet shapes and volumes over time revealed distinctly different evaporation rates for different types of droplets. Our results indicate that the subfunctionalization of the tarsal secretion is facilitated by morphologically distinct components, which are likely a result of different proportions of components within the emulsion. Understanding these components and their functions may aid in gaining insights for developing adaptive and multifunctional biomimetic adhesive systems.

Keywords: Phasmatodea; tarsal secretion; evaporation rate; adhesion; cryo-scanning electron microscopy

Citation: Thomas, J.; Gorb, S.N.; Büscher, T.H. Characterization of Morphologically Distinct Components in the Tarsal Secretion of *Medauroidea extradentata* (Phasmatodea) Using Cryo-Scanning Electron Microscopy. *Biomimetics* **2023**, *8*, 439. https://doi.org/10.3390/biomimetics8050439

Academic Editor: Hermann Ehrlich

Received: 22 August 2023
Revised: 14 September 2023
Accepted: 18 September 2023
Published: 20 September 2023

1. Introduction

Attachment to the substrate is an important phenomenon influencing the everyday lives of most insects, as it is used to accomplish different tasks, such as locomotion [1], resisting predators [2], supporting copulation [3], etc. Different attachment devices have evolved, which all follow various combinations of certain basic principles, to fulfil the function of attachment [4–6].

To generate attachment, most insect tarsi utilize adhesion supported by lateral shear (friction) [7]. Two types of tarsal attachment systems emerged in insects: hairy ones and smooth ones [6,8,9]. Both maximize the contact area with the substrate to increase the contribution of different physical forces (e.g., Van der Waals forces, capillary interactions, and viscous forces) and, consequently, adhesion [7,10–12]. In a smooth attachment system, the contact surface maximization is caused by the soft material of adhesive pads. It is hierarchically organized internally towards the pad surface with progressively splitting fibers or integral foams, resulting in a rather smooth and flexible surface, and enabling the replication of the substrate profile and the maximization of the actual contact area [7,13] To support the performance of the adhesive system, most insects utilize wet adhesion, meaning they secrete a tarsal fluid into the contact interface [7].

Besides insects, other animal groups also produce a fluid onto the substrate to support the adhesion process [14–16]. However, the mechanisms can largely differ from those employed by insects. Tree frogs similarly, but convergently, produce an adhesive liquid in their mucus glands to strengthen their attachment [17,18]. Geckos do not produce large amounts of tarsal secretion, but it has been shown that a nanometer thin lipid film covers the surface of their adhesive setae [19]. Especially in marine environments, other depletion mechanisms are found: Echinoderms, for example, secrete a two-phase secretion, where the first phase generates the adhesion and the second phase dissolves the first phase [20–23]. Freshwater polyps (*Hydra* spp.) can achieve temporary adhesion based on their adhesive secretion, which consists mainly of fibers [24]. The fluid depletion of stick and leaf insects, in contrast, works differently: it is most likely a passive delivery mechanism of tarsal fluid from the adhesive pads onto the substrate that is facilitated by pressure and intermolecular forces [12].

A histological investigation of the smooth tarsal adhesive organs of *Gromphadorhina portentosa* (Blattodea) revealed that the tarsal secretion is produced by exocrine cells into a storage volume underneath the outer part of the endocuticle. It is then released through channels onto the substrate [25,26]. The chemical compositions and functions of the tarsal fluids of insects have been investigated in several studies during the past decades (e.g., [2,27–35]). A chemical analysis of the tarsal secretion was conducted for representatives of different insect groups, such as Diptera [36], Hymenoptera [37,38], Coleoptera [39–46], Orthoptera [28,47], and Blattodea [34]. These analyses show that the chemical composition of the secretion differs between insect groups. The components found in the tarsal secretion include a water-soluble part and a lipid-soluble part. The water-soluble substances include alcohols, glucose and other saccharides, amino acids, unipolar carbohydrates, polar proteins, and peptides. The lipid-soluble substances include hydrocarbons, fatty acids (saturated and unsaturated with a chain length between C16 and C20 in both free and glyceride forms), and true waxes [28,34,42,47]. Based on the chemical analysis conducted by Vötsch et al., (2002), it was concluded that the tarsal secretion of *Locusta migratoria* (Orthoptera) is a highly viscous emulsion consisting of lipid droplets in a water-like solution [28]. The identified differences in the composition suggest that the functions of the tarsal fluids and their mixtures can also differ in detail.

Experiments investigating the functions of tarsal secretions in several insect groups demonstrated that the fluid could support three main functions as follows:

(I) It can increase the adhesion to a broad range of substrates. Experiments on the attachment performance of Phasmatodea [48] and Coleoptera [49], where the volume of the tarsal secretion was diminished through consecutive steps or by porous substrates, showed that the attachment forces were enhanced on smooth surfaces, but were reduced on rough surfaces, indicating that the fluid is a crucial part of attachment generation on rough surfaces [48,49]. This effect was additionally supported by experiments on the bioinspired micropatterned samples [50]. These results indicate that the secretion can fill the asperities of non-smooth substrates, thus increasing the real contact area and thereby the attachment forces [32,48,51–53]. The immersion of nanometric beads in the accumulated tarsal secretions of the beetle *Coccinella septempunctata* and the fly *Calliphora vicina* indicated different viscosities of 21.8 and 10.9 mPa × s, respectively, showing that the physical properties of the fluid diverge between the species [33]. The presence of the liquid in contact is expected to provide capillary forces that increase adhesion [4,7]. Additionally, the high viscosity of the fluid likely implements viscous forces and thereby increases attachment [54].

(II) It contributes to decontamination. Contaminating the adhesive pads of the stick insect *Carausius morosus* with polystyrene beads and manipulating the amount of adhesive fluid showed that a high amount of fluid led to a faster recovery rate of adhesion than a low fluid amount. Thus, it is an important part of the self-cleaning mechanism of smooth adhesive pads [30,31].

(III) It can compensate for different surface chemistry of substrates. Chemical analyses of the fluids allow for the interpretation of the interaction with different surfaces. Due to the presence of two phases (water-soluble and lipid-soluble phases), the emulsion should improve the attachment to hydrophilic and hydrophobic surfaces, as it acts as a coupling agent between the pad and substrates with different free surface energies [7,28,52,55].

Some of the components of the tarsal secretion resemble those found on the surface of the insect cuticle, potentially helping to reduce the evaporation rate of water through the adhesive pad and assisting in communication [38,56]. An investigation of the ultrastructure and frictional properties of the smooth pad of *Tettigonia viridissima* (Orthoptera) revealed that the fluid within the pad contributes to its viscoelastic behavior and the frictional forces subjected to the substrate [57,58]. As highlighted by these experiments and chemical analyses, it is evident that the tarsal secretion supports and affects locomotion and attachment and therefore insect behavior. It is also apparent that these results show that the physical and chemical properties of the tarsal fluids differ greatly between species.

Despite considerable insights into the compositions and functions of tarsal secretions, approaches to investigate the details of fluid depletion in insects are scarce. Particularly, details of the interactions of the tarsal fluids with the substrates and between different components of the fluid on the surface of the attachment pads and on the substrate remain unexplored. Insects with large smooth attachment pads possess large areas that need to be covered by these secretions and are particularly prone to contamination.

We therefore analyzed the morphological characteristics of the footprint residues of the stick insect *Medauroidea extradentata*. Stick insects are among the largest insects [59] and they possess smooth adhesive pads [13,60–62], which are rather voluminous [55,63], and therefore should produce a significant volume of tarsal secretion. Stick insects have two types of smooth adhesive pads: the arolium and the euplantulae [13,64,65]. The arolium is situated on the pretarsus between the two claws and is mainly used to generate an adhesion force (force perpendicular to the surface), whereas the euplantulae are situated on the tarsomeres and contribute to friction (force horizontal to the surface) (Figure 1) [6,62,65–67]. We used cryo-scanning electron microscopy to analyze frozen footprints at a high magnification in their quasi-native (frozen) states. In addition, white light interferometry was used to measure the change in the volume of individual liquid components over time and quantify their evaporation rates. Through this combination of approaches, we aimed to investigate the structural and physical properties of the footprints left by both types of smooth attachment pads of this species. The findings may provide useful information (1) to understand adhesion in stick insects and (2) to enhance advances in the field of biomimetic multifunctional adhesives.

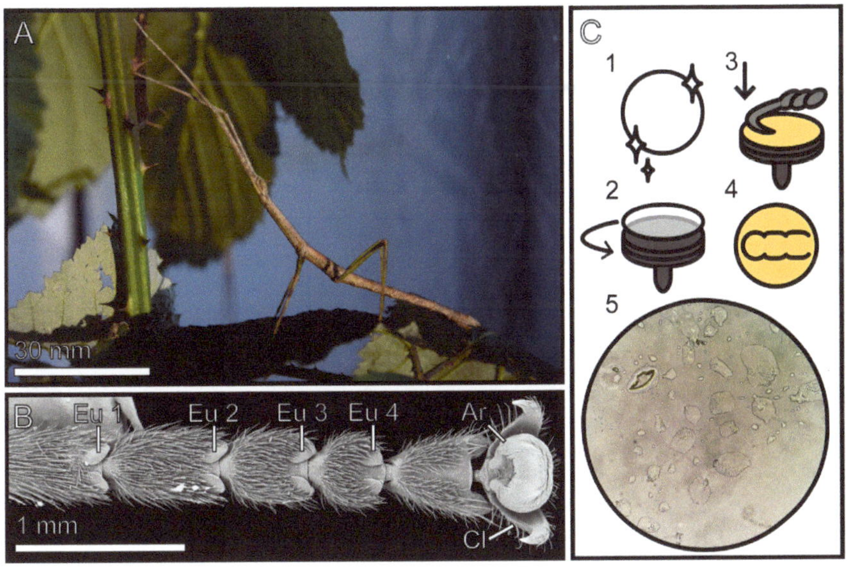

Figure 1. Experimental setup. (**A**) Adult female *Medauroidea extradentata*. (**B**) Overview of the tarsus of *M. extradentata*; Ar = arolium, Cl = claw, Eu = euplantulae. (**C**) The sequence of footprint generation: round glass coverslip is cleaned (1), glass cover slip is mounted on a cryo-SEM stub (2), cover slip is sputtered, tarsus of living insect is positioned on top, and pressure is applied (arrow) (3), footprint remains on the glass (4), footprint under a light microscope (5). (**C**) Reproduced with permission from Thomas et al. (2023) [55]. Copyright: The Company of Biologists.

2. Materials and Methods

2.1. Animals

We used the phasmid species *Medauroidea extradentata* (Brunner von Wattenwyl, 1907) (Figure 1A) because of the presence of a broad range of data on the functional morphology and biomechanics of its tarsal attachment system [13,55,62,68].

The morphology of arolium and euplantulae represents the most common and least derived setup among phasmids with smooth adhesive microstructures on both attachment pads, without micro-ornamentation [63] (Figure 1B). Individuals were obtained from the laboratory cultures of the Department of Functional Morphology and Biomechanics (Kiel University, Kiel, Germany). The insects were fed with blackberry leaves ad libitium and kept in a regular day and night cycle. Only adult individuals with clean and intact legs were selected. The insects were kept with blackberry leaves in clean hard plastic boxes to reduce contamination of the adhesive pads.

2.2. Footprint Collection

Microscope slides (76 × 26 mm) and glass coverslips (12 mm) (Thermo scientific, Budapest, Hungary) were used as sampling substrates for investigation using white light interferometry (WLI) and cryo-scanning electron microscopy (cryo-SEM). The glass surfaces were thoroughly cleaned with the following protocol prior to sampling footprints (Figure 1C1): (1) 15 min in an ultrasonic bath with distilled water and soap (neutral intensive cleaner); (2) 15 min in an ultrasonic bath with distilled water; (3) 15 min in an ultrasonic bath with 100% pure ethanol; and (4) 1 h in vacuum in a desiccator.

To obtain a footprint, the insects were first anaesthetized with CO_2 for 20 s. The tarsus was placed on a carefully cleaned glass slide or glass coverslip within a marked area. Glass slides for cryo-SEM investigation were previously sputter-coated with a 20 nm layer of gold–palladium (Figure 1C2). A second cleaned glass slide was placed on the dorsal side of the tarsus and pressed for 5 s with even pressure, and the leg was simultaneously pulled to generate some shear forces (Figure 1C3). Lastly, the second glass slide and the foot were carefully removed, and alteration of the footprint was avoided (Figure 1C4). The footprints were immediately used for investigation in WLI and cryo-SEM (Figure 1C5). The glass slides were stored in a closed glass chamber at 20.6–22.9 °C room temperature and 43.2–51.4% ambient humidity (measured within the chamber).

2.3. Cryo-Scanning Electron Microscopy

Fresh footprints were sampled on cleaned glass coverslips that were previously sputter-coated with 20 nm gold–palladium (Figure 1C).

The glass coverslips with the fluid footprints were mounted on aluminum stubs and carefully immersed in liquid nitrogen for 5 s. The footprint was then transferred into the cryo-preparation chamber (Gatan ALTO-2500 cryo-preparation system, Gatan, Abingdon, UK) at −140 °C of the SEM Hitachi S-4800 (HitachiHigh-Technologies, Tokyo, Japan). The frozen footprints were then observed in the SEM at −120 °C at an accelerating voltage of 3 kV without sputter coating. Subsequently, each sample was sputter-coated with gold–palladium (layer thickness 10 nm) in the preparation chamber at −140 °C and observed again at −120 °C with 3 kV accelerating voltage. Sputter coating was used to enhance the visualization of the surface structure of the footprint components. For some frozen samples, sublimation (freezing-drying) at −80 °C in the prechamber was performed prior to observations. Contrast adjustment and image cropping were performed using the software Photoshop CS6 (Adobe Systems Inc., San Jose, CA, USA).

2.4. White Light Interferometry (WLI)

The glass slides with the footprints were examined in the white light interferometer New View 6000 (Zygo, Darmstadt, Germany) and analyzed using the software MetroPro (Zygo, Middlefield, CT, USA). The glass slides were placed under the WLI and measured at 20.6–22.9 °C room temperature and 61–64.3% ambient humidity. Areas of interest, where enough of the fluid accumulated to form a measurable droplet, were selected with the build-in-mask function (Figure 1C5). The droplet volume and its change over time (rate of evaporation) were measured. The droplet volume was measured at the beginning and afterwards and was remeasured every day for at least 14 days. If the drops still showed measurable volume after 14 days, the measurements were continued. Each measurement series consisted of three measurements, which were performed at an interval of about three minutes; for the analysis, the mean value of the three measurements was subsequently used. In the case of strong changes in volume, the respective mask was adjusted accordingly. In some rare cases, it was observed that water vapor likely accumulated in the footprints and thereby increased their volumes. Footprints with an accumulation of water were excluded from the analysis.

To determine the evaporation rate of the droplets, the initial volume (day 0) was considered to be 100%, and the change in volume percentage over minutes (%/min) and days (%/day) was determined (for raw data see Supplementary Data S1). The data were statistically analyzed using R (R version 4.2.3, R Core Team, Vienna, Austria). For statistical analysis, the evaporation rates were compared with a Kruskal–Wallis one-way analysis of variance (ANOVA) on ranks, followed by Dunn's post hoc test, since they were not normally distributed (Shapiro–Wilk test) and showed no homoscedasticity (Levene's test)

2.5. Temperature and Ambient Humidity Measurements

Temperature and the ambient humidity were measured with a Tinytag Plus 2 TGP—4500 (Gemini Data Loggers, Chichester, UK) and analyzed using the software Tiny-Tag Explorer 6.0 (Gemini Data Loggers, Chichester, UK). For the measurements in the closed glass chamber, 250 measurements at intervals of 30 min were conducted, and for the measurements at the WLI 200, measurements at intervals of 1 min were conducted.

3. Results

3.1. Analysis of Frozen Footprints

We analyzed the appearance of the fluid and solid residuals resulting from the contact of the tarsus with the glass surface. Cryo-SEM enabled us to visualize the components in their frozen state at −120 °C with a high magnification. Immediate freezing with liquid nitrogen after deposition allowed us to investigate the footprints in the condition just after fluid depletion. The micrographs, hence, show a temporary impression of the footprint at the time of its application (Figure 2).

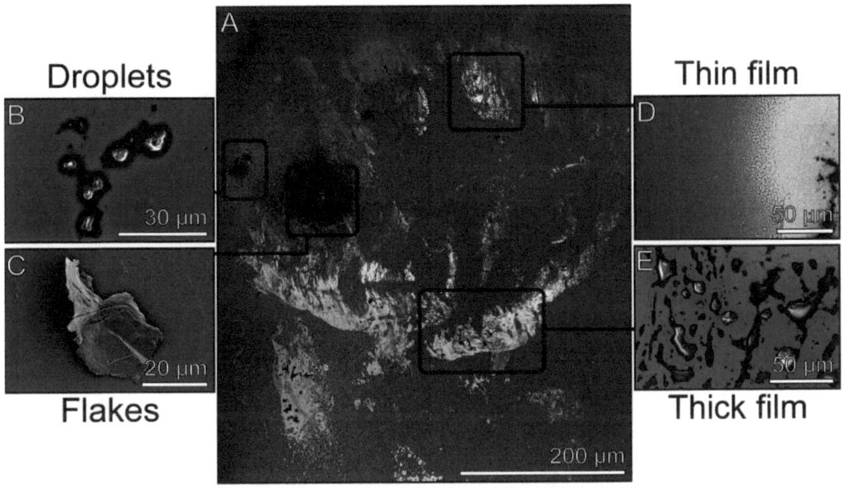

Figure 2. Footprint overview containing the four main components. (**A**) Arolium footprint of *M. extradentata*. Examples of droplets (**B**), flakes (**C**), thin films (**D**), and thick films (**E**).

The grey scale on the cryo-SEM images is influenced by two factors: the distance of the footprint to the detector and the electron density of the secretion. Both factors are affected by the thickness and conductivity of the secretion itself. The electron density is additionally influenced by the fluid's composition. Accordingly, thin liquid layers with a low electron density are displayed brighter than the background, as well as the thick layers with a high electron density that appear dark (Figure 2). After sputter coating, the differences in electron density (conductivity) vanished due to the coverage by the gold–palladium sputtering (see Supplementary Figure S1). An observation of the frozen footprints revealed distinct components, which differed in their morphology and the site of occurrence. These can generally be divided into four groups that can be distinguished based on their shape, size, surface structure, and site of occurrence. In Figure 2A, an overview of the imprint of an arolium of *M. extradentata* is shown with representations of all four components (Figure 2B–E). The footprint components include droplets (Figure 2B), flakes (Figure 2C), thin films (Figure 2D), and thick films (Figure 2E). These components, their exact distribution within a footprint, as well as the characteristics shared between the groups, will be described in detail below.

3.2. Distribution of the Tarsal Secretion and Solid Bodies within the Footprints

Footprints are characterized as all structures that are found in the vicinity of the application site of the adhesive pads and are morphologically different from the sputtered glass surface. They include the putative components of the secretion that support attachment, known as solid bodies, which presumably originate from the solidified tarsal secretion and environmental contaminations. The attachment pads of the tarsus and pretarsus leave distinct imprints in terms of their position and shape (Figure 3) in most cases, which enable differentiation of the origin of the residuals. The arolium usually leaves large, single-surface impressions, which comply with the shape of the adhesive pad with diameters between 500 µm and 1 mm (Figure 3A,B). The imprints of the euplantulae generally possess diameters between 100 and 200 µm and are situated 100–200 µm apart from one another (Figure 3C,D). Additional imprints were often observed between the two pad imprints that had an elongated form with lengths of around 400 µm (Figure 3D).

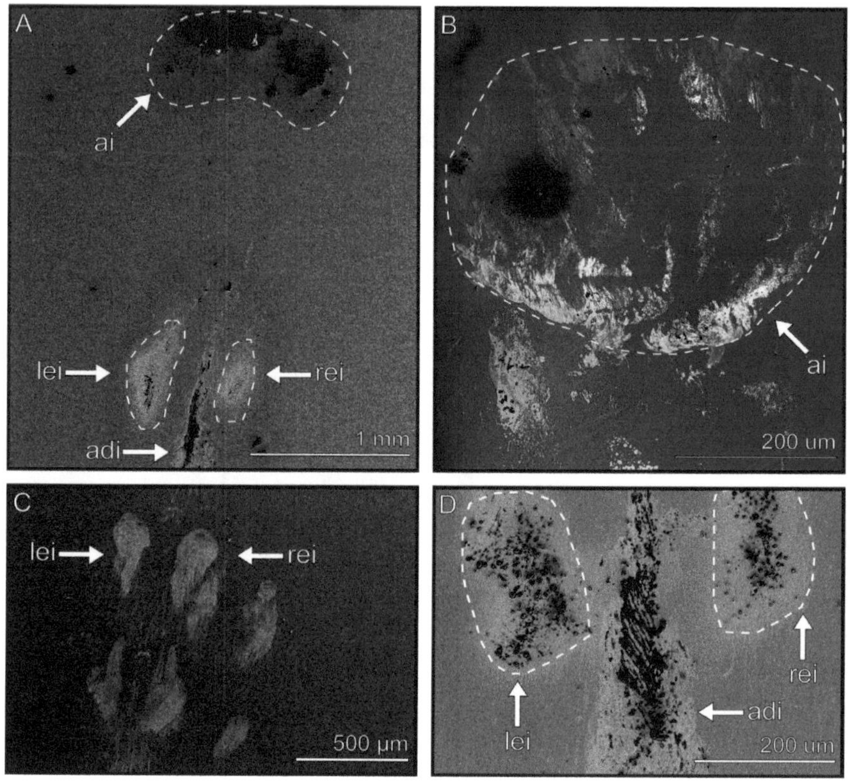

Figure 3. *M. extradentata* footprints from different pads. (**A**) Arolium imprint visible in the top half and imprint of first euplantulae in the lower half, with an additional imprint between them. (**B**) Arolium imprint. (**C**) Two euplantulae imprints below one another. (**D**) Euplantulae imprints with an additional tarsal secretion imprint between them. Outlines highlight the edges of the imprint. ai = arolium imprint, lei = left euplantulae imprint, rei = right euplantulae imprint, adi = additional imprint.

A footprint generally consists of a mixture of all four morphologically distinct components as well as contaminants (Figure 2). Usually, the thin film is the most common component, followed by thick films and droplets, while the flakes are the least frequently observed. Contaminants are usually covered by a film, which, in the case of small contaminants, causes them to aggregate with others to form large clusters (see below). The largest amount of residues were found at the edge of the adhesive pads' imprints, which was especially noticeable in the footprint areas left by the arolium (Figure 3A,B). It was observed that some imprints can consist of only one or two components (Figure 3D imprints mainly consist of thin and thick films).

3.3. Droplets

Droplet components have a round, compact, and voluminous shape (Figure 4). Droplets were observed as single components (2–10 μm) (Figure 4C,F), as well as in larger complexes (30–100 μm in diameter) (Figure 4A,B). Different surface structures were observed on droplets. These can be either smooth, rough, or covered with nano-droplets (Figure 4C,E (smooth), F (rough), and D (nano-droplets)). When sputtered, no layered surface is visible in contrast to the other footprint components (see Supplementary Figure S1). These components were found throughout the whole footprint and were unrelated to the position of the attachment pad.

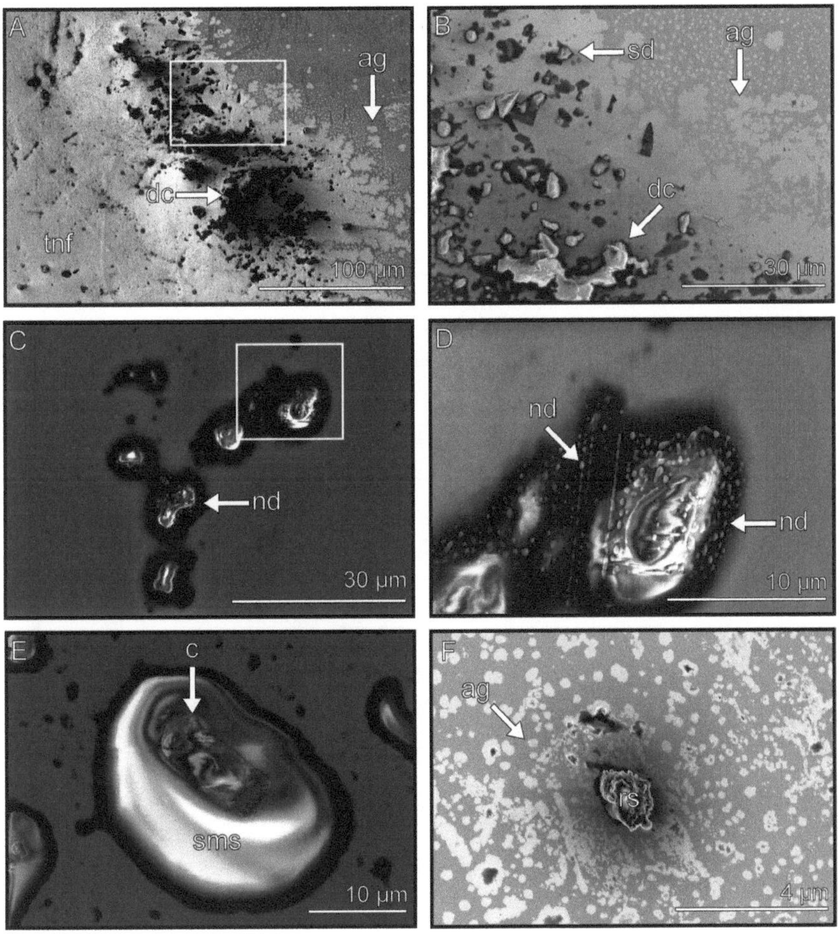

Figure 4. Droplets in the secretion. (**A**) Overview of an imprint, which mainly consists of droplets and thin films. The box displays the location that image (**B**) originates from. (**B**) Droplets can occur as single droplets, or they can accumulate into complexes. (**C**) Droplets with nano-droplets on and around them. The box displays the location where image (**D**) originates from. (**D**) Single droplet with magnified view on the nano-droplets. (**E**) Single droplet with a smooth surface structure and an absorbed contamination particle. (**F**) Single droplet with a rough surface and aggregated thin film around it. sd = single droplet, dc = droplet complex, ag = thin film aggregate, nd = nano-droplets, c = contamination, sms = smooth surface, rs = rough surface, tnf = thin film.

3.4. Flakes

Flakes are components that can be observed individually (Figure 5A) and have a length between a few μm (Figure 5F) and 50 μm (Figure 5B). They have compressed (Figure 5C) or elongated shapes (Figure 5D). After sputtering, the flakes show a structured surface consisting of several parallel thin layers with a thickness of a few nanometers, which do not have a uniform shape (Figure 5E,F). Flakes were more frequently found in the arolium imprints than in the euplantulae imprints. In the imprints of both types of attachment structures, they were deposited at the edges of the majority of observed footprints (Figure 5A). Flakes were less often observed compared to the other tarsal footprint components.

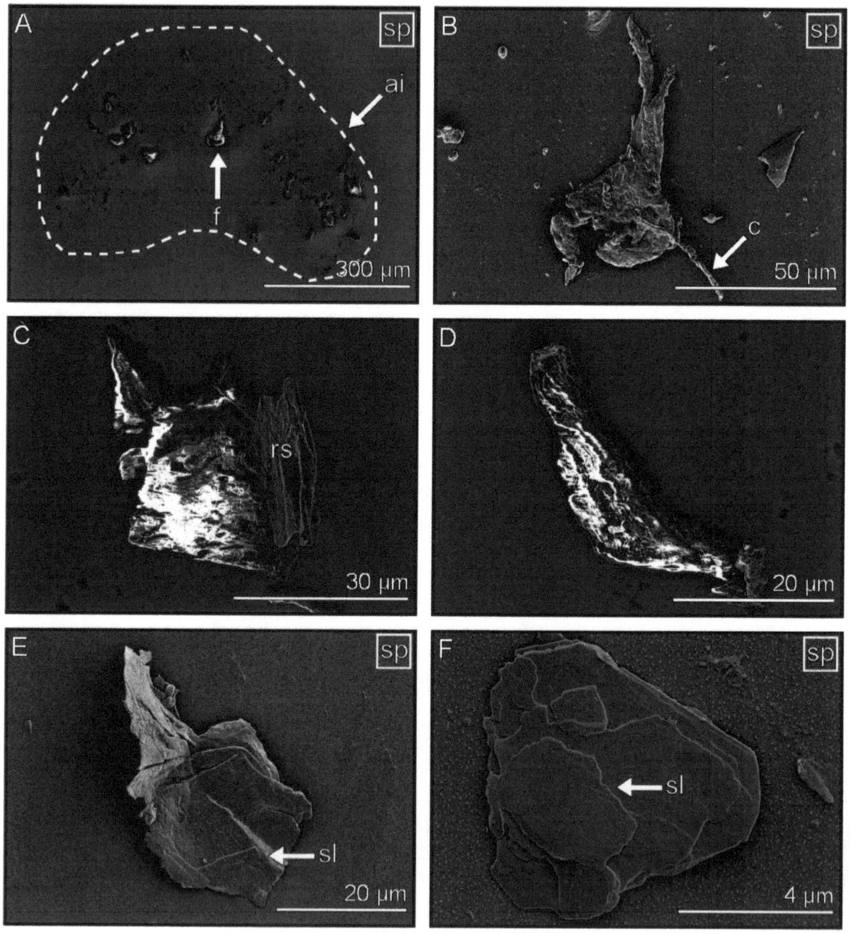

Figure 5. Flakes in the tarsal secretion. (**A**) Distribution pattern of the flakes in an arolium imprint (**B–F**) Exemplary shapes, sizes, and surface structures of the flakes. (**B**) Elongated sputtered flake with adsorbed contamination. (**C**) Compressed unsputtered flake. (**D**) Elongated unsputtered flake. (**E**) Sputtered flake with layered surface structure. (**F**) Small and sputtered flake with layered surface structure. Outlines highlight the edges of the imprint. Images marked with "sp" show sputtered samples. ai = arolium imprint, f = flake, c = contamination, sl = surface layer, rs = rough surface structure.

3.5. Thin Films

The components that are displayed brightly in the cryo-SEM images, either because of their low volume or low electron density, and that cover large surface areas are classified as thin films (Figure 6). These components were observed in each imprint of the arolium and euplantulae (Figure 6B,C). The structures of the thin films differed depending on whether they were situated close to the other components or not. When the films were near the other components, we observed that the thin films formed a structure that covers a large area (Figure 6A–C). With an increasing distance to the other components, the thin films increasingly aggregated into smaller circles down to a few nanometers (Figure 6C,D,F). The thin films covered the largest areas in the observed footprints, with areas between 100 μm and 1 mm in diameter (Figure 6A,B). The surface structure was not discernible without sputtering (Figure 6E,F). Sputter coating of the thin films covered them completely and made them invisible, as the thickness of the coating was likely higher than the films themselves.

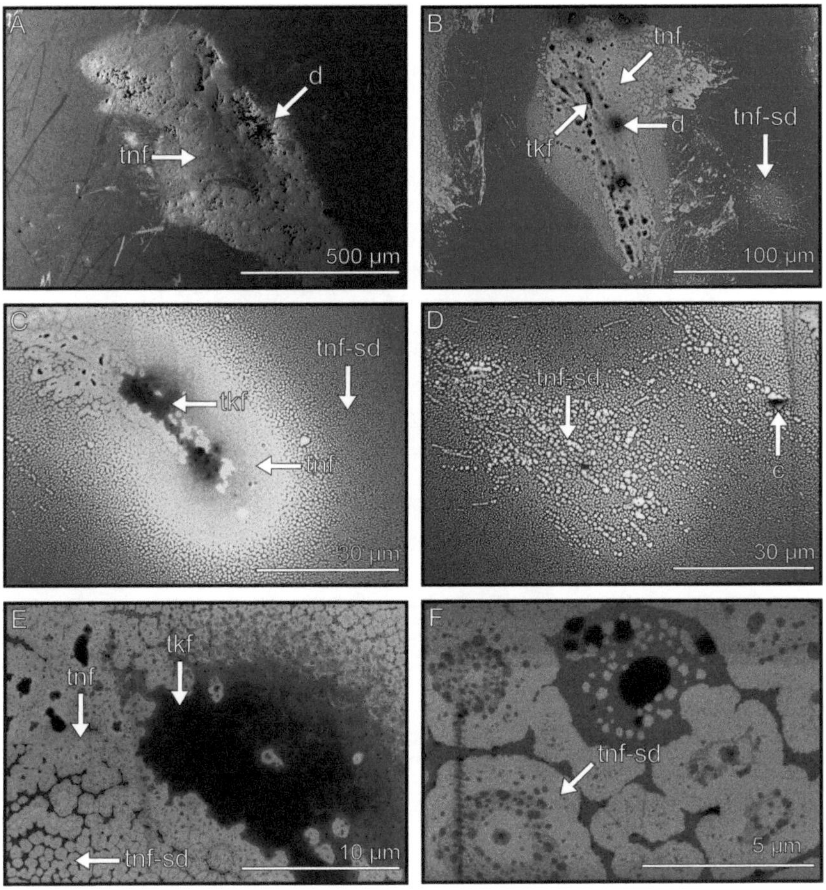

Figure 6. Thin films in the tarsal secretion. (**A**) Overview of a large imprint with a high proportion of the thin film. (**B**) Imprint with thin film and thick film components. The thin film forms small aggregates in the periphery of the other components. (**C**) When the thin film is closer to other components, it forms a uniform surface, whereas when it is distant to other components, it forms progressively smaller aggregates. (**D**) In absence of other components, small round aggregates are formed. (**E**,**F**) Higher magnification of a uniform thin film (**E**) and a small aggregate (**F**). tnf = thin film, d = droplet, tkf = thick film, tnf-sd = thin film small droplet, c = contamination.

3.6. Thick Films

Thick films are components that are dark in the cryo-SEM images and thus have either a large volume and/or a high electron density (Figure 7A). They were found in the majority of footprint samples studied. In addition, these components exhibit a smooth surface structure before sputtering (Figure 7C) and form a coherent coverage of up to 100 μm in diameter, which resembles a thick liquid film (Figure 7B). These films can be uniform or show some gaps (Figure 7D,E). Individual isolated portions, possessing the above-mentioned appearance, were also observed and are classified as part of the thick film. These single units possessed no gaps and covered areas of only a few μm (Figure 7F). When sputtered, some thick films showed rough or granular surface structures (Figures 7B and 8C,F).

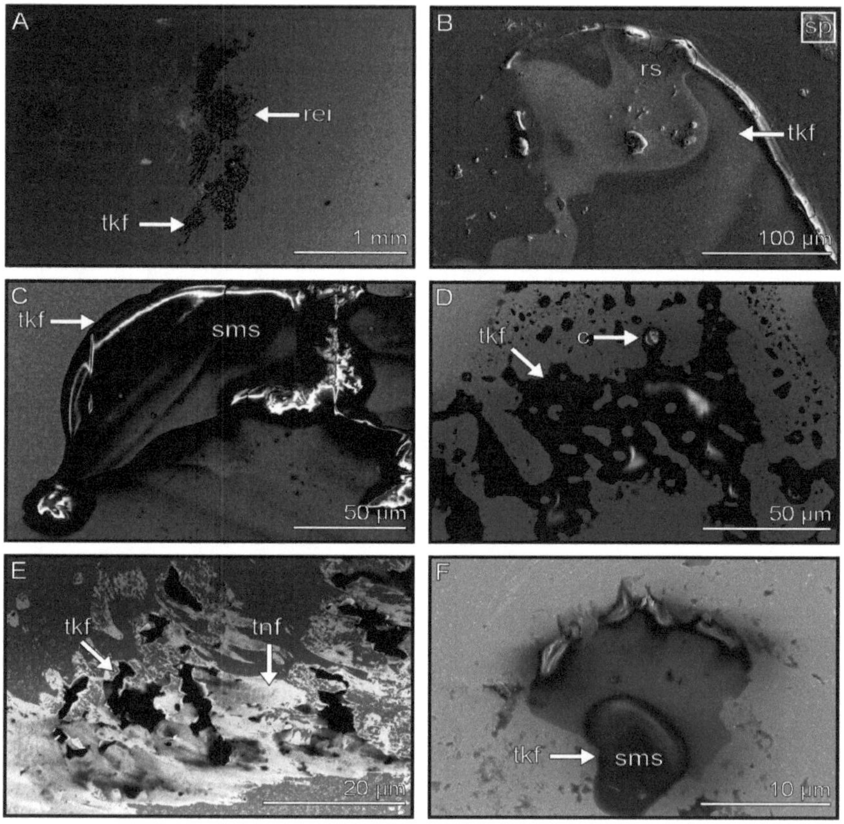

Figure 7. Thick films in the tarsal secretion. (**A**) Imprint mainly consists of thick film components. (**B**) Uniform sputtered thick films with rough surfaces. (**C**) Uniform imprint with a smooth surface structure. (**D**) Thick film imprint with gaps and a smooth surface. (**E**) Transition between uniform thick film and thin film residuals. (**F**) Thick film with smooth surface structure. Images marked with "sp" show sputtered samples. rei = right euplantulae imprint, tkf = thick film, sms = smooth surface structure, c = contamination, tnf = thin film, rs = rough surface structure.

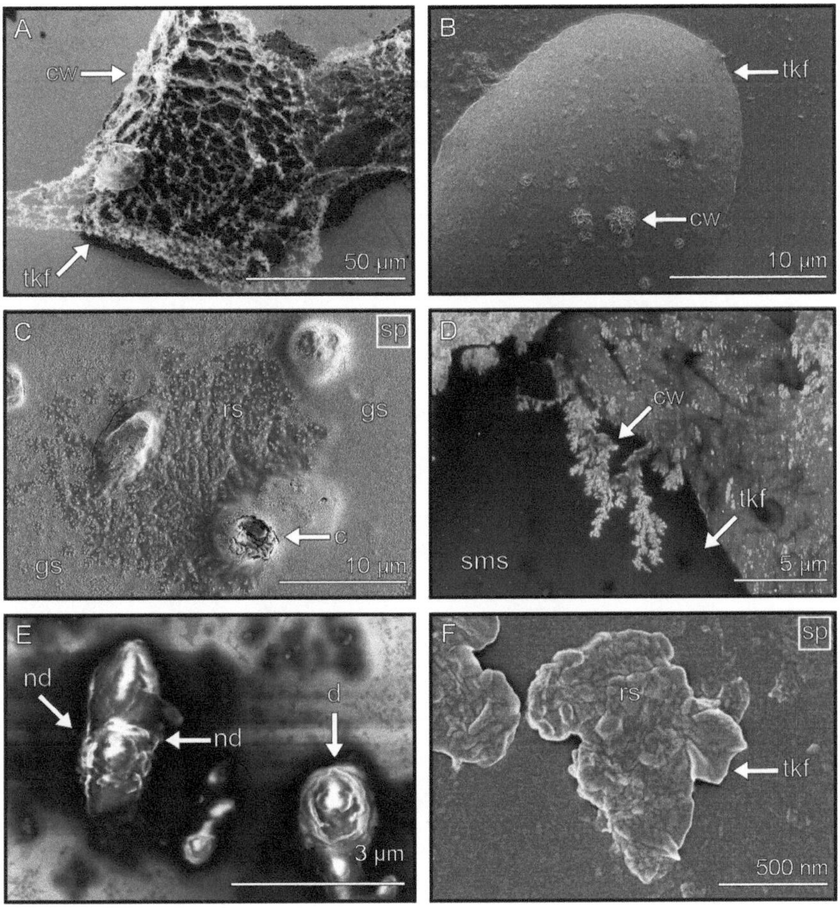

Figure 8. Different surface structures of tarsal secretion components. (**A**) Thick film component covered by a mesh of frozen water. (**B**) Thick film component with multiple single patches consisting of ice crystals. (**C**) Close-up of a sputtered thick film with rough and granular surface and enclosed contamination. (**D**) Thick film with ice formation on its smooth surface. (**E**) Droplet component with nano-droplets. (**F**) Rough surface of sputtered thick film. Images marked with "sp" show sputtered samples. cw = crystallized water (ice), tkf = thick film, rs = rough surface structure, gs = granular surface structure, smooth surface structure, c = contamination, nd = nano droplets, d = droplet.

In some samples, iced water was observed close to the thick film components. The crystalline water could envelop the thick film (Figure 8A), form individual circles on the surface (Figure 8B), or cover the surface as a network (Figure 8D). Thick films were observed in both the arolium and euplantula imprints. While they seemed to be evenly distributed in the euplantulae (Figure 7A), they were found more frequent at the edges in the arolium impressions (Figure 3B).

3.7. Contaminations

A wide variety of contaminations were found in the phasmid footprints, ranging from large contaminations (around 100 μm in width) (Figure 9A) to small 3 μm wide particles (Figure 9B). Usually, they were covered by the tarsal secretion (Figure 9C,D). The contaminations were mostly observed at the edge of the impressions. We detected that large contaminations were often isolated from the others (Figure 9A), whereas small contaminations were combined into large clusters by the tarsal secretion (Figure 9B–D). In addition to the contamination being coated by the adhesive secretion, it was also observed that they adhered to the surface of residuals (Figure 9E,F).

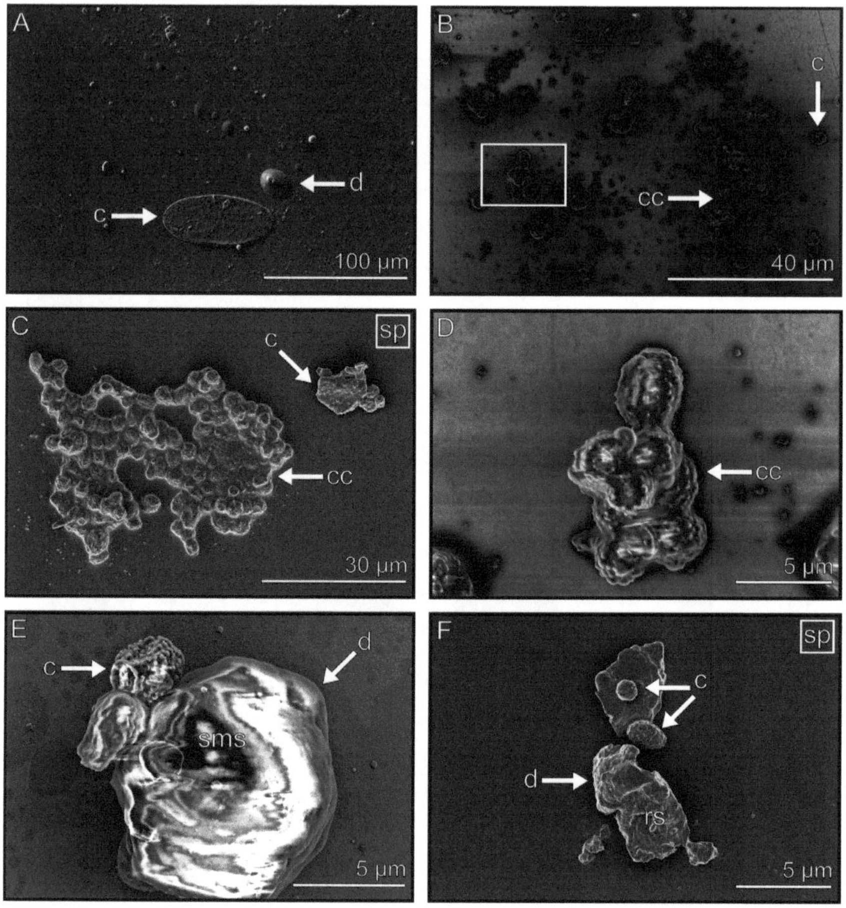

Figure 9. Fluid interactions with contaminants. (**A**) Large contamination covered and surrounded by tarsal secretion. (**B**,**C**) Multiple contaminants clustered by the tarsal secretion (unsputtered (**B**) and sputtered (**C**)). Box in (**B**) displays the magnified region of (**D**). (**D**) Close-up of contaminants clustered by tarsal secretion. Single contamination adhered to the surface of a droplet (unsputtered (**E**) and sputtered (**F**)). Images marked with "sp" show sputtered samples. c = contamination, d = droplet, cc = contamination cluster, sms = smooth surface structure, rs = rough surface structure.

3.8. Evaporation Rates

A measurement of the change in volume over time (evaporation rate) of 68 fluid droplets from 25 footprints was performed over a period of up to 75 days (Figure 10). The change in the droplet volume over the period of measured days and the evaporation rate (in %/min) of the three droplet types are visualized in Figure 11A,B.

The droplets revealed evaporation rates with different slopes. Based on their evaporation behavior and rates, we distinguished three different types of droplets indicated by different colors in Figure 11A. These three types are (1) the non-evaporating droplets (yellow), whose volume did not change over the entire measured period; (2) the slowly evaporating droplets (blue), which showed a slow evaporation rate over the measured period; and (3) the fast-evaporating droplets (red), which displayed a fast evaporation rate and completely evaporated after a few days, or at maximum, 32 days.

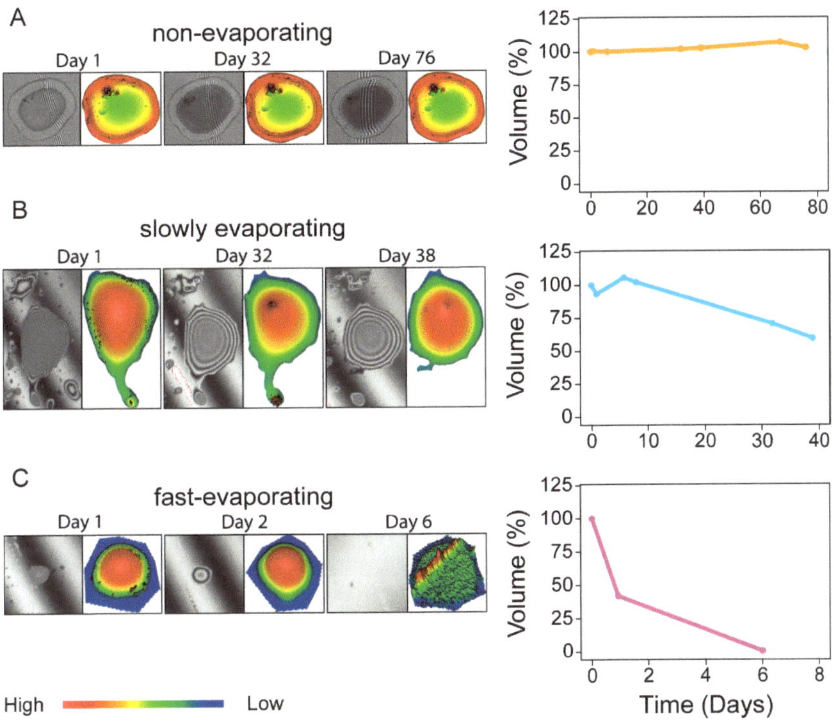

Figure 10. WLI measurements of droplet evaporations. Examples of non-evaporating (yellow) (**A**), slowly evaporating (blue) (**B**), and fast-evaporating (pink) droplets (**C**). On the left side, droplets are shown at three different measurement days for each droplet type. For every droplet, a microscopy image of the droplet taken with the WLI microscope (left) and the corresponding 3D heatmap showing its volume (right) are given. Graphs show measurement curves of the corresponding curves. The colors of the 3D images represent the relative height of the droplets: red = highest part; blue = lowest part.

The non-evaporating droplets showed an evaporation rate, which is the change in the droplet volume over the measured days, that was lower than 0.34%/day. In total, 14 droplets were assigned to this type, which showed mean evaporation rates of 0.0531 ± 0.283%/day and 0.0000369 ± 0.000196%/min (Figures 10A and 11).

The slowly evaporating type included all droplets with an evaporation rate between 0.34%/day and 3.2%/day. The group contained 12 droplets and displayed mean evaporation rates of 1.075 ± 0.479%/day and 0.000747 ± 0.000333%/min (Figures 10B and 11).

The fast-evaporating type showed a faster volume loss and included all droplets with an evaporation rate higher than 3.2%/day. This type was present in 44 droplets and exhibited mean evaporation rates of 8.629 ± 7.503%/day and 0.00599 ± 0.00521%/min (Figures 10C and 11).

A statistical comparison between the evaporation rates of the three types showed that the evaporation rate of the fast-evaporating type was significantly higher than that of the other types (Dunn's post hoc test, $p < 0.001$), whereas there was no significant difference between the evaporation rates of the slowly evaporating and non-evaporating types (Dunn's post hoc test, $p = 0.326$) (Kruskal–Wallis one-way ANOVA on ranks, $H = 50.912$, d.f. = 2, N (fast-evaporating) = 44, N (slowly evaporating) = 12, N (non-evaporating) = 14).

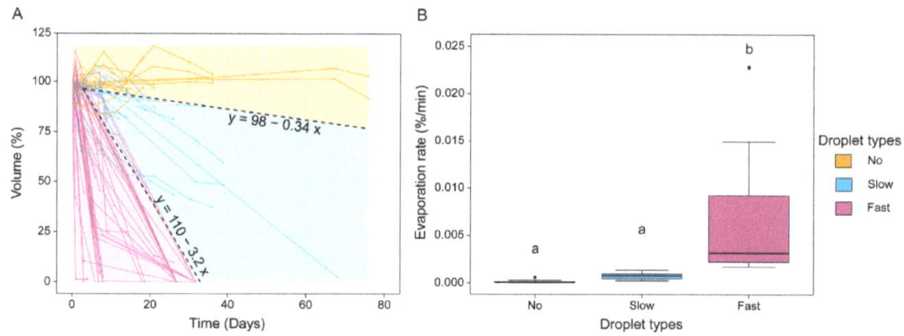

Figure 11. Change in droplet volume (in %) during 78 days of experiment (**A**) and evaporation rates (in %/min) (**B**). (**A**) Change in droplet volume. Droplet types are color-coded (non-evaporating droplets = yellow; slowly evaporating droplets = blue; fast-evaporating droplets = pink) and the regression lines of the boundaries to the three evaporation rate types are represented with corresponding linear regression equations. (**B**) Evaporation rates (in %/min) of non-evaporating droplets (yellow, $n = 14$), slowly evaporating droplets (blue, $n = 12$), and fast-evaporating (pink, $n = 44$) droplets. The values correspond to the mean evaporation rate (in %/min) of each individual droplet. Groups with different lowercase letters are statistically different (Kruskal–Wallis one-way ANOVA on ranks, $p < 0.001$ with Dunn's post hoc test, $p < 0.05$). Boxes indicate the 25th and 75th percentiles, whiskers are the 10th and 90th percentiles, and the line within the boxes shows the median.

3.9. Light Microscopy Observations

Randomly selected glass slides with deposited footprints were observed and filmed at different time intervals under an inverted microscope. During the observation, a fine needle was carefully pulled through individual droplets and the different behaviors of the droplets were observed and filmed (Supplementary Videos S1 and S2). Two different droplet responses were observed: (1) droplets that appeared to be liquid and were split into smaller droplets by the needle (Supplementary Video S1) and (2) droplets that appeared to be more viscous, whereby the needle scratched their surface (Supplementary Video S2).

4. Discussion

The investigations of the tarsal secretion of *Medauroidea extradentata* using cryo-SEM and WLI revealed that the fluid contains morphologically more diverse components than previously assumed. Previous research on the morphological and physical properties of the tarsal secretion of species with large smooth attachment pads is rare due to its higher viscosity when compared to the tarsal secretion of flies and beetles [33,69]. Peisker et al. (2014) were able to measure the viscosity of the tarsal secretion of flies and beetles by using the Brownian motion of micro-beads within the fluid [33]. Their lower viscosity enabled them to accumulate enough fluid, but this method could not be applied to the tarsal secretion of phasmids. Firstly, due to the high viscosity, it was not possible to collect a large amount of tarsal fluid using self-pulled glass needles and a micromanipulator, and secondly, the tarsal fluid was distributed over the surface in numerous fine droplets, and thus, there was no sufficient initial volume (own observations). Nevertheless, some results regarding the physical properties of the tarsal secretion of Phasmatodea were recently provided in [35]. They measured the contact angle (°) and dewetting speed (µm/s) of the tarsal secretion of phasmids with different body sizes and showed that the surface tension and viscosity of the fluid on glass are independent of the body size [35]. The high variance of the contact angle and dewetting speed suggests that the tarsal secretion is not perfectly homogenous and likely consists of multiple physically distinct components, which is confirmed by our results.

We discovered that the fluid consists of at least four morphologically different parts (droplets, flakes, thin films, and thick films (Figure 2)) including liquid components with different physical properties (non-evaporating, slowly evaporating, and fast-evaporating droplets) (Figures 10 and 11). These various constituents could explain the diverse properties of the tarsal secretions in the smooth attachment devices of Ensifera [58] that likely have similar secretions to that of the representatives of Phasmatodea. However, differences in the composition and quantity of the components may account for the differences in the fluid properties of different species. The complexity of biological tarsal fluid makes it difficult to reproduce the properties of this secretion using artificial-hydrocarbon-based components because the artificial fluid only mimics a part of this complex mixture, which might be surface-specific. Additionally, the mimicking quality of the other physical properties of natural secretions in biomimetic adhesive fluids remains unknown [70].

Sometimes, an additional imprint with a similar composition as those from the pads appeared between the imprints of the two euplantulae (Figure 3A,D). This is an interesting observation, which deserves mentioning here. Since there are fields of setae situated between them (Figure 1B), there are two possible origins of this imprint. First, the fluid potentially originates from the adhesive pad and is transported onto the hairs. Second, the hairs themselves secrete the fluid and therefore contribute to attachment.

4.1. Possible Origin of the Flake Component

The cryo-SEM allowed us to observe the tarsal fluid at a high resolution and thus identify four morphologically distinct structures (Figure 2). The stage temperature of −120 °C affected the behavior of the secretion, causing all of the previously liquid components to solidify. The solidification allowed us to identify the structural characteristics of each component. In order to obtain information about the actual behavior of the tarsal fluid, we included light microscopy observations and measurements of evaporation rates using the WLI (Figures 10 and 11). The light microscopy observations revealed that the tarsal secretion consists of different parts with varying viscosity, with few droplets hardening after time passes. The varying viscosities were detected by drawing a fine needle through individual droplets (see light microscopy observations). In some droplets, a scratching of the surface was observed, indicating a hardening (see Supplementary Videos S1 and S2) (Figure 1C). These observations were supported using the WLI, as measurements of the volume of individual droplets over time revealed three different evaporation rates (Figure 10).

Although no chemical analyses of the tarsal fluid of *M. extradentata* are readily available, of the composition of the fluids of other insect species allows us to draw assumptions on the composition based on the morphological observations (Figure 10).

Previous research on the adhesive fluid of insects with smooth pads (stick insects, cockroaches, and ants) also showed that it is a two-phase microemulsion, which consists of a volatile hydrophilic phase and a non-volatile hydrophobic phase [12,71,72].

The fast evaporation rate could be a result of the fraction of the tarsal fluid that has a potentially high volatile content (e.g., short-chained hydrocarbons and alcohols). A slow evaporation rate can be indicative of a droplet type that contains a higher proportion of non-volatile components (e.g., long-chained hydrocarbons and fatty acids) and a lower proportion of volatile components. The non-evaporating part of the tarsal liquid might consist of hardening non-volatile components (Figures 10 and 11) [27,28,34,39,40,42,47,73].

Similar measurements were conducted by Peisker and Gorb (2011), where they measured the evaporation rates of individual tarsal adhesive secretion droplets of the hairy attachment systems of the fly *Calliphora vicina* and the beetle *Coccinella septempuctata* with an atomic force microscope over a time span of 60 min [73]. Within this time span, they discovered fast evaporation for the tarsal secretion of the fly and a comparably slower evaporation for the tarsal secretion of the beetle. The main difference in our findings for the secretion on the smooth attachment devices of *M. extradentata* is the presence of a range of different evaporation rates including both fast and slow evaporation rates in different droplets from the same footprint. As Peisker and Gorb measured the evaporation in the first 60 min, and we observed the evaporation for up to 80 days with larger time spans between measurements, a more precise comparison within the same time scale would be necessary for these examples. However, the tarsal secretions from the hairy attachment systems investigated therein seem to consist of more uniform droplets compared to the smooth attachment system of *M. extradentata*.

Our microscopy observations and experiments reported above allow us to make predictions about the material properties and behavior of the four tarsal secretion components. We may hypothesize that one part of the tarsal secretion being fluid and solidifying over time in the light microscope is the same component that shows no evaporation and the flake-like appearance in the cryo-SEM. The morphological indications for this hypothesis are the particular layered surface structure of sputtered flakes resembling the structure of dried fluid (Figure 5B,E,F), and the position of the flakes on the edge of the imprints, since this is the site where the majority of the other contaminants was located (Figure 5A). Chemical analyses of the tarsal secretion of other insects also found lipid-soluble components that could solidify over time [28,34]. Also, flake-like structures were previously detected in the footprints of *Locusta migratoria* [28]. Therefore, we may assume that the flakes are hardened and accumulate old parts of tarsal secretion, which are removed via the passive self-cleaning mechanism to the margin of the pad. The occurrence of flakes indicates that the adhesive fluid is at least partially composed of non-volatile components.

4.2. Self-Cleaning Mechanism

The passive self-cleaning mechanism of smooth attachment devices is important for attachment maintenance, as it removes contaminants that would reduce adhesive performance [30,31]. Contaminants reduce attachment by increasing the distance between the adhesive pad and the substrate, which, in turn, reduces the actual surface area that is available for contact with the substrate [31]. Two different kinds of contaminants were observed in the footprints: the old, hardened tarsal secretion in the form of flakes (Figure 5) and the contamination from the environment (Figure 9).

Clemente and colleagues described the effect of the tarsal secretion for self-cleaning in the stick insect *Carausius morosus*. They discovered that a high amount of tarsal fluid increases the recovery rate of the adhesion and hypothesized that this is due to the liquid filling the gaps [30,31]. We can confirm with our cryo-SEM data that the tarsal fluid supports the self-cleaning mechanism.

We observed that all contaminants were either covered or surrounded by the tarsal secretion (Figure 9A (surrounded) and B–D (covered)). Small particles are agglomerated together via the adhesive secretion [74], reducing the ratio of volume to surface area (Figure 9B–D), thereby enabling an agglomerate to be removed easier. Due to the convex shape of the adhesive pads, as well as the pressure of the newly produced tarsal secretion both types of contaminants are transported further to the edge in subsequent steps during locomotion and are finally deposited on the substrate via the shearing motions of the tarsus (Figures 3A,B and 5A).

4.3. Attachment

Various experiments showed that the tarsal secretion has different effects on the attachment force generation, ranging from implementing viscous and capillary forces [4,33] to leveling the asperities on the substrate surfaces [48,50]. Our observations on the morphology of the secretion components provide support for the understanding of the following effects of the secretion in *M. extradentata*.

4.4. Viscosity and Capillary Forces

Multiple experiments indicate that the action of the capillary and viscous forces is important to generate adhesion in wet contacts [32,33,35,54,75]. We observed different morphological manifestations of the same fluid, which must vary in viscosity, and thus may affect the viscous and capillary forces to different extents.

The thick film components display morphological characteristics that can be attributed to those of viscous fluids, such as covering a large surface area (Figure 5A,E) and possessing a large volume (Figure 5 (dark grey scale)). We also measured droplets exhibiting a slow evaporation rate and appearing to be more viscous than others (Supplementary Video S2). Besides viscosity, the surface tension influences the liquids' interaction with the substrate and can contribute to them having similar appearances. However, viscosity and surface tension are somewhat related [76,77].

Therefore, the thick films are likely rather viscous and/or possess a comparably high surface tension, which potentially aids in implementing the viscous force at the tarsus–substrate interface.

The droplet components show different morphology than the films, as they form individual roughly round shapes that can accumulate into big complexes. We found droplets in the footprints exhibiting a fast evaporation rate (mean evaporation rate of $8.629 \pm 7.503\%/\text{day}$ and $0.00599 \pm 0.00521\%/\text{min}$) (Figure 10C). Due to the spectrum of different morphologies and evaporation rates occurring in the same secretion, the adhesive fluid can likely adapt to different substrate qualities and thus effectively combine capillary and viscous forces, enhancing attachment performance.

4.5. Free Surface Energy

For insects, the adaptation to different free surface energies of the substrates is challenging, as this is a factor that can influence locomotion, which, for example, plants utilize to either repel or capture insects [78–80]. A chemical analysis of the tarsal fluids of insects has shown that they are composed of hydrophilic and hydrophobic components in an emulsion that is capable of adapting to different free surface energies [28,34]. Although our morphological observations are not sufficient to predict the chemical composition of the footprints, the different morphological appearances of the fluids in the same footprint allow us to speculate about the physical properties of the components.

Nano-droplets were found along with the larger droplets (Figures 4D and 8E), supporting the previous findings of a highly viscous emulsion consisting of lipid droplets in a water-based solution [28]. Different ice crystal shapes were observed on the surfaces of the thick film components (Figure 8A,B,D), which were likely formed due to the freezing of the water that was present within these components.

The droplet and thick film components show a range of surface structures, which can be smooth, rough, or granular (Figure 4C,F (droplets) and Figure 8C,F (thick film)). This suggests that the chemical composition of these components could differ and that, overall, the different components consist of a mixture of the same compounds, which assemble into the morphology we observed. Further evidence for this hypothesis could be found in the specificity of evaporation rates (Figure 11A). Assuming that the different components consist of a mixture of the same ingredients with varying proportions within the composition, it appears likely that the broad spectrum of evaporation rates found in the droplets is a result of the volatile components' decreasing proportions in droplets with slower evaporation rates. This would also explain why the evaporation rates of the non-evaporating and slowly evaporating droplets did not differ statistically (Figure 10B) in spite of different evaporation behaviors and visual differences under light microscopes (our own observations). Further evidence of the ability of the tarsal secretion to support the attachment on surfaces with different surface energies was shown in *M. extradentata*. In these experiments, *M. extradentata* was able to adhere to a highly hydrophobic surface (Polytetrafluoroethylene (PTFE)) even underwater [55].

Since different compositions of the adhesive fluid should respond differently to the free surface energy of the substrate, detailed analyses of the chemical composition are required for a deeper understanding of the functional role of the fluid in the tarsal attachment system of the stick insect.

4.6. Leveling Substrate Asperities

An insect's adhesive secretion is often mentioned as having a function in leveling the surface roughness in the adhesive contact [32,51,81,82]. Rough surfaces cause a reduced contact area between the adhesive pad and the substrate. Tarsal adhesive secretions fill up gaps in the roughness, increasing the contact area and thus the attachment forces [48,49,53]. To fulfil this task, such secretion needs to cover large areas to be available in sufficient amounts, and their viscosity must be adapted to the corresponding roughness hierarchy. The presumable fluid components observed in the cryo-SEM showed different dewetting morphologies. Thin film components cover a wide gap-less area with a thin film (thin films cover 1 mm in diameter (Figure 6A)). Thick film components cover the surface with a patchy thick film (thick films cover 2 mm in diameter (Figure 7A,E)), and the droplet components form single droplets, which can accumulate into larger complexes (droplets cover ~200 μm in diameter (Figure 4)).

As the surfaces of natural substrates have fractal roughness at different hierarchical levels [83–85], a mixture of fluids with a range in viscosity would be helpful to quickly adapt to a range of roughness at once. Thick films are probably more viscous and, hence, are able to fill large gaps of coarse roughness. Thin films potentially have a lower viscosity, which is judged on their low volume and wide spreading on the substrates, and likely more readily fill gaps of finer roughness. The low volume is visible in the lower height of the latter fluids in imprints, which disappear completely when covered by a 10 nm Au-Pa layer when sputter-coated (see Supplementary Figure S1). Another indication that the tarsal fluid of *M. extradentata* can have an influence on the leveling of the substrate asperities is the performance of *M. extradentata* on substrates with varying roughness, which was reported in previous experiments [62,68].

5. Conclusions

The cryo-SEM enabled an examination of the pad fluid in its frozen state immediately after deposition. We identified four morphologically different components that originate from the same tarsal secretion. The measurements of the evaporation rate of individual droplets indicate that the liquid consists of a spectrum of slowly evaporating to fast-evaporating components. These observations suggest that the tarsal fluid is a mixture of volatile and non-volatile components that, working in concert, extend the properties of the adhesive secretion.

Parts of the adhesive secretion can harden over time into flakes and thus contaminate the adhesive area of the attachment pads. Contaminations can be glued together via new adhesive fluid and can be passively removed via tarsal movement during locomotion. Due to the presence of morphologically and physically different components, the adhesive fluid can support different phases of attachment, including contact generation, contact maintenance, and contact breakage, for example, by filling varying degrees of roughness and generating capillary and/or viscous forces.

With this study, we show how the morphologically and physically diverse tarsal secretion of *Medauroidea extradentata* could potentially contribute to the range of functions. These results allow for several possible ideas to be generated for further investigations. A detailed chemical analysis of the adhesive secretion would aid in making a correlation between the morphological features and chemical composition. Histological studies could provide insights into the structure and distribution of the exocrine cells that are involved in the production of the secretion. Studies of the composition of the tarsal fluid of different ecologically specialized taxa can aid in understanding the adaptability of tarsal secretions. These insights could be valuable for the development of novel biomimetic adhesive fluids.

Supplementary Materials: The following supporting information can be downloaded at: https://www.mdpi.com/article/10.3390/biomimetics8050439/s1, Figure S1: Unsputtered and sputtered structures, Data S1: raw data of the evaporation measurements, Video S1: video of liquid droplet, Video S2: video of viscous droplet.

Author Contributions: Conceptualization: S.N.G., J.T. and T.H.B.; Methodology: J.T., S.N.G. and T.H.B.; Validation: J.T. and T.H.B.; Formal Analysis: J.T.; Investigation: J.T.; Resources: S.N.G.; Data Curation: J.T. and T.H.B.; Writing—Original Draft: J.T.; Writing—Review and Editing: S.N.G. and T.H.B.; Visualization: J.T.; Supervision: T.H.B.; Project Administration: S.N.G. and T.H.B.; Funding Acquisition: S.N.G. All authors have read and agreed to the published version of the manuscript.

Funding: This work was supported the German Research Foundation (DFG) (grant GO 995/34-1).

Institutional Review Board Statement: Not applicable.

Data Availability Statement: The raw data of the evaporation measurements (Supplementary Data S1) and an additional SEM picture (Supplementary Figure S1) and light microscopy videos (Supplementary Videos S1 and S2) are available as Supplementary Materials.

Acknowledgments: We thank Esther Appel and Alexander Kovalev (Department of Functional Morphology and Biomechanics, Kiel University, Germany) for their technical assistance.

Conflicts of Interest: The authors declare no conflict of interest.

References

1. Creton, C.; Gorb, S. Sticky Feet: From Animals to Materials. *MRS Bull.* **2007**, *32*, 466–472. [CrossRef]
2. Eisner, T.; Aneshansley, D.J. Defense by foot adhesion in a beetle (*Hemisphaerota cyanea*). *Proc. Natl. Acad. Sci. USA* **2000**, *97*, 6568–6573. [CrossRef] [PubMed]
3. Voigt, D.; Schuppert, J.M.; Dattinger, S.; Gorb, S.N. Sexual dimorphism in the attachment ability of the Colorado potato beetle *Leptinotarsa decemlineata* (Coleoptera: Chrysomelidae) to rough substrates. *J. Insect Physiol.* **2008**, *54*, 765–776. [CrossRef] [PubMed]
4. Ditsche, P.; Summers, A.P. Aquatic versus terrestrial attachment: Water makes a difference. *Beilstein J. Nanotechnol.* **2014**, *5*, 2424–2439. [CrossRef] [PubMed]
5. Büscher, T.H.; Gorb, S.N. Physical constraints lead to parallel evolution of micro- and nanostructures of animal adhesive pads: A review. *Beilstein J. Nanotechnol.* **2021**, *12*, 725–743. [CrossRef]
6. Büscher, T.H.; Gorb, S.N. Convergent Evolution of Animal Adhesive Pads. In *Convergent Evolution*; Springer: Cham, Switzerland, 2023; pp. 257–287. [CrossRef]
7. Gorb, S.N. Smooth Attachment Devices in Insects: Functional Morphology and Biomechanics. *Adv. Insect Physiol.* **2007**, *34*, 81–115. [CrossRef]
8. Gorb, S.N.; Beutel, R.G. Evolution of locomotory attachment pads of hexapods. *Naturwissenschaften* **2001**, *88*, 530–534. [CrossRef]
9. Federle, W. Why are so many adhesive pads hairy? *J. Exp. Biol.* **2006**, *209*, 2611–2621. [CrossRef]
10. Autumn, K.; Sitti, M.; Liang, Y.A.; Peattie, A.M.; Hansen, W.R.; Sponberg, S.; Kenny, T.W.; Fearing, R.; Israelachvili, J.N.; Full, R.J. Evidence for van der Waals adhesion in gecko setae. *Proc. Natl. Acad. Sci. USA* **2002**, *99*, 12252–12256. [CrossRef]
11. Dirks, J.-H.; Federle, W. Fluid-based adhesion in insects—Principles and challenges. *Soft Matter* **2011**, *7*, 11047–11053. [CrossRef]
12. Dirks, J.-H.; Federle, W. Mechanisms of fluid production in smooth adhesive pads of insects. *J. R. Soc. Interface* **2011**, *8*, 952–960. [CrossRef]
13. Busshardt, P.; Wolf, H.; Gorb, S.N. Adhesive and frictional properties of tarsal attachment pads in two species of stick insects (Phasmatodea) with smooth and nubby euplantulae. *Zoology* **2012**, *115*, 135–141. [CrossRef] [PubMed]
14. Smith, A.M. *Biological Adhesives*; Springer: Berlin/Heidelberg, Germany, 2006.
15. Davey, P.A.; Power, A.M.; Santos, R.; Bertemes, P.; Ladurner, P.; Palmowski, P.; Clarke, J.; Flammang, P.; Lengerer, B.; Hennebert, E.; et al. Omics-based molecular analyses of adhesion by aquatic invertebrates. *Biol. Rev. Camb. Philos. Soc.* **2021**, *96*, 1051–1075. [CrossRef]
16. Delroisse, J.; Kang, V.; Gouveneaux, A.; Santos, R.; Flammang, P. Convergent Evolution of Attachment Mechanisms in Aquatic Animals. In *Convergent Evolution*; Springer: Cham, Switzerland, 2023; pp. 523–557. [CrossRef]
17. Federle, W.; Barnes, W.J.P.; Baumgartner, W.; Drechsler, P.; Smith, J.M. Wet but not slippery: Boundary friction in tree frog adhesive toe pads. *J. R. Soc. Interface* **2006**, *3*, 689–697. [CrossRef] [PubMed]
18. Langowski, J.K.A.; Dodou, D.; Kamperman, M.; van Leeuwen, J.L. Tree frog attachment: Mechanisms, challenges, and perspectives. *Front. Zool.* **2018**, *15*, 32. [CrossRef]
19. Rasmussen, M.H.; Holler, K.R.; Baio, J.E.; Jaye, C.; Fischer, D.A.; Gorb, S.N.; Weidner, T. Evidence that gecko setae are coated with an ordered nanometre-thin lipid film. *Biol. Lett.* **2022**, *18*, 20220093. [CrossRef] [PubMed]
20. Biederman-Thorson, M.A. *Biological Mechanisms of Attachment: The Comparative Morphology and Bioengineering of Organs for Linkage, Suction, and Adhesion*; Springer: Berlin/Heidelberg, Germany, 1974.
21. Flammang, P.; Michel, A.; Cauwenberge, A.V.; Alexandre, H.; Jangoux, M. A study of the temporary adhesion of the podia in the sea star *asterias rubens* (Echinodermata, asteroidea) through their footprints. *J. Exp. Biol.* **1998**, *201 Pt 16*, 2383–2395. [CrossRef]
22. Flammang, P. Adhesive Secretions in Echinoderms: An Overview. In *Biological Adhesives: With 15 Tables*; Smith, A.M., Callow, J.A., Eds.; Springer: Berlin/Heidelberg, Germany, 2006; pp. 183–206.

23. Hennebert, E.; Maldonado, B.; Ladurner, P.; Flammang, P.; Santos, R. Experimental strategies for the identification and characterization of adhesive proteins in animals: A review. *Interface Focus* **2015**, *5*, 20140064. [CrossRef]
24. Seabra, S.; Zenleser, T.; Grosbusch, A.L.; Hobmayer, B.; Lengerer, B. The Involvement of Cell-Type-Specific Glycans in *Hydra* Temporary Adhesion Revealed by a Lectin Screen. *Biomimetics* **2022**, *7*, 166. [CrossRef]
25. Betz, O. Adhesive Exocrine Glands in Insects: Morphology, Ultrastructure, and Adhesive Secretion. In *Biological Adhesive Systems*; Springer: Vienna, Austria, 2010; pp. 111–152. [CrossRef]
26. Schmitt, C.; Betz, O. Morphology and ultrastructure of the tarsal adhesive organs of the Madagascar hissing cockroach Gromphadorhina portentosa. *Cell Tissue Res.* **2017**, *370*, 243–265. [CrossRef]
27. Gorb, S.S.N. *Attachment Devices of Insect Cuticle*; Springer: Dordrecht, The Netherlands, 2001.
28. Vötsch, W.; Nicholson, G.; Müller, R.; Stierhof, Y.-D.; Gorb, S.; Schwarz, U. Chemical composition of the attachment pad secretion of the locust *Locusta migratoria*. *Insect Biochem. Mol. Biol.* **2002**, *32*, 1605–1613. [CrossRef] [PubMed]
29. Abou, B.; Gay, C.; Laurent, B.; Cardoso, O.; Voigt, D.; Peisker, H.; Gorb, S. Extensive collection of femtolitre pad secretion droplets in the beetle *Leptinotarsa decemlineata* allows nanolitre microrheology. *J. R. Soc. Interface* **2010**, *7*, 1745–1752. [CrossRef] [PubMed]
30. Clemente, C.J.; Bullock, J.M.R.; Beale, A.; Federle, W. Evidence for self-cleaning in fluid-based smooth and hairy adhesive systems of insects. *J. Exp. Biol.* **2010**, *213*, 635–642. [CrossRef] [PubMed]
31. Clemente, C.J.; Federle, W. Mechanisms of self-cleaning in fluid-based smooth adhesive pads of insects. *Bioinspir. Biomim.* **2012**, *7*, 046001. [CrossRef]
32. Dirks, J.-H. Adhesion in Insects. In *Encyclopedia of Nanotechnology*; Bhushan, B., Ed.; Springer: Dordrecht, The Netherlands, 2014; pp. 1–10.
33. Peisker, H.; Heepe, L.; Kovalev, A.E.; Gorb, S.N. Comparative study of the fluid viscosity in tarsal hairy attachment systems of flies and beetles. *J. R. Soc. Interface* **2014**, *11*, 20140752. [CrossRef]
34. Betz, O.; Albert, K.; Boley, M.; Frenzel, M.; Gerhardt, H.; Grunwald, I.; Hartwig, A.; Kleemeier, M.; Maurer, A.; Neuenfeldt, M.; et al. Struktur und Funktion des tarsalen Haftsystems der Madagaskar-Fauchschabe *Grompadorhina portentosa* (Blattodea). *Mitt. Dtsch. Ges. Allg. Angew. Ent.* **2018**, *21*, 159–164.
35. Kaimaki, D.-M.; Andrew, C.N.S.; Attipoe, A.E.L.; Labonte, D. The physical properties of the stick insect pad secretion are independent of body size. *J. R. Soc. Interface* **2022**, *19*, 20220212. [CrossRef]
36. Bauchhen, E. Die Pulvillen von *Calliphora erythrocephala* (Diptera, Brachycera) als Adhsionsorgane. *Zoomorphology* **1979**, *93*, 99–123. [CrossRef]
37. Schmitt, U. Hydrocarbons in tarsal glands of *Bombus terrestris*. *Experientia* **1990**, *46*, 1080–1082. [CrossRef]
38. Eltz, T. Tracing pollinator footprints on natural flowers. *J. Chem. Ecol.* **2006**, *32*, 907–915. [CrossRef]
39. Ishii, S. Adhesion of a Leaf Feeding Ladybird *Epilachna vigintioctomaculta* (Coleoptera: Coccinellidae) on a Vertically Smooth Surface. *Appl. Entomol. Zool.* **1987**, *22*, 222–228. [CrossRef]
40. Kosaki, A.; Yamaoka, R. Chemical Composition of Footprints and Cuticula Lipids of Three Species of Lady Beetles. *Jpn. J. Appl. Entomol. Zool.* **1996**, *40*, 47–53. [CrossRef]
41. Betz, O. Structure of the tarsi in some *Stenus* species (Coleoptera, Staphylinidae): External morphology, ultrastructure, and tarsal secretion. *J. Morphol.* **2003**, *255*, 24–43. [CrossRef] [PubMed]
42. Geiselhardt, S.F.; Geiselhardt, S.; Peschke, K. Comparison of tarsal and cuticular chemistry in the leaf beetle *Gastrophysa viridula* (Coleoptera: Chrysomelidae) and an evaluation of solid-phase microextraction and solvent extraction techniques. *Chemoecology* **2009**, *19*, 185–193. [CrossRef]
43. Attygale, A.B.; Aneshansley, D.J.; Meinwald, J.; Eisner, T. Defense by foot adhesion in a chrysomelid beetle (*Hemisphaerota cyanea*): Characterization of the adhesive oil. *Zoology* **2000**, *103*, 1–6.
44. Geiselhardt, S.F.; Federle, W.; Prüm, B.; Geiselhardt, S.; Lamm, S.; Peschke, K. Impact of chemical manipulation of tarsal liquids on attachment in the Colorado potato beetle, *Leptinotarsa decemlineata*. *J. Insect Physiol.* **2010**, *56*, 398–404. [CrossRef]
45. Geiselhardt, S.F.; Lamm, S.; Gack, C.; Peschke, K. Interaction of liquid epicuticular hydrocarbons and tarsal adhesive secretion in *Leptinotarsa decemlineata* Say (Coleoptera: Chrysomelidae). *J. Comp. Physiol. A* **2010**, *196*, 369–378. [CrossRef]
46. Geiselhardt, S.F.; Geiselhardt, S.; Peschke, K. Congruence of epicuticular hydrocarbons and tarsal secretions as a principle in beetles. *Chemoecology* **2011**, *21*, 181–186. [CrossRef]
47. Reitz, M.; Gerhardt, H.; Schmitt, C.; Betz, O.; Albert, K.; Lämmerhofer, M. Analysis of chemical profiles of insect adhesion secretions by gas chromatography–mass spectrometry. *Anal. Chim. Acta* **2015**, *854*, 47–60. [CrossRef]
48. Drechsler, P.; Federle, W. Biomechanics of smooth adhesive pads in insects: Influence of tarsal secretion on attachment performance. *J. Comp. Physiol. A Neuroethol. Sens. Neural Behav. Physiol.* **2006**, *192*, 1213–1222. [CrossRef]
49. Gorb, E.V.; Hosoda, N.; Miksch, C.; Gorb, S.N. Slippery pores: Anti-adhesive effect of nanoporous substrates on the beetle attachment system. *J. R. Soc. Interface* **2010**, *7*, 1571–1579. [CrossRef] [PubMed]
50. Kovalev, A.E.; Varenberg, M.; Gorb, S.N. Wet versus dry adhesion of biomimetic mushroom-shaped microstructures. *Soft Matter* **2012**, *8*, 7560–7566. [CrossRef]
51. Barnes, W.J.; Smith, J.; Oines, C.; Mundl, R. Bionics and wet grip. *Tire Technol. Int.* **2002**, *2002*, 56–60.
52. Dirks, J.-H.; Clemente, C.J.; Federle, W. Insect tricks: Two-phasic foot pad secretion prevents slipping. *J. R. Soc. Interface* **2010**, *7*, 587–593. [CrossRef] [PubMed]

53. Bullock, J.M.R.; Drechsler, P.; Federle, W. Comparison of smooth and hairy attachment pads in insects: Friction, adhesion and mechanisms for direction-dependence. *J. Exp. Biol.* **2008**, *211*, 3333–3343. [CrossRef]
54. Stefan, J. Versuche über die scheinbare Adhäsion. *Ann. Phys. Chem.* **1875**, *230*, 316–318. [CrossRef]
55. Thomas, J.; Gorb, S.N.; Büscher, T.H. Influence of surface free energy of the substrate and flooded water on the attachment performance of stick insects (Phasmatodea) with different adhesive surface microstructures. *J. Exp. Biol.* **2023**, *226*, jeb244295. [CrossRef]
56. Lockey, K.H. Lipids of the insect cuticle: Origin, composition and function. *Comp. Biochem. Physiol. Part B Comp. Biochem.* **1988**, *89*, 595–645. [CrossRef]
57. Gorb, S.; Jiao, Y.; Scherge, M. Ultrastructural architecture and mechanical properties of attachment pads in *Tettigonia viridissima* (Orthoptera Tettigoniidae). *J. Comp. Physiol. A* **2000**, *186*, 821–831. [CrossRef]
58. Gorb, S.; Scherge, M. Biological microtribology: Anisotropy in frictional forces of orthopteran attachment pads reflects the ultrastructure of a highly deformable material. *Proc. Biol. Sci.* **2000**, *267*, 1239–1244. [CrossRef]
59. Phasmida Species File Online. Available online: http://phasmida.speciesfile.org/HomePage/Phasmida/HomePage.aspx (accessed on 11 August 2023).
60. Büscher, T.H.; Buckley, T.R.; Grohmann, C.; Gorb, S.N.; Bradler, S. The Evolution of Tarsal Adhesive Microstructures in Stick and Leaf Insects (Phasmatodea). *Front. Ecol. Evol.* **2018**, *6*, 69. [CrossRef]
61. Büscher, T.H.; Kryuchkov, M.; Katanaev, V.L.; Gorb, S.N. Versatility of Turing patterns potentiates rapid evolution in tarsal attachment microstructures of stick and leaf insects (Phasmatodea). *J. R. Soc. Interface* **2018**, *15*, 20180281. [CrossRef]
62. Büscher, T.H.; Gorb, S.N. Complementary effect of attachment devices in stick insects (Phasmatodea). *J. Exp. Biol.* **2019**, *222*, jeb209833. [CrossRef]
63. Büscher, T.H.; Grohmann, C.; Bradler, S.; Gorb, S.N. *Tarsal Attachment Pads in Phasmatodia (Hexapoda: Insecta)*; Schweizerbart Science Publishers: Stuttgart, Germany, 2019.
64. Labonte, D.; Federle, W. Functionally different pads on the same foot allow control of attachment: Stick insects have load-sensitive "heel" pads for friction and shear-sensitive "toe" pads for adhesion. *PLoS ONE* **2013**, *8*, e81943. [CrossRef] [PubMed]
65. Büscher, T.H.; Becker, M.; Gorb, S.N. Attachment performance of stick insects (Phasmatodea) on convex substrates. *J. Exp. Biol.* **2020**, *223*, jeb226514. [CrossRef]
66. Labonte, D.; Williams, J.A.; Federle, W. Surface contact and design of fibrillar 'friction pads' in stick insects (*Carausius morosus*): Mechanisms for large friction coefficients and negligible adhesion. *J. R. Soc. Interface* **2014**, *11*, 20140034. [CrossRef]
67. Winand, J.; Gorb, S.N.; Büscher, T.H. Gripping performance in the stick insect *Sungaya inexpectata* in dependence on the pretarsal architecture. *J. Comp. Physiol. A* **2023**, *209*, 313–323. [CrossRef] [PubMed]
68. Burack, J.; Gorb, S.N.; Büscher, T.H. Attachment Performance of Stick Insects (Phasmatodea) on Plant Leaves with Different Surface Characteristics. *Insects* **2022**, *13*, 952. [CrossRef]
69. Peattie, A.M.; Dirks, J.-H.; Henriques, S.; Federle, W. Arachnids secrete a fluid over their adhesive pads. *PLoS ONE* **2011**, *6*, e20485. [CrossRef]
70. Fowler, J.E.; Gorb, S.; Baio, J.E. Multi-Technique Investigation of a Biomimetic Insect Tarsal Adhesive Fluid. *Front. Mech. Eng.* **2021**, *7*, 681120. [CrossRef]
71. Dirks, J.-H. Physical principles of fluid-mediated insect attachment—Shouldn't insects slip? *Beilstein J. Nanotechnol.* **2014**, *5*, 1160–1166. [CrossRef] [PubMed]
72. Federle, W.; Riehle, M.; Curtis, A.S.G.; Full, R.J. An integrative study of insect adhesion: Mechanics and wet adhesion of pretarsal pads in ants. *Integr. Comp. Biol.* **2002**, *42*, 1100–1106. [CrossRef] [PubMed]
73. Peisker, H.; Gorb, S.N. Evaporation dynamics of tarsal liquid footprints in flies (*Calliphora vicina*)and beetles (*Coccinella septempunctata*). *J. Exp. Biol.* **2012**, *215*, 1266–1271. [CrossRef] [PubMed]
74. Chen, H.; Liu, W.; Li, S. Random loose packing of small particles with liquid cohesion. *AIChE J.* **2019**, *65*, 500–511. [CrossRef]
75. Kendall, K. *Molecular Adhesion and Its Applications: The Sticky Universe*; Kluwer Academic Publishers: Boston, MA, USA, 2004.
76. Pelofsky, A.H. Surface Tension-Viscosity Relation for Liquids. *J. Chem. Eng. Data* **1966**, *11*, 394–397. [CrossRef]
77. Zheng, M.; Tian, J.; Mulero, Á. New correlations between viscosity and surface tension for saturated normal fluids. *Fluid Phase Equilibria* **2013**, *360*, 298–304. [CrossRef]
78. Gorb, E.; Gorb, S. Effects of surface topography and chemistry of *Rumex obtusifolius* leaves on the attachment of the beetle *Gastrophysa viridula*. *Entomol. Exp. Appl.* **2009**, *130*, 222–228. [CrossRef]
79. Gorb, E.V.; Gorb, S.N. Anti-adhesive effects of plant wax coverage on insect attachment. *J. Exp. Bot.* **2017**, *68*, 5323–5337. [CrossRef]
80. Labonte, D.; Robinson, A.; Bauer, U.; Federle, W. Disentangling the role of surface topography and intrinsic wettability in the prey capture mechanism of Nepenthes pitcher plants. *Acta Biomater.* **2021**, *119*, 225–233. [CrossRef]
81. Bullock, J.M.R.; Federle, W. The effect of surface roughness on claw and adhesive hair performance in the dock beetle *Gastrophysa viridula*. *Insect Sci.* **2011**, *18*, 298–304. [CrossRef]
82. England, M.W.; Sato, T.; Yagihashi, M.; Hozumi, A.; Gorb, S.N.; Gorb, E.V. Surface roughness rather than surface chemistry essentially affects insect adhesion. *Beilstein J. Nanotechnol.* **2016**, *7*, 1471–1479. [CrossRef] [PubMed]
83. Gorb, S.N. Biological Fibrillar Adhesives: Functional Principles and Biomimetic Applications. In *Handbook of Adhesion Technology*; Da Silva, L.F.M., Öchsner, A., Adams, R.D., Eds.; Springer: Berlin/Heidelberg, Germany; Cham, Switzerland, 2011; pp. 1409–1436.

84. Büscher, T.H.; Quigley, E.; Gorb, S.N. Adhesion Performance in the Eggs of the Philippine Leaf Insect *Phyllium philippinicum* (Phasmatodea: Phylliidae). *Insects* **2020**, *11*, 400. [CrossRef] [PubMed]
85. Salerno, G.; Rebora, M.; Piersanti, S.; Büscher, T.H.; Gorb, E.V.; Gorb, S.N. Oviposition site selection and attachment ability of *Propylea quatuordecimpunctata* and *Harmonia axyridis* from the egg to the adult stage. *Physiol. Entomol.* **2022**, *47*, 20–37. [CrossRef]

biomimetics

MDPI

Article

Effect of the Mechanical Properties of Soft Counter-Faces on the Adhesive Capacity of Mushroom-Shaped Biomimetic Microstructures

May Gonen [1] and Haytam Kasem [1,2,*]

1 Faculty of Mechanical Engineering, The Technion-Israel Institute of Technology, Haifa 32000, Israel
2 Department of Mechanical Engineering, Azrieli College of Engineering, Jerusalem 9103501, Israel
* Correspondence: haytamka@jce.ac.il

Abstract: The effects of mechanical properties and contact environment conditions on the adhesiveness of the biomimetic adhesive mushroom-shaped micro-structure have been experimentally investigated. The idea is based on the adhesive micro-structures and surfaces inspired by nature after observing the abilities of some animals. Applications are proposed in various fields of engineering and technology. However, to enable unconventional uses of these biomimetic adhesion surfaces, such as in the biomedical field, it is necessary to adjust and optimize their tribological properties (friction, adhesion, and peeling strength) in contact with soft substrates that can simulate the mechanical features of biological tissues. Our work explores the effect of the combinations of the various parameters on the strength of adhesion. Under dry contact conditions, soft counter-faces lead to lower adhesion than hard counter-faces, whereas under wet conditions, soft counter-faces lead to higher adhesion than harder counter-faces.

Keywords: biomimetic; mushroom-shaped microstructure; adhesion; substrate; mechanical proprieties

Citation: Gonen, M.; Kasem, H. Effect of the Mechanical Properties of Soft Counter-Faces on the Adhesive Capacity of Mushroom-Shaped Biomimetic Microstructures. *Biomimetics* **2023**, *8*, 327. https://doi.org/10.3390/biomimetics8030327

Academic Editors: Stanislav N. Gorb and Thies Henning Büscher

Received: 2 July 2023
Revised: 16 July 2023
Accepted: 17 July 2023
Published: 24 July 2023

1. Introduction

Over the few last decades, the field of adhesive sciences has evolved due to the growing need for reversible and rapid adhesive systems in various fields of technology [1–3]. These developments were inspired by the biological adhesive systems found in several species of insects, reptiles, and spiders, which have developed unique biological attachment systems during their natural evolution. These systems allow them to grip and run on the wide range of horizontal, vertical, rough, and smooth surfaces that they encounter in their living environments [4–6]. Systems based on permanent or long-term adhesion mainly rely on mushroom-shaped micro-structures, while systems involved in short-term temporary adhesion mainly rely on the spatula shape of individual contacts [1].

One of the dry biomimetic adhesives developed for real use is based on mushroom-shaped contact elements [4]. Inspired by the sticky hairs found in male beetles from the Chrysomelidae family, this microstructure does not present a hierarchical geometry like that found in the biological attachment system of the gecko; rather, it is simply a surface covered with mushroom-shaped microstructures. This attachment system is suitable for creating a long-term passive adhesive force on smooth substrates with almost no pre-load. The potential of these biomimetic adhesives was first verified using a robot-type device walking on smooth vertical surfaces that used this microstructure [7]. The tribological performances of mushroom-shaped adhesive microstructures attracted the attention of many researchers during the last decade. Research was conducted to investigate their various properties and the influence of different operational parameters. Pre-load and contamination have been studied by Gorb et al. [4]. They compared the adhesive properties of a biomimetic mushroom-shaped fibrillar microstructure to that of a control flat surface that was made of the same material and operated under the same operational conditions, and they proved that the adhesive features of the structured surface were more than twice as effective as those of thee flat surface.

The influence of pre-load on the adhesion was studied by Varenberg et al. [8]. They showed that the mushroom-shaped geometry of fibrillar contact elements was responsible for a stable adhesive attachment. This type of contact element promotes the fast and simple generation of reliable adhesion. The mushroom-shaped geometry seems to transform fibrillar contact elements into passive suction devices and makes them tolerant to overload, thus enhancing their robustness and stability [9]. Additional parameters have been studied, such as the rule of hierarchy. These works show that the adhesion enhancements are the result of increased surface conformation [10]. In addition, it is reported that ambient pressure and suction may contribute up to 10% of the pull-off force measured on the structured surfaces at high velocities [9], while oil lubrication (wet adhesion) involves both capillarity- and viscosity-dependent forces [11]. As for performance under different contact conditions, reversible adhesion has previously been achieved using a mushroom-shaped microstructure, which is inspired by the beetle's microbial structure, submerged underwater [12]. Surfaces with a defined structure have a 25% increase in adhesion when immersed in water compared to a dry surface. Adhesion of a mushroom-shaped microstructure via underwater contact is 20 times more effective than that of a flat surface made of the same material. The Van der Waals interaction that creates adherence is greatly enhanced via the suction effect that occurs in underwater interaction. The resulting higher adherence of the substance encourages possible applications in biomedical technologies, as well as a variety of applications in which mushroom-shaped microstructures are submerged in fluid environments. Thus, it is important to note that in a wide variety of engineering applications, biomimetic mushroom-shaped adhesive microstructures are usually used in dry contact, whereas natural methods usually contains fluid [13].

In fields related to engineering, such as machines and robots, the tribological performances (adhesion, friction, and peeling strength) of biomimetic adhesive microstructures are often evaluated using smooth and hard counter-faces (in general glass). However, as shown by research into biological or medical applications, it is becoming necessary to ensure that counter-faces made of soft materials are as close as possible to the mechanical properties of biological tissues. In the light of the above issues, the present study aims to experimentally investigate the influence of the mechanical properties of different soft and hard counter-faces on the adhesive strength of biomimetic mushroom-shaped micro-structures. Their adhesive capacity will also be evaluated under different contact conditions. The inspiration for modeling our soft counter-faces emanated from a study that investigated the effect of surface micro-structures on the friction and lubrication properties of the tongue-based tribological system model [14]. Given that the mushroom-shaped contact elements are commonly developed to achieve passive long-term adhesion, while they often fail to generate friction resistance, in this study, only their adhesive properties will be investigated.

2. Materials and Methods

2.1. Mushroom-Shaped Microstructure and Flat Reference Samples

In this study, the mushroom-shaped microstructure tape used was manufactured by Gottlieb binder GmbH (Holzgerlingen, Germany) [15]. The manufacturing process consisted of pouring two-compound polymerizing poly(vinylsiloxane) (PVS; Coltene Whaledent AG, Altsatten, Switzerland) to 0.3-millmeter thick cast tape with Young's modulus at around 3 MPa [16], and the tape that contained the microstructures was then released from the negative template. The use of such a soft elastomer helped us to obtain very compliant structures that increase adhesive performance. Indeed, the compliant structures barely store elastic repulsive energy and, therefore, easily follow the roughness of the counter-face with which they are in contact, thus increasing the intimate contact area and the resulting adhesive forces. The obtained mushroom-shaped microstructure consisted of hexagonally packed pillars of about 100 μm in height, bearing terminal contact plates of about 40 μm in diameter, and an areal density of the terminal contact plates of around 40%, according to our calculations. The backside of the cast micro-structured tape was used as a smooth reference surface, as it was made of the same material.

To prepare samples to fit the customized adhesion test-rig used in this study, the adopted concept placed six small cylinders (∅ 2 mm and 1 mm in height) on the same sample, while all cylinders faces, whether they had either mushroom-shaped or smooth flat faces, were aligned on the same plane. To complete this step, the process reported in reference [17] was used. A specific aluminum template was manufactured via a CNC process. The mold contained a socket with six round holes in the bottom, which were arranged symmetrically to achieve, as far as possible, an equal load distribution, which is a necessary condition in tribological characterization. The process consisted of placing the aluminum mold onto a glass panel, before inserting the six cylindrical models into the holes of the biomimetic microstructures, meaning that the tested side was in contact with the glass to enable flattening, thus verifying that the cylinders were aligned in the same plane. Next, a PVS fixative elastomer was placed on top of the aluminum mold to unite the six different cylinders into one model. A second flat glass was used to flatten the PVS to achieve a uniform thickness of the final model. Finally, after polymerization, the resulting model was released from the aluminum mold (Figure 1).

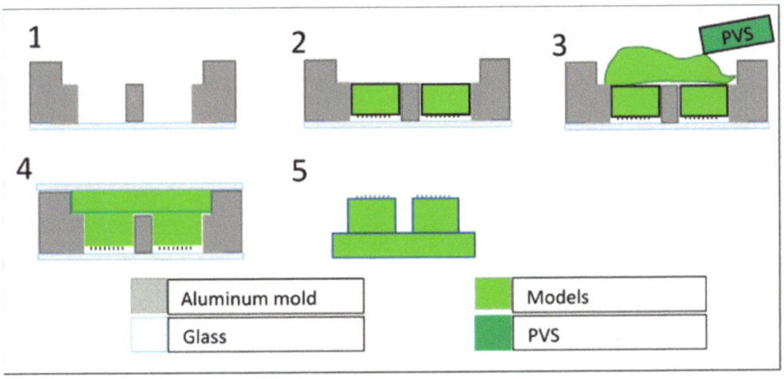

Figure 1. Schematic description of the process of integration of the small biomimetic cylinders into a single model. (1) the specific mold positioned on a smooth and clean glass; (2) small cylinders containing the mushroom elements inserted inside the specific mold with textures facing the glass; (3) a small quantity of PVS gently poured over the backside, which once solidified fixes their position together; (4) a cover glass used to remove extra PVS and to unify the shape and thickness of the final sample; (5) release of the final combined sample after PVS solidification.

Following the preparation process described above, two different samples were prepared and tested: a flat reference sample in which all six sub-contract points were cylinders with flat smooth surfaces (see Figure 2a), and a mushroom-shaped microstructure sample in which all six sub-contract points were cylinders with mushroom-shaped microstructures (see Figure 2b). It is important to note that all cylinders (micro-mushroom-shaped and flat) were randomly placed inside of the aluminum mold due to their isotropic property.

2.2. Counterface (Substrate)

In contrast to previous works, in which only one hard material was used for the counter-face, general glass, or Epoxy [18–20], in the present study, the adhesion experiments were performed on three different counter-faces, which were duplicated from the same smooth surface (Microscope slide) using three different materials that had different mechanical properties, i.e., PVS, SILFLO©, and Epoxy used in this study as a reference counter-face material. All counter-face specimens ($25 \times 20 \times 1$ mm^3 in size) were cast via replicating the same surface (Microscope slide) using a two-step molding technique [21]. PVS (Poly-vinyl siloxane), which is an addition–reaction silicone elastomer, has a Young's modulus of E_P = 3.12 MPa once polymerized [13]. SILFLO© is a brand of silicone impression material that consists of a base and catalyst in a putty consistency. It is a soft material that has a Young's modulus of about Es = 1.5 MPa. This material is mainly used to simulate mechanical proprieties of biological tissues, such as the tongue [14]. Epoxy is a hard resin used to manufacture adhesives, coatings, and other products and materials It has a Young's modulus about E_E = 3.1 GPa [8]. Epoxy is also used to cast different counter-faces (substrates) in previous studies [19,20].

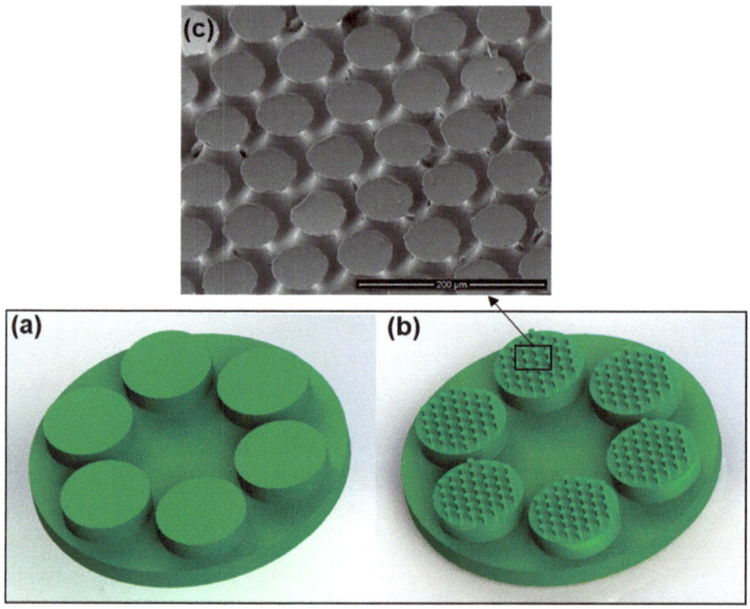

Figure 2. Illustration of the (**a**) flat control sample, (**b**) mushroom-shaped biomimetic microstructure sample, and (**c**) SEM image of the mushroom-shaped pillars.

These counter-face specimens were fully characterized using 3D optical profilometer (Figure 3) Wyko NT1100 (Veeco, Tucson, AZ, USA). The counter-faces were examined three times using different areas on the surfaces. The main roughness parameters measured, for which average values are shown in Table 1, are as follows:

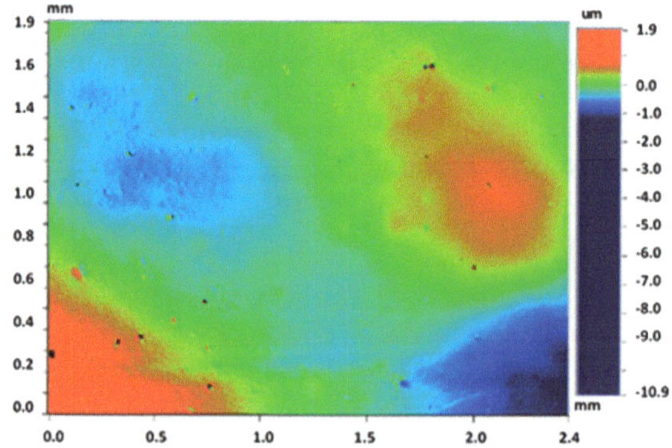

Figure 3. 3D optical profilometer image of the tested counter-face.

Table 1. Average values of the main roughness parameters obtained using four measurements at different zones on each tested counter-face.

Surface Material	*Ra* um	*Rq* um	*Rz* um	*Rpk* nm	*Rvk* nm	Wettability Angle
PVS	0.12	0.15	2.22	142.52	201.19	114.5
SILFLO©	0.59	0.85	13.78	1246.42	1022.09	108.05
EPOXY	0.27	0.35	2.95	366.30	474.30	97.8

Ra—the average roughness calculated over the entire measured array;

Rq—the root-mean-squared roughness calculated over the entire measured array;

Rz—the average of the ten greatest peak-to-valley separations;

Rpk—reduced peak height, i.e., the top portion of the surface that can be worn away during the run-in period;

Rvk—reduced valley depth, i.e., the lowest portion of the surface that might retain lubricant during wet contact.

The contact wettability angle was measured via water contact angle measurement. The measurements were conducted with a droplet of double-distilled water (DDW) using an Easy Drop contact angle goniometer (FM40Mk2, Krüss GmbH, Hamburg, Germany) at room temperature and ambient humidity. The contact wettability angle characterized the properties of the surfaces in terms of hydrophilicity ($\theta < 90°$) or hydrophobicity ($\theta > 90°$).

We noted that the sample replicated in this study was obtained in a previous work [19] (Microscope slide), while the roughness parameters obtained are very similar and are within the measurement error range. Therefore, it can be concluded that the samples were properly prepared in the present study.

2.3. Experimental Procedure

Adhesion experiments were performed via a customized tribometer that was developed at the Laboratory of Tribology and Microstructures of the Azrieli College of Engineering, Jerusalem (JCE). A full description of the used device is given in [17]. Based on a moving horizontal counter-face, this tribometer allowed us to evaluate the tribological properties (friction, adhesion, and peeling) of different materials, including textured surfaces, under dry or wet contact conditions according to needs. Using this tribometer, the drive unit consisted of three translation stages (two motorized and one manual) to adjust the contact location and apply the loads between the friction pair components. The measurement unit consisted of two load cells (FUTEK's FSH00092-LSB200) used to measure force variations at a high resolution (0.1 mN) in both the normal and tangential directions. The operating and control software were written in a LabVIEW environment. The measurements were sampled via a multifunctional data acquisition board Lab-PC- NI USB-6211 (National Instruments Co., Austin, TX, USA) and processed using the LabVIEW 2017 software package (National Instruments Corporation, 11500 N. Mopac Expressway, Austin, TX, 78759, USA). The current study used a passive self-aligning system to ensure full flat-on-flat contact between the mating surfaces during the adhesion experiments.

The samples, i.e., flat sample (FS) or mushroom-shaped microstructure sample (MSMS), were mounted on the holder and connected to the self-alignment system. Next, the selected counter-face specimen, which was already glued to a microscope slide glass, was mounted on the moving holder attached to the translating stage. The fixation screws were then reinforced to prevent any unwanted movement (see Figure 4).

Once the samples were mounted, the measurement and self-alignment systems were calibrated by resetting the load cells to eliminate the effect of mass gravity. It is important to note that the same calibration was performed after each sample replacement. Adhesion tests were conducted as follows: The counter-face specimen was brought into contact with the patterned microstructure samples at a pre-defined speed, leading to a gradual increase in the normal load P until the pre-defined value was reached. Next, the translation stage was withdrawn in the normal opposite direction at a pre-defined constant velocity, while the load cell measured the generated pull-off force. The maximal adhesion force at the separation point was recorded for each test. Each sample or configuration was tested four times, from which tests the average value of the maximal adhesion force, as well as the standard deviation, was calculated.

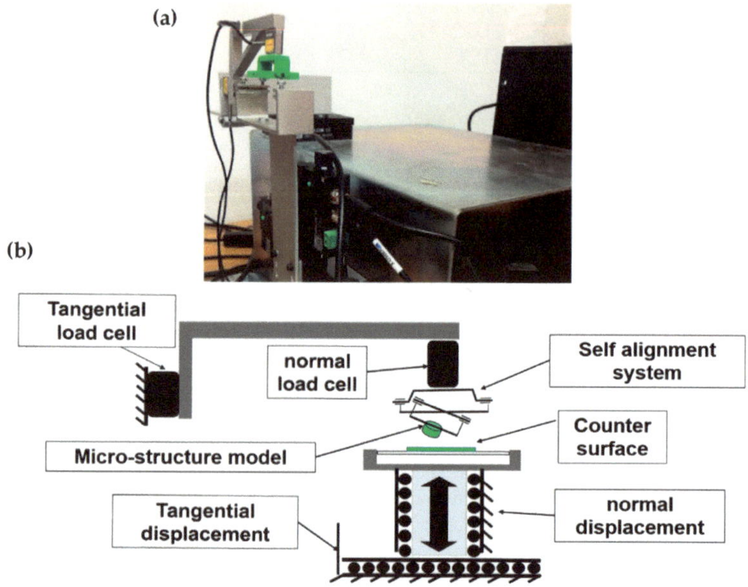

Figure 4. (**a**) General view and (**b**) schematic illustration of the customized tribometer.

The adhesion strength presented in the graphs of the experimental results was computed by dividing the measured adhesion force by the total contact area of the sample (6 small cylinders, see Figure 2). Next, the obtained value was normalized over the nominal aspect ratio of contact surface η (Equation (1)). This value was equal to 1 for the smooth control reference model and 0.4 for the mushroom-shaped microstructure model.

$$\eta = \frac{S}{\Delta S} \cdot 100\ [\%] \tag{1}$$

where S is the relative area of the mushroom, and ΔS is the total equilateral area.

All experiments were performed under the same ambient condition at a room temperature of 23 °C ± 1 °C and a relative humidity of 45% ± 5%.

2.4. Contact Environment

In the present study, the adhesion experiments were performed under three different environment contact conditions, i.e., dry (in the air), distilled water, and glycerol. To retain the liquid on the counter-face (substrate) for the experiments performed under distilled water and glycerol, a PVS belt of 1 mm height was glued onto the contour of the counter-face (see Figure 5).

2.5. Tests Operational Conditions

Each model was tested by applying normal loads of 100, 200, 300, 500, 700 and 1000 mN to cover the load range inferred in previous studies [19,20] (giving nominal contact pressures on the mushroom-shaped microstructure: 0.013 to 0.13 MPa). The loading and unloading speed was 0.5 mm/s.

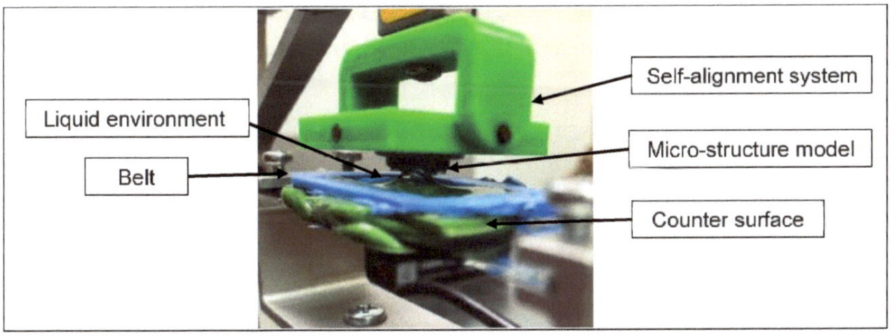

Figure 5. Glass slide microscope with a PVS belt during an adhesion experiment under contact when submerged with glycerol.

3. Results and Discussion

Figure 6 presents the typical behavior of an adhesive contact, which occurs when the normal load is displayed as a function of the vertical displacement Hmm (distance between the model and the substrate "counterface"). This behavior can be divided into five characteristic stages: In stage (1), models approach the opposite models before making contact with each other, and in stage (2), the models (smooth control reference model or mushroom-shaped microstructure model) come into contact with the counter-face. In step (3), the system is loaded until it reaches the desired value of the normal pre-load, and in step (4), the resistance to detachment is measured as a function of displacement in the opposite direction at a pre-determined separation speed. When there is no adhesion, disconnection occurs almost immediately. However, when the contact is adhesive, disconnection does not occur immediately. The force continues to decline in the negative stage due to resistance to disconnection. Full disconnection occurs at point $P_{a,max}$ (5), which corresponds to the maximum adhesion force measured for each test. Finally, the system stabilizes after slight fluctuations.

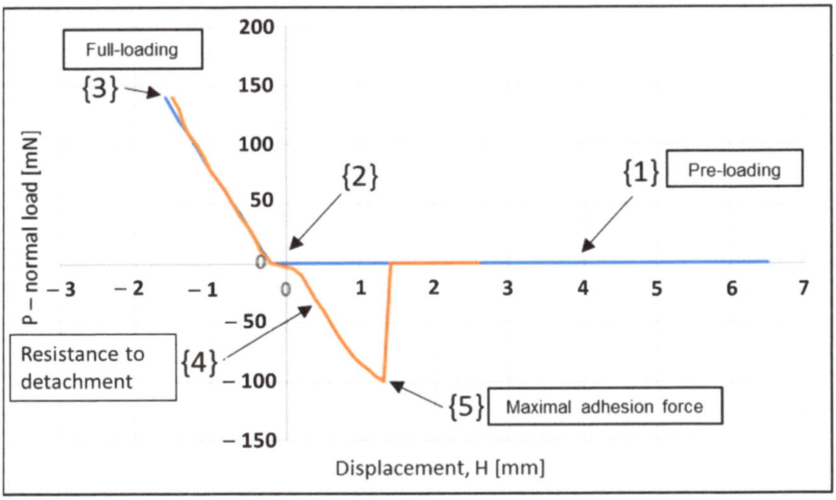

Figure 6. Force–distance curve showing the typical behavior during adhesion test. Loading (blue line) and unloading (orange line).

As mentioned above, in this study, we investigated the influence of counter-face material under different environmental conditions.

3.1. Dry Contact Condition

Figure 7 displays the average values of the maximal adhesion strength obtained using the three counter-face materials (epoxy, PVS, and SILFLO©) that have almost the same surface roughness (Ra around 0.1–0.5 μm, replicated from a microscope slide) under dry contact conditions. The maximal adhesion strength is displayed as a function of the applied normal pre-load. Data presented in (a) were obtained via a smooth control reference model, while data presented in (b) were obtained via a mushroom-shaped microstructure model tested under the same operational conditions.

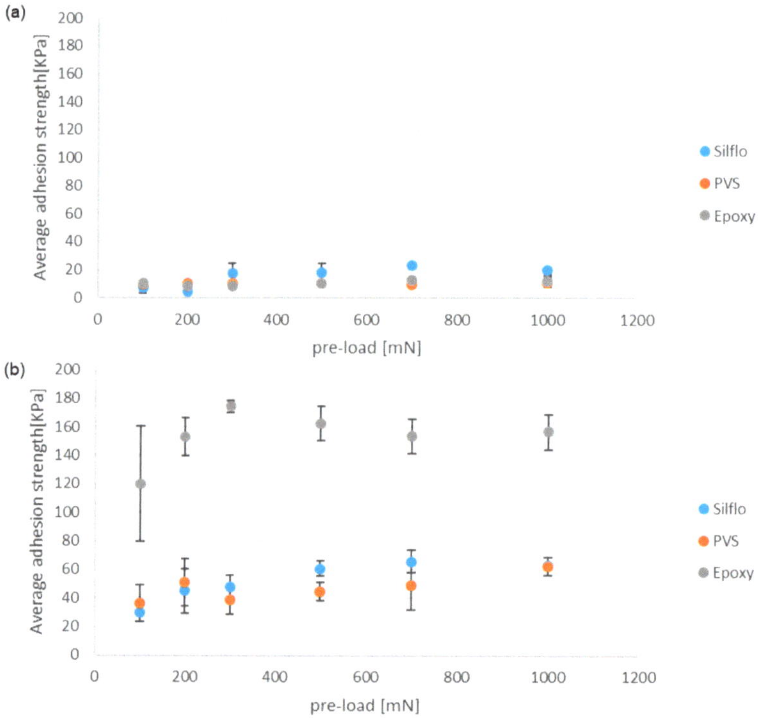

Figure 7. The average values of the maximal adhesion force under dry contact conditions. (**a**) The smooth control reference model and (**b**) the mushroom-shaped microstructures. All counter-faces are replicated using the same microscope slide.

The performance of mushroom-shaped microstructures can be seen in Figure 7b. The trend lines suggest that the maximum adhesion force appears to be unaffected by the value of the initial normal pre-load. This behavior has already been reported in the literature [20] Indeed, a certain minimum pre-load value is required to form the maximum contact area between the mushroom-shaped microstructures and the opposite counter-face, beyond which no additional contact area can be achieved. As for the current results, it is likely that the minimum pre-load applied is higher than the requested minimum preload. Under dry contact conditions, the hard epoxy counter-face gives the highest adhesion strength, while the softer materials (PVS and SILFLO©) give smaller adhesion strengths that are close to each other, with a slight advantage for SILFLO©. These results tend to highlight that, at least in the case of dry contact conditions, the high mechanical properties of the substrate (counter-face) do not affect negatively the adhesion force when in contact with biomimetic adhesive elements such as mushroom-shaped microstructures. This result can be explained by the stress concentration distribution on each mushroom cap [21]. When separating a soft micro-mushroom element from a softer counter-face, stress concentration occurs at the edge of the mushroom cap, leading to initial detachment from the side toward the center of the mushroom. This behavior leads to relatively fast detachment, reducing the measured adhesive force [21]. This behavior can be approximated using the model of mushroom-shaped pillar with a thick plate described in [22], in which the interfacial stress singularity appears at the cap's edge, probably due to the diminution of the stiffness ratio between the mushroom caps and the counter-face. In contrast, when separating a soft mushroom cap model with optimal plate thickness from a rigid counter-face, the stress concentration occurs in the center of the mushroom [23]. In this case, the initial detachment between the mushroom caps and the rigid counter-face begins at the center of the pillar and propagates under the shape of circumferential peeling, progressively increasing the peeling line which, when coupled with the resulting artificial suction effect, contributes positively to increasing adhesion force [12]. Figure 7a is related to the reference smooth control model. For all three materials, the adhesion strength is lower than that of the mushroom samples (Figure 7b). SILFLO©, however, gives slightly higher adhesion than the two other materials, especially when the applied normal pre-load is higher than 300 mN. In Figure 7, it can be seen that under dry conditions, there is a very limited influence of the pre-load, as was previously reported in [4].

3.2. *Wet Contact Environment—Distilled Water*

Figure 8 displays the average values of the maximal adhesion strength for (a) the smooth reference and (b) mushroom-shaped microstructure samples under water-wet conditions. The maximum adhesion strength is displayed as a function of the applied normal pre-load for the three counter-face materials (Epoxy, PVS, and SILFLO©), which have the same surface roughness. The contact was completely submerged within distilled water during the adhesion test (see illustration in Figure 5).

When submerged with distilled water, the hard Epoxy counter-face shows almost non-adhesive behavior relative to the smooth reference (a), while there is a very low value for the textured sample (b). However, concerning soft material, SILFLO© presents the highest adhesion (on average 8 to 10 times higher than those of the other two materials). Humidity-related effects on adhesion can be explained based on the capillary forces due to the formation of liquid bridges [24]. Hence, the low elasticity modulus, when combined with possible capillarity forces, seems to be the cause of its high adhesion capacity within distilled water. The low elasticity modules of the SILFLO© counter-face, when combined with the high flexibility of mushroom caps under water, might accentuate the effect of artificial suction, as reported in [8,25,26], hence the increasing adhesion force. The slightly higher elasticity modulus and greater hydrophobicity of PVS than SILFLO© seems to be the cause of its lower adhesion strength.

3.3. *Wet Contact Environment—Glycerol*

Figure 9 displays the average values of the maximal adhesion strength for (a) smooth reference and (b) mushroom-shaped microstructure samples tested under Glycerol wet condition (contact submerged with glycerol during the adhesion test). The maximum adhesion strength is displayed as a function of the applied normal pre-load for the three counter-face materials (epoxy, PVS, and SILFLO©), which have the same surface roughness.

In the case of mushroom-shaped microstructure sample (b), when tested under small pre-loads, the hard epoxy counter-face generates an adhesion strength smaller than those of the two other soft materials (PVS and SILFLO©). This behavior can be explained based on the fact that under small pre-loads, the hard epoxy does not deform enough to generate sufficient contact surface with the mating mushroom-shaped micro-structures. It is also possible that glycerol contained air bubbles that were retained between the mating surfaces under small pre-loads. In contrast, the PVS and SILFLO© counter-faces deformed more noticeably under the same normal pre-load, which is something that, along with the deformation of the patterned mushroom-shaped microstructure, contributes to the increase in the real contact area, leading to the generation of higher adhesion force. While PVS gives better results (highest adhesion), the difference with SILFLO© is insignificant. SILFLO© still presents the best enhancement ratio between the smooth reference and mushroom-shaped microstructure samples.

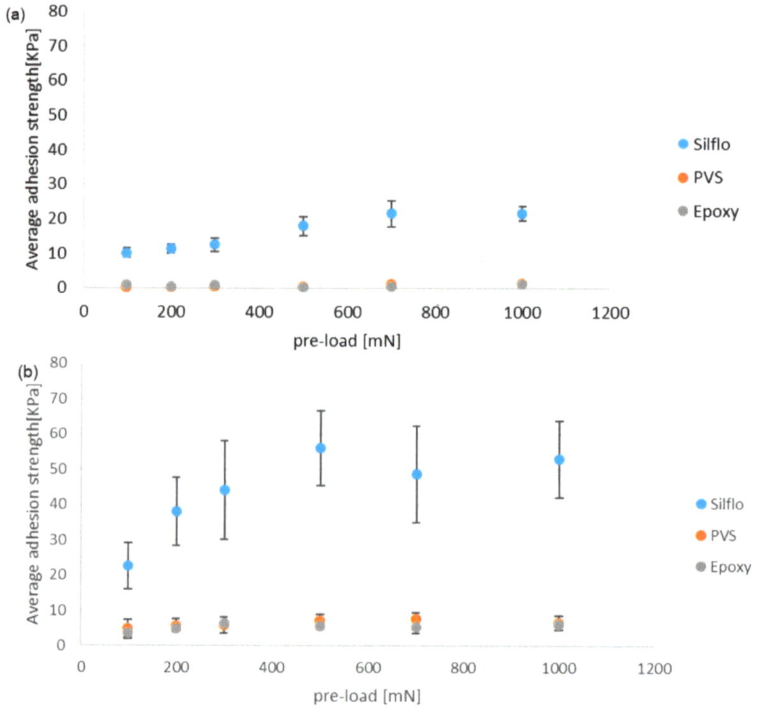

Figure 8. Average values of the maximal adhesion strength as a function of the applied normal pre-load under wet contact conditions with distilled water for two models: (**a**) a smooth control reference model and (**b**) mushroom-shaped microstructures. All counter-faces are replicated using a microscope slide.

It is important to note that the present work does not take in consideration the influence of the detachment velocity, which can affect the behavior of the interface. In another recently published work [27], it was shown that the adhesion properties of biomimetic mushroom- and spatula-like elements were affected by the detachment velocity, and three different regimes were reported: (i) a quasi-static range, in which no clear dependent was obtained; (ii) an intermediate range, in which the maximum adhesion force at detachment increased in line with the detachment velocity; and (iii) an upper limit, which represents a velocity beyond which the pull-off force no longer depends on the detachment velocity.

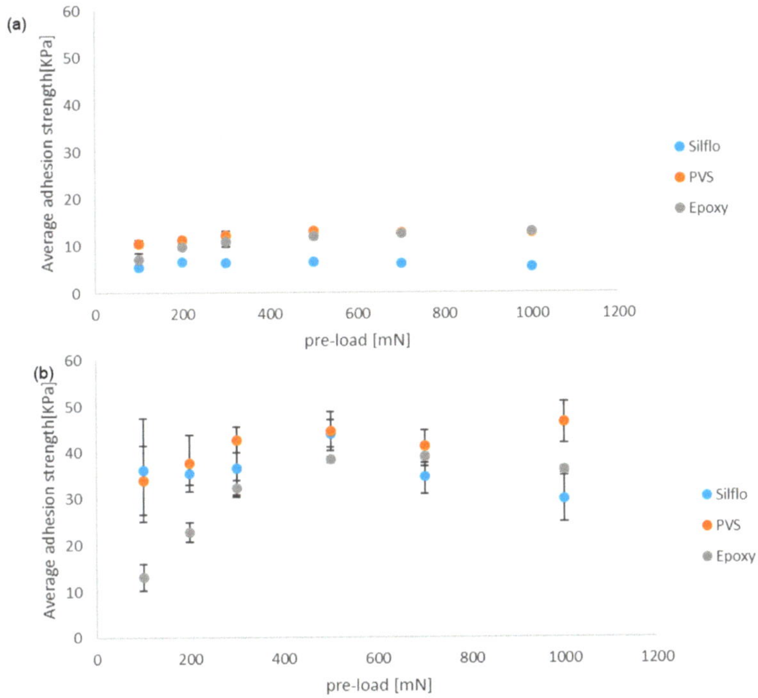

Figure 9. Average value of the maximal adhesion strength as a function of the applied normal pre-load under wet contact conditions with glycerol for two models: (**a**) a smooth control reference model and (**b**) a mushroom-shaped microstructure. All counter-faces are replicated using a microscope slide.

4. Conclusions

The present work experimentally investigates the influence of mechanical proprieties of substrates (counter-faces) when in contact with mushroom-shaped biomimetic adhesive microstructures. The biomimetic mushroom-shaped microstructure tape was made of poly(vinylsiloxane) (PVS) and manufactured by Gottlieb Binder GmbH (Holzgerlingen, Germany) [15]. The counter-faces were cast via replication with three different materials i.e., (i) PVS (Poly-vinyl siloxane); (ii) SILFLO©, which is a brand of silicone impression material; and (iii) a hard Epoxy. The adhesive properties under different contact conditions were investigated using a customized test-rig. The results of this work will help us to identify the key mechanical proprieties responsible for the observed variation in pull-off adhesion force. The effects of counter-face mechanical properties on the adhesion of mushroom-shaped biomimetic microstructures were experimentally investigated under different environmental and operational conditions. The following conclusions were drawn:

- In smooth and rigid counter-faces tested under dry contact conditions, mushroom-shaped micro-structures generated almost 6 times more adhesion strength than a smooth control reference. This result is in full agreement with other results reported in the literature [20], although different test-rigs and samples shapes were used.
- Under dry contact conditions, soft counter-faces led to lower adhesion than hard counter-faces. This different behavior seemed to be related to the change in the interfacial stress distribution [21].
- Under wet conditions, soft counter-faces led to higher adhesion than hard counter-faces. This result can be explained by both additional capillary forces due to the formation of liquid bridges and, possibly, more suction effect favored by the elastic deformation of the mushroom cap and counter-face [24].

In summary, the adaptation and proper use of the adhesive capabilities of biomimetic adhesive microstructures can advance studies in the field of adhesion and promote adhesion to soft surfaces in dry and wet environments. An example of a potential application is the field of biomedicine.

Author Contributions: M.G. contributed to the design and implementation of the research, analysis of the results, and writing the first draft of the manuscript. H.K. supervised the research and contributed to the conceptualization of the tests, analysis of the results, and writing, reviewing, and editing the paper. All authors have read and agreed to the published version of the manuscript.

Funding: This research received no external funding

Institutional Review Board Statement: Not applicable

Data Availability Statement: Not applicable

Acknowledgments: The authors would like to thank Yuri Kligerman for providing valuable insights and suggestions. Special thanks are also extended to David Badler for his helpful discussion and providing help with the tribometer system.

Conflicts of Interest: The authors declare no conflict of interest.

References

1. Gorb, S.N.; Varenberg, M. Mushroom-shaped geometry of contact elements in biological adhesive systems. *J. Adhes. Sci. Technol.* **2007**, *21*, 1175–1183. [CrossRef]
2. Parness, A.; Soto, D.; Esparza, N.; Gravish, N.; Wilkinson, M.; Autumn, K.; Cutkosky, M. A microfabricated wedge-shaped adhesive array displaying gecko-like dynamic adhesion, directionality and long lifetime. *J. R. Soc. Interface* **2009**, *6*, 1223–1232. [CrossRef] [PubMed]
3. Jagota, A.; Hui, C.-Y. Adhesion, friction, and compliance of bio-mimetic and bio-inspired structured interfaces. *Mater. Sci. Eng. R Rep.* **2011**, *72*, 253–292. [CrossRef]
4. Gorb, S.; Varenberg, M.; Peressadko, A.; Tuma, J. Biomimetic mushroom-shaped fibrillar adhesive microstructure. *J. R. Soc. Interface* **2007**, *4*, 271–275. [CrossRef] [PubMed]
5. Aksak, B.; Murphy, M.P.; Sitti, M. Adhesion of Biologically Inspired Vertical and Angled Polymer Microfiber Arrays. *Langmuir* **2007**, *23*, 3322–3332. [CrossRef] [PubMed]
6. Smith, A.M.; Callow, J.A. (Eds.) *Biological Adhesives*; Springer: Berlin/Heidelberg, Germany, 2006. [CrossRef]
7. Santos, D.; Spenko, M.; Parness, A.; Kim, S.; Cutkosky, M. Directional adhesion for climbing: Theoretical and practical considerations. *J. Adhes. Sci. Technol.* **2007**, *21*, 1317–1341. [CrossRef]
8. Carbas, R.J.C.; da Silva, L.F.M.; Marques, E.A.S.; Lopes, A.M. Effect of post-cure on the glass transition temperature and mechanical properties of epoxy adhesives. *J. Adhes. Sci. Technol.* **2013**, *27*, 2542–2557. [CrossRef]
9. Heepe, L.; Varenberg, M.; Itovich, Y.; Gorb, S.N. Suction component in adhesion of mushroom-shaped microstructure. *J. R. Soc. Interface* **2011**, *8*, 585–589. [CrossRef]
10. Murphy, M.P.; Kim, S.; Sitti, M. Enhanced Adhesion by Gecko-Inspired Hierarchical Fibrillar Adhesives. *ACS Appl. Mater. Interfaces* **2009**, *1*, 849–855. [CrossRef]
11. Meng, F.; Liu, Q.; Wang, X.; Tan, D.; Xue, L.; Barnes, W.J.P. Tree frog adhesion biomimetics: Opportunities for the development of new, smart adhesives that adhere under wet conditions. Philosophical Transactions of the Royal Society A: Mathematical. *Phys. Eng. Sci.* **2019**, *377*, 20190131. [CrossRef]
12. Varenberg, M.; Gorb, S. A beetle-inspired solution for underwater adhesion. *J. R. Soc. Interface* **2008**, *5*, 383–385. [CrossRef]
13. Kovalev, A.E.; Varenberg, M.; Gorb, S.N. Wet versus dry adhesion of biomimetic mushroom-shaped microstructures. *Soft Matter* **2012**, *8*, 7560. [CrossRef]
14. Ranc, H.; Servais, C.; Chauvy, P.-F.; Debaud, S.; Mischler, S. Effect of surface structure on frictional behaviour of a tongue/palate tribological system. *Tribol. Int.* **2006**, *39*, 1518–1526. [CrossRef]
15. Tuma, J. Patent "Process for creating adhesion elements on a substrate material". WO2005087033A1, 22 September 2005.
16. Peressadko, A.; Gorb, S.N. When less is more: Experimental evidence for tenacity enhancement by division of contact area. *J. Adhes.* **2004**, *80*, 247–261. [CrossRef]
17. Badler, D.; Kasem, H. Synergetic effect of the simultaneous use of different biomimetic adhesive micro-structures on tribological performances. *Biotribology* **2020**, *22*, 100124. [CrossRef]
18. Gorb, S.N. Visualisation of Native Surfaces by Two-Step Molding. *Microsc. Today* **2007**, *15*, 44–47. [CrossRef]
19. Kasem, H.; Cohen, Y. Effect of counterface roughness on the friction of bionic wall-shaped microstructures for gecko-like attachments. *Bioinspir. Biomim.* **2017**, *12*, 046010. [CrossRef]
20. Kasem, H.; Varenberg, M. Effect of counterface roughness on adhesion of mushroom-shaped microstructure. *J. R. Soc. Interface* **2013**, *10*, 20130620. [CrossRef]

21. Voigt, D.; Schuppert, J.M.; Dattinger, S.; Gorb, S.N. Sexual dimorphism in the attachment ability of the Colorado potato beetle Leptinotarsa decemlineata (Coleoptera: Chrysomelidae) to rough substrates. *J. Insect Physiol.* **2008**, *54*, 765–776. [CrossRef]
22. Carbone, G.; Pierroac, E.; Gorb, S.N. Origin of the superior adhesive performance of mushroom-shaped microstructured surfaces. *Soft Matter* **2011**, *7*, 5545. [CrossRef]
23. Afferrante, L.; Carbone, G. The Mechanisms of Detachment of Mushroom-Shaped Micro-Pillars: From Defect Propagation to Membrane Peeling. *Macromol. React. Eng.* **2013**, *7*, 609–615. [CrossRef]
24. Heepe, L.; Wolff, J.O.; Gorb, S.N. Influence of ambient humidity on the attachment ability of ladybird beetles (*Coccinella septempunctata*). *Beilstein J. Nanotechnol.* **2016**, *7*, 1322–1329. [CrossRef] [PubMed]
25. Persson, B.N.J. On the mechanism of adhesion in biological systems. *J. Chem. Phys.* **2003**, *118*, 7614. [CrossRef]
26. Kasem, H.; Tsipenyuk, A.; Varenberg, M. Biomimetic wall-shaped hierarchical microstructure for gecko-like attachment. *Soft Matter* **2015**, *11*, 2909–2915. [CrossRef]
27. Badler, D.; Goltsberg, R.; Ammar, A.A.; Kasem, H. Experimental study of adhesion, friction, and peeling of biomimetic combined micro-mushroom and micro-spatulae textures. *Tribol. Int.* **2023**, *186*, 108609. [CrossRef]

biomimetics

Article

The Influence of Temperature on Anisotropic Wettability Revealed by Friction Force Measurement

Zhen Lin [1,†], Kangjian Xiao [1,†], Lijun Li [1], Yurong Zhang [2], Xiaolong Zhang [1], Daobing Chen [2,*] and Longjian Xue [1,2,*]

[1] School of Power and Mechanical Engineering, Wuhan University, South Donghu Road 8, Wuhan 430072, China
[2] The Institute of Technological Science, Wuhan University, South Donghu Road 8, Wuhan 430072, China
* Correspondence: chendb22@whu.edu.cn (D.C.); xuelongjian@whu.edu.cn (L.X.)
† These authors contribute equally to this work.

Abstract: Anisotropic surfaces with special wettability under various temperatures are of both fundamental interest and practical importance in many fields. However, little attention has been paid to the surfaces at temperatures between room temperature and the boiling point of water, which is partially due to the lack of a suitable characterization technique. Here, using the MPCP (monitoring of the position of the capillary's projection) technique, the influence of the temperature on the friction of a water droplet on the graphene-PDMS (GP) micropillar array (GP-MA) is investigated. The friction forces in the orthogonal directions and the anisotropy in the friction decrease when the GP-MA surface is heated up, based on the photothermal effect of graphene. The friction forces also decrease along the pre-stretching direction but increase in the orthogonal direction when the stretching is increased. The change in the contact area, the Marangoni flow inside a droplet, and the mass reduction are responsible for the temperature dependence. The findings strengthen our fundamental understanding of the dynamics of drop friction at high temperatures and could pave the way for the design of new functional surfaces with special wettabilities.

Keywords: anisotropic surface; liquid-solid friction; graphene; wettability; bioinspired

Citation: Lin, Z.; Xiao, K.; Li, L.; Zhang, Y.; Zhang, X.; Chen, D.; Xue, L. The Influence of Temperature on Anisotropic Wettability Revealed by Friction Force Measurement. *Biomimetics* **2023**, *8*, 180. https://doi.org/10.3390/biomimetics8020180

Academic Editors: Stanislav N. Gorb and Thies Henning Büscher

Received: 31 March 2023
Revised: 21 April 2023
Accepted: 21 April 2023
Published: 25 April 2023

1. Introduction

Anisotropic wetting originating from anisotropic surface structure or arrangement provides vital functions for many plants and animals to survive in nature [1–5]. For instance, with the asymmetric hierarchical topography of the peristome surface, the pitcher plant is able to achieve directional liquid transport to accumulate nectar and water and even to form a slippery liquid film for trapping insects [6]. Anisotropic wettability also enables fish to reduce drag [7], water striders to walk on water [8], and beetles to capture water in the desert [9]. In addition, the wettability can change once the temperature of the surface or the environment changes, which has attracted increasing attention in recent years [10–12] For instance, with the one-dimensional distribution of the mastoid structure in the parallel direction of leaf veins and uneven distribution in the vertical direction [3], water droplets accumulate on rice leaf when the temperature is low in the morning and roll along the veins as the temperature increases. The study of the influence of the temperature on the anisotropy surface wettability is thus of great importance in fundamental research and engineering fields, such as water accumulation, spray cooling, droplet transport, drag reduction, and agricultural spray [13–17].

Under various temperatures, the surface anisotropy can be easily regulated. At room temperature, the anisotropic adhesion of a water droplet on surfaces composed of an array of triangular pillars or stripes can be regulated in situ by the mechanical stretching of the elastic substrate [18–21]. Making use of the shape memory effect of the materials, smart surfaces with reversible isotropic/anisotropic wettability achieved the control of droplet motion by mechanical force or other stimulations [22–24]. With the photothermal effect, the temperature of a water droplet can be regulated locally, which breaks up the wetting symmetry of the droplet and thus manipulates the droplet motion [25–29]. Adding polypyrrole nanoparticles into a water droplet, Wooh et al. [25] were able to drive the droplet on a lubricant-impregnated surface and superamphiphobic surface with the focused irradiation of near-infrared light on the droplet. Once the temperature of the substrate could be heated up by light irradiation, the droplet temperature and therefore the droplet motion could be finely regulated. When the infrared-light irradiation was focused on the substrate at one side of a droplet, a temperature difference was generated between the two sides of the droplet, which resulted in the unbalanced surface tension and Marangoni force, driving the droplet towards the side without light irradiation [26]. When the surface temperature was close to the Leidenfrost point, the droplet levitated on a vapor layer; the movements of a droplet on surfaces have been intensively investigated [10,11,13,30]. The existence of a vapor layer would result in a negligible normal adhesion or lateral friction of the water droplet on the surface, which, however, would cause a high thermal resistance. Wang et al. [13] fabricated a micropillar surface with gradient periods and thus the coefficient of heat transfer and realized the directional transport of a high-temperature (close to Leidenfrost point) droplet towards the region with a higher heat-transfer coefficient. Liu et al. [11] found an interesting phenomenon in which a droplet showed a steerable bouncing on heated concentric microgrooves arrays under different temperatures, which is believed to originate from the synergistic action of the surface structure and boiling states. That is, the motion of a water droplet could be manipulated by controlling the temperature (close to the Leidenfrost point) and the topography substrate surface. On the other hand, the motion of a droplet on asymmetric surfaces with a temperature higher than room temperature but lower than the boiling temperature has required more investigation, though the phenomenon is quite common in our daily life. We assume the lack of a suitable technique to characterize the droplet motion within such a temperature window may partially be responsible for this situation.

Here, we investigate the influence of the temperature on the friction of a water droplet on a graphene-PDMS (GP) micropillar array (GP-MA) by MPCP (monitoring of the position of the capillary's projection) technique [20]. The temperature of the GP-MA is regulated by the photothermal effect of graphene in the micropillars. While the water contact angle (CA) and sliding angle (SA) show a negligible dependence on the temperature (between room temperature and boiling point of water) or the surface geometries, the friction measurements reveal a clear change when the period and temperature of the GP-MA are changed. As the temperature increases, the friction force and the anisotropy along the orthogonal directions decrease. The results offer us the chance to better explore the surfaces with anisotropic liquid-solid friction.

2. Materials and Methods

2.1. Materials and Preparation

The PDMS elastomer kit (Sylgard 184) was purchased from Dow Corning (Midland, MI, USA). Graphene sheets were purchased from Aladdin (Shanghai, China).

The fabrication of the GP-MA samples contained the preparation of GP micropillars (demolded from PDMS mold) and pre-stretched PDMS film. A PDMS mold containing an array of micro-holes with 7×7 mm^2 area, 50 μm in diameter, 20 μm in depth, and 70 μm in period was acquired via a conventional soft lithography technique. The PDMS precursor was prepared by mixing the base prepolymer and the cross linker in a weight ratio of 10:1. Graphene was added to the as-prepared PDMS precursor at a concentration of 0.8 wt%, according to our previous study, to form the GP precursor [31]. After stirring for 30 min, the GP precursor was degassed in a desiccator for 10 min and was filled into the PDMS mold to prepare the GP micropillars. The PDMS film was prepared by filling the as-prepared PDMS precursor into a glass chamber template, followed by curing at 90 °C for 1 h. After peeling from the template, the fully cured PDMS film with a thickness of 0.5 mm was cut into a rectangular shape (30 × 10 mm) and mechanically stretched to the predefined degrees. The pre-stretched PDMS film tightly covered the GP precursor-filled mold, followed by a curing at 90 °C for 1 h. After the demolding and relaxation of stress, the GP-MA sample was successfully fabricated.

2.2. Characterization

The morphologies of the graphene were examined by a field emission scanning electron microscope (MIRA 3 LMH, Tescan AG, Brno, Czech Republic) and an atomic force microscope (AFM, Nano Wizard 4, JPK Inc., Germany) in tapping mode (QI mode, scan rate = 5 Hz). The Raman spectroscopy was carried out by a laser micro-Raman spectrometer (Renishaw, English, Sheffield, UK) with an excitation wavelength of 532 nm.

The microstructures of the GP-MA were observed by dark-field optical microscopy (ECLIPSE Ci-L, Tokyo, Japan). The 3D structure of the GP-MA and the roughness of the micropillar top were characterized by a white light interference 3D profiler (New View TM 9000, ZYGO, Middlefield, CA, USA). The contact states of the water droplet on the GP-MA were observed by inverted optic microscopy (ECLIPSE MA100N, Nikon, Tokyo, Japan).

The water contact angle (CA) and slide angle (SA) were measured on a droplet shape analysis (OCA25, Dataphysics, Hamburg, Germany). The volume of the water droplet for the CA and SA measurements was 4 and 8 μL, respectively. The CA and SA were measured at least five times, and the mean values were calculated.

The liquid-solid friction force was tested by the MPCP technique, as established in our previous work [20]. Before the measurement, the GP-MA was mounted on the motor stage and brought into contact with a water droplet of 6 μL. The droplet remained adhered to the capillary with a diameter of 0.3 mm throughout the measuring process. The droplet was driven at a constant speed of 0.3 mm/s relative to the steady capillary, and the displacement of the capillary (D) was monitored and recorded simultaneously. The liquid-solid friction force (F) of the droplet on the surface can be described as

$$F = kD,$$

where k is the spring constant of the capillary. The friction force on one sample was tested no less than five times, and the mean value was calculated.

The temperature control was achieved with a 365 nm UV light source (XC-102, IGEtec. China) with an irradiation area of 20 × 20 mm^2. The infrared images and temperature data of the samples were acquired by an infrared thermal imaging camera (TiX640 60Hz, Fluke, Everett, WA, USA). A constant irradiation was applied to ensure a stable temperature of the GP-MA during the measurements. The mass of the droplet was monitored by an electronic scale (ME204/02, Mettler-Toledo, Greifensee, Switzerland).

3. Results and Discussion

3.1. Geometry of the GP-MA

The GP-MA, which is composed of a PDMS backing layer and GP micropillars, was successfully prepared following the procedure adopted from our previous work (Figure 1a) [18]. Graphene was added in order to offer GP-MA with a photothermal effect making use of the strong capability of graphene to absorb light with wavelengths across the entire spectrum. The Raman spectrum confirmed the state of graphene rather than graphite (Figure 1b). The G peak at ~1581 cm^{-1} represented the E_{2g} phonon at the Brillouin zone center, and the G′ peak at ~2698 cm^{-1} originated from the double resonance Raman process in sp^2 carbon. The D peak at ~1352 cm^{-1} gave evidence of the presence of defects while the intensity ratio between the D and G peaks (~0.15) indicated a small number of defects in the graphene sheets [32,33]. The AFM characterization showed that the thickness of the graphene sheets was 2~4 nm (Figure 1c), suggesting that the graphene sheets had two to four layers [34]. Meanwhile, the graphene sheets had a lateral size of ~1 μm or less The small lateral size together with the crumpled state of the graphene sheets (Figure 1d) facilitated the dispersion of the graphene sheets in the PDMS matrix and provided a large strain deformation of GP [35]. A 0.8 wt% concentration of graphene was chosen, as the composite has a similar elastic modulus to that of pure PDMS [36,37], which would allow the GP micropillars to be deformed together with the supporting layer. Moreover, the dark field illumination indicated that graphene sheets solely and homogeneously dispersed in the micropillars without diffusion to the backing layer (Figure 2a).

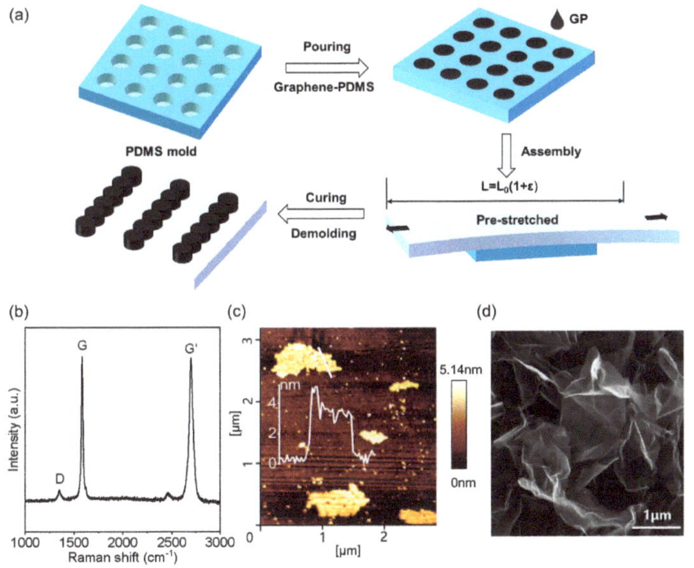

Figure 1. Fabrication of the graphene-PDMS (GP) micropillar array (GP-MA) and the morphologies of the graphene sheets. (**a**) Schematic illustration of the fabrication process; (**b**) Raman spectrum, (**c**) AFM, and (**d**) SEM of the graphene sheets. The inset in (**c**) shows the thickness profile of the graphene sheet.

The resulting GP-MA without pre-elongation ($\varepsilon = 0$) faithfully replicated the geometry of the micro-holes in the template, showing 50 μm in diameter, 20 μm in height, and 70 μm in period (Figure 2b,c). Here, the direction of pre-stretching was defined as the x direction, while the orthogonal direction of the pre-stretching was defined as the y direction. To quantitatively characterize the geometry of the resulting GP-MA, the corresponding diameter and periodic distances in the x and y directions are noted as D_x, D_y, P_x, and P_y, respectively. With the increase in the pre-elongation ($\varepsilon > 0$), the D_x and P_x decreased gradually due to the following relaxation of the pre-stretching. Meanwhile, the D_y and P_y gradually increased due to the Poisson's ratio effect (Figure 2b,c). That is, as the ε increased, the anisotropy of the GP-MA surface increased. For instance, an ε of 40% decreased D_x and P_x to 45.58 ± 0.52 μm and 55.16 ± 0.77 μm, respectively; meanwhile, it increased D_y and P_y to 53.91 ± 0.89 μm and 78.96 ± 0.82 μm, respectively. When the ε reached 80%, the micropillars presented a "side by side" state in the x direction, as D_x at 40.25 ± 1.17 μm and P_x at 40.91 ± 1.29 μm were quite close. On the other hand, the D_y and P_y reached 56.73 ± 0.55 μm and 88.54 ± 1.01 μm, respectively.

As the GP micropillars have the same elastic modulus as the backing layer, the release of the pre-elongation would cause the synchronized deformation of the GP micropillars with the backing layer. Therefore, the GP micropillars showed elliptical shapes with the supporting layer pre-stretched (Figure 2a, such as an ε of 40% and 80%). Moreover, the release of pre-elongation squeezed the micropillar top towards the center along the x direction and stretched the center toward the two sides in the y direction (Figure 2d). As a result, the micropillar top was deformed into a saddle shape (Figure 2e). For instance, at $\varepsilon = 80\%$, the top of GP micropillars presented an arched shape (high in the middle and short on both sides) along the x direction but a curved shape (short in the middle and high on both sides) along the y direction. Along with the deformation of the micropillar top, the roughness of the micropillar top increased from 0.16 ± 0.01 μm to 0.82 ± 0.11 μm when the pre-elongation was increased (Figure 2e). Therefore, with the increase in ε, the anisotropy of the GP-MA also increased in the macro- and microscale.

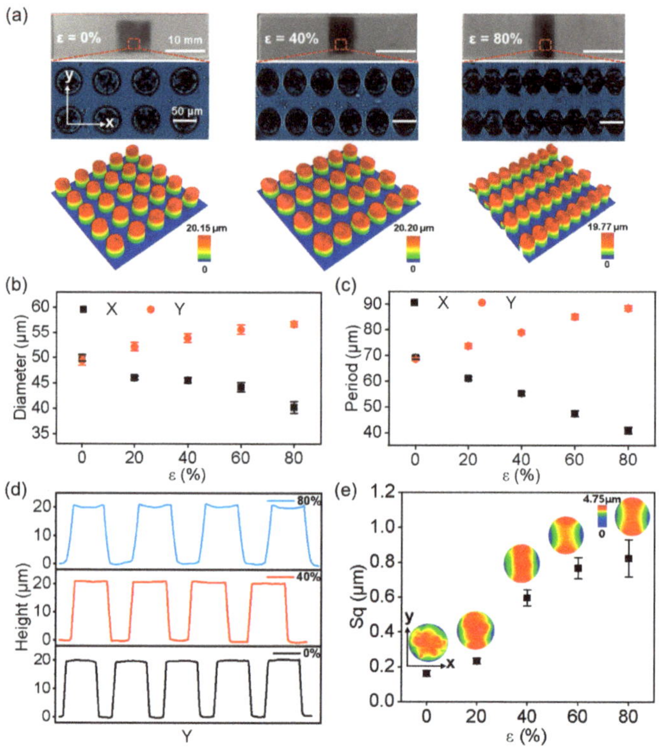

Figure 2. Geometry of the GP-MA. (**a**) The optical image and 3D morphology of the GP-MA with elongations of 0, 40%, and 80%; (**b**) diameter and (**c**) period of the GP-MA along the x and y directions under various elongations; (**d**) typical profile of the GP-MA along the y direction with elongations of 0, 40%, and 80%; (**e**) the root mean square roughness (Sq) of the micropillar top under various elongations. The inset in (**e**) shows the 3D morphology of the micropillar top under various elongations. Each data point in (**b**,**c**,**e**) represents the mean value of at least five measurements. Standard deviations are indicated by error bars.

3.2. Interfacial Interaction of a Water Droplet on the GP-MA at Room Temperature

The wettability of the GP-MA was evaluated by traditional CA and SA measurements. Generally, the wettability of the GP-MA is determined by the surface geometry. When the ε was 0, the GP-MA had a CA of 145.8 ± 1.9 and 145.9 ± 1.3° in the x and y directions, respectively (Figure 3a). The negligible difference in the CA in the orthogonal directions suggested that the GP-MA surface was isotropic, which is reasonable, as the period and diameter in the two directions were the same (Figure 2b–d). When the ε was increased from 0 to 80%, the CA slightly increased to 148.4 ± 3.4° in the x direction and decreased to 132.0 ± 2.7° in the y direction. The changes in the CA were 2.6° and 13.9°, which meant a difference of 1.8% and 9.5%, respectively. When the backing layer was pre-stretched, i.e., an ε up to 60%, the period and the micropillars were anisotropic (Figure 2b,c,e); however, the CAs in the two directions were almost the same. With an ε of 80%, the P_y increased to 88.54 ± 1.01 μm, which was quite large compared to the size of the droplet, and a partial penetration of the droplet into the array occurred, resulting in a slight decrease in the CA. Thus, the difference in the CA in the two directions reached 16.4°, which meant a difference of 11.0%. On the other hand, it has been reported that on superhydrophobic surfaces, the uncertainty of one pixel at the diffuse edge and baseline could introduce substantial systematic errors in the CA from 1° to more than 10° [38]. Similarly, the SAs in two directions also showed no difference in that the water droplet did not fall even when the sample was turned upside-down in either the x or y direction (Figure 3b) [18]. That is, the CA and SA measurements could not distinguish the difference between the GP-MA surfaces with various elongations and could not reveal the anisotropy of the GP-MA surfaces.

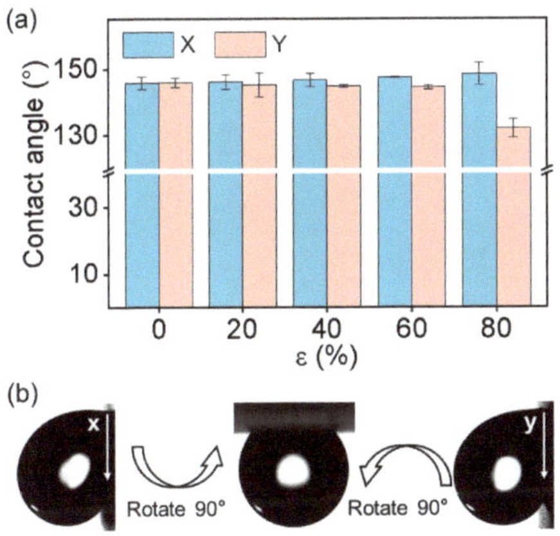

Figure 3. The wettability characterized by the CA and SA measurements. (**a**) The CA of the GP-MA with various elongations; (**b**) typical SA image of the GP-MA. Each data point in (**a**) represents the mean value of at least five measurements. Standard deviations are indicated by error bars.

The liquid-solid friction of a water droplet on the GP-MA was then determined to examine the surface. Similar to the solid-solid friction, a liquid-solid friction curve also has three sections: static friction, kinetic friction, and the transition zone from static friction to kinetic friction (Figure 4a) [39]. The peak value during the static friction period is considered as the static friction force (F_S), and the mean value during the kinetic friction is calculated to be the kinetic friction force (F_K) [19]. As with the solid-solid friction, F_S is larger than F_K. The friction forces along the x and y directions are then noted as $F_{S,x}$, $F_{K,x}$, $F_{S,y}$, and $F_{K,y}$, respectively, for convenience.

When the ε was increased from 0 to 80%, the $F_{K,x}$ decreased from 41.82 ± 3.48 to 29.80 ± 2.8 μN, with a decrease of 28.7%, while the $F_{K,y}$ increased from 41.66 ± 5.53 to 51.10 ± 2.68 μN, showing an increase of 22.6% (Figure 4b). Compared with the CA measurements when the ε increased from 0 to 80%, the differences in the $F_{K,x}$ and $F_{K,y}$ were 15.9 and 2.5 times larger, respectively. When the elongation reached 80%, the differences of the F_K (F_S) in two directions reached 71.5% (67.9%), which was more than six times the difference in the CA. That is, the friction measurement can clearly reveal the anisotropy of the surface.

The contact geometry is responsible for the anisotropy friction. At the initial stage, the contact area was a circle (Figure 4c-i), and it changed to an ellipse along the moving direction of the droplet (Figure 4c-ii). Since the friction force is proportional to width of the contact area (short axis of the ellipse), the mean number of micropillars along the short axis (n) were then counted (Figure 4c-iii). A larger n means a longer pinning front of the droplet during the lateral movement and thus a larger F_K. When the droplet slid in the x direction, the $F_{K,x}$ decreased from 41.82 ± 3.48 to 29.80 ± 2.8 μN when the ε increased from 0 to 80%, showing a linear dependence on n (decreased from 13.5 to 11.5, Figure 4d). Similarly, the

$F_{K,y}$ also showed a linear dependence on n (increased from 13.5 to 31.0, Figure 4e), while the dependence on n was much weaker than for $F_{K,x}$. In other words, the regulation of the micropillar arrangement by stress is much more efficient along the x direction. While we could not directly observe the three phase contact line on each micropillar top due to the limitation of our device, we propose an easier movement of the droplet along the x direction than along y direction because of the saddle-shaped micropillar top. That is, along with the period of GP micropillars, the anisotropic micropillar top could also contribute to the anisotropic friction. Once again, the liquid-solid friction strongly suggests the anisotropy of the surface.

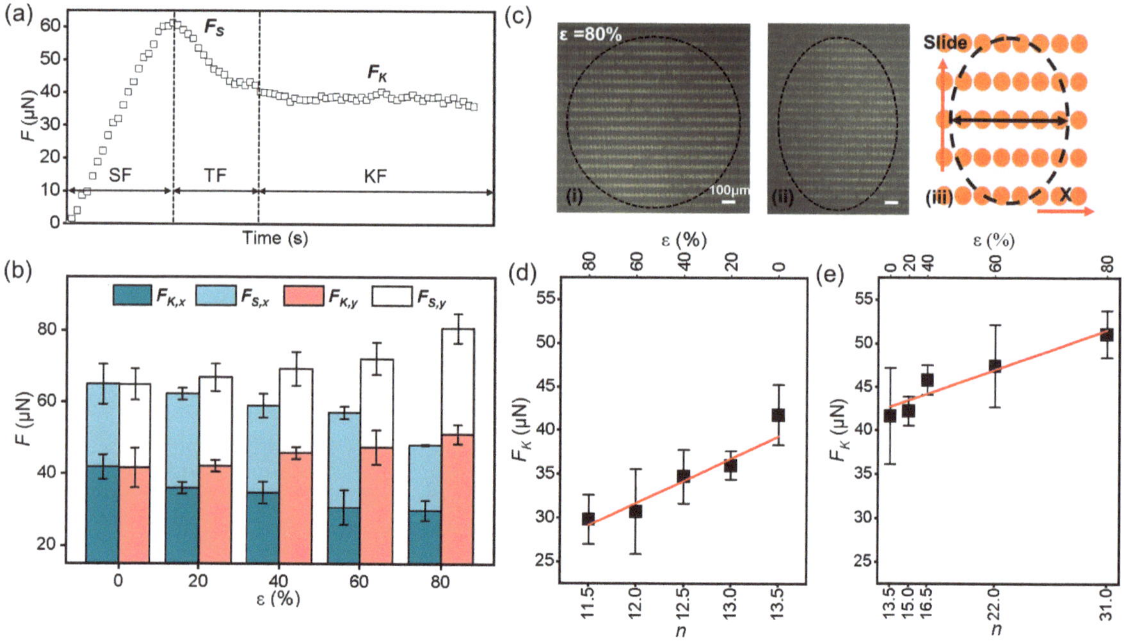

Figure 4. The liquid-solid friction measurement on the GP-MA at room temperature. (**a**) Typical curve showing the region of the static friction (SF), the kinetic friction (KF), and the transition zone of friction (TF); (**b**) the static friction force (F_S) and kinetic friction force (F_K) along two directions on the GP-MA under various elongations; (**c**) the contact interface of the droplet at the initial (i) and KF (ii) states and the corresponding schematic illustration (iii) on the GP-MA with an elongation of 80%; the black dashed lines represent the perimeter of the contact area; the black line at (iii) indicates the short axes of the contact area, where the number of micropillars were counted; (**d**,**e**) influence of the micropillar number on the F_K along the direction of (**d**) the x and (**e**) y under various elongations. Each data point in (**b**,**d**,**e**) represents the mean value of at least five measurements. Standard deviations are indicated by error bars.

3.3. Liquid-Solid Friction Measurement on the GP-MA at Elevated Temperatures

Due to the ability of graphene to absorb light and its photothermal effect [40,41], the temperature of the GP-MA surface could be effectively heated up by UV irradiation remotely. For instance, under a UV irradiation of 30 mW/cm^2, the GP-MA surface achieved a homogeneous temperature of ~80 °C after 80 s (Figure 5a). By changing the light intensity between 10 and 40 mW/cm^2, the temperature between 43.9 and 106.5 °C was easily realized (Figure 5b). Stable temperatures ranging from 40 to 80 °C could then be easily and remotely regulated by controlling the light intensity for the following investigation. As the graphene sheets were dispersed solely in the GP micropillars, the temperature of the GP-MA increased while the backing layer remained at room temperature. As a result, the diameter and period of the GP-MA remained the same (D_x = 49.42 ± 0.54 μm, D_y = 49.31 ± 0.42 μm, P_x = 69.17 ± 0.53 μm, and P_y = 69.49 ± 0.31 μm) when the surface temperature was heated up to 80 °C, which was beneficial for the following tests (Figure 5c). When the droplet slid on the heated GP-MA, the bottom part of the droplet would thus be heated up, causing a Marangoni flow inside the droplet (Figure 5d).

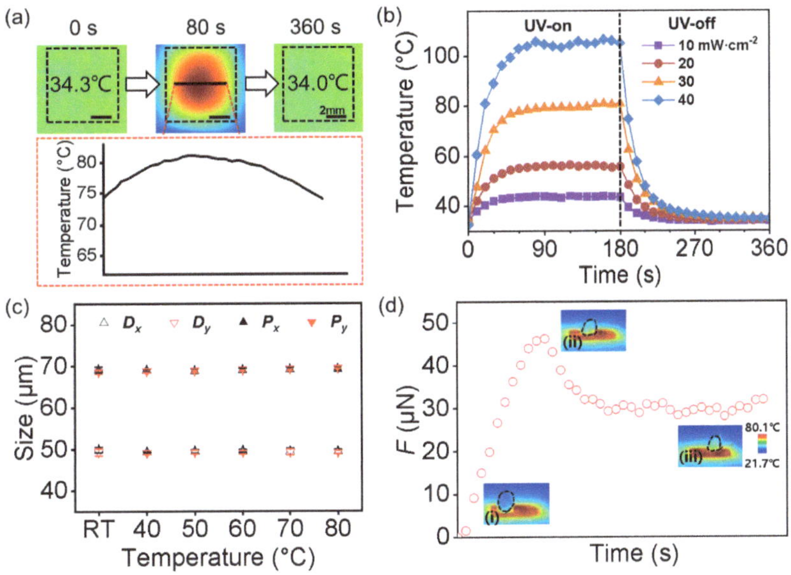

Figure 5. The photothermal effect in the GP-MA. (**a**) Typical infrared image of the GP-MA under UV irradiation at different time. The black line is a typical temperature profile across the GP-MA; (**b**) temperature change of the GP-MA under various UV-light intensities; (**c**) the structure parameters (D_x, D_y, P_x, and P_y) of the GP-MA with an ε of 0 at various temperatures; (**d**) typical measuring curve for the friction and infrared images of a water droplet at the initial (i), SF (ii) and KF (iii) states on the GP-MA at 80 °C. Each data point in (**c**) represents the mean value of at least five measurements. Standard deviations are indicated by error bars.

When the temperature of the GP-MA was increased, both the F_S and F_K decreased in both directions. For the sample with ε = 0, $F_{K,x}$ and $F_{K,y}$ decreased to 30.90 ± 4.9 and 29.94 ± 6.4 μN, with a decrease of 26.1% and 28.1%, respectively, when the temperature increased from room temperature (RT) to 80 °C (Figure 6a). Considering the statistics, there was no difference in the orthogonal directions, showing an anisotropy ($\Delta F_K = |F_{K,x} - F_{K,y}|$) close to 0. The same dependence of friction forces on the temperature was also demonstrated on the GP-MA with an ε of 40% (Figure 6b). Different from the GP-MA with ε = 0, there were clear differences in the F_S and F_K along the orthogonal directions at all the temperatures tested. Moreover, ΔF_K decreased from 11.09 to 6.74 μN as the temperature increased from RT to 80 °C. That is, the anisotropy changed following the surface temperature.

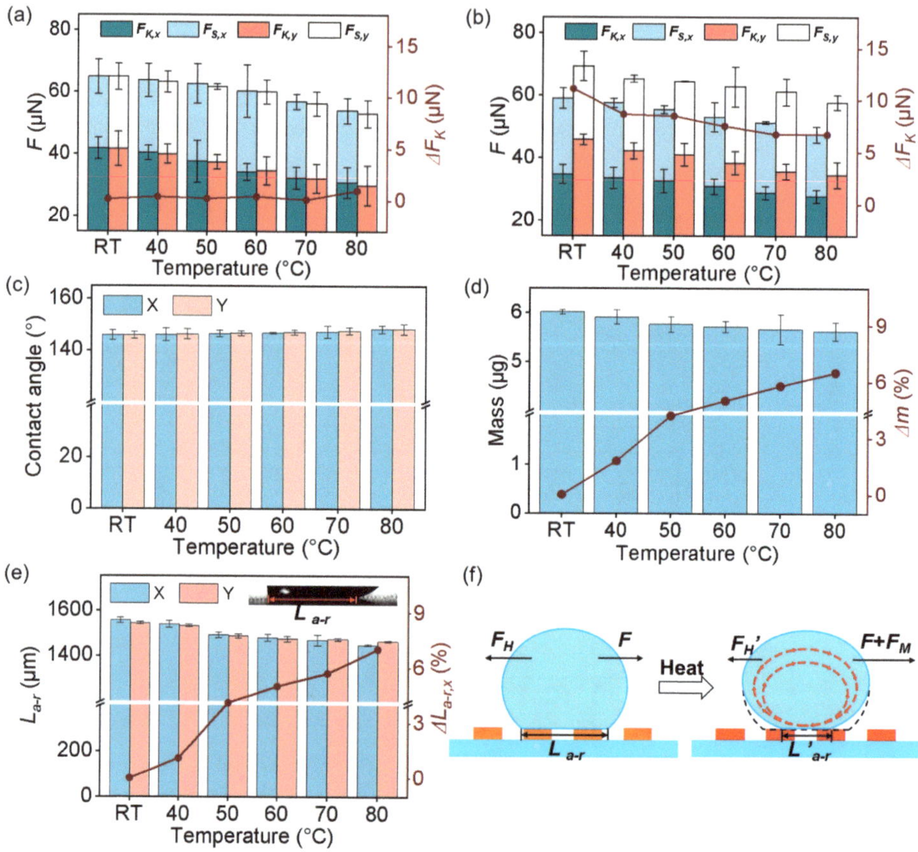

Figure 6. The liquid-solid friction measurement on the GP-MA at elevated temperatures. (**a**,**b**) The friction forces along the x and y directions and the corresponding ΔF_K on the GP-MA with an ε of (**a**) 0 and (**b**) 40% at various temperatures; (**c**) the CA along the x and y directions on the GP-MA with an ε of 0 at various temperatures; (**d**) mass, the proportion of the mass decrease (Δm) and (**e**) length between the advancing and receding fronts (L_{a-r}) and the changing proportion of L_{a-r} in the x direction ($\Delta L_{a-r,x}$) of a droplet during the friction measurements on the GP-MA with an ε of 0 at various temperatures; (**f**) the proposed mechanism for the decrease in the friction force at increased temperature. Each data point in (**a**–**e**) represents the mean value of at least five measurements Standard deviations are indicated by error bars.

To understand the temperature dependence of the friction and anisotropy, considering the stability of the structure (Figure 5c), the contact angle (Figure 6c), the mass change (Figure 6d), and the length between the advancing and receding fronts (L_{a-r}) during friction (Figure 6e) were investigated. At temperatures ranging from RT to 80 °C, the CA of the GP-MA with an ε of 0 ranged from 148.4 ± 2.0° to 148.3 ± 1.4° in both directions (Figure 6c). As a difference of 1.7% in the CA was detected at various temperatures, the influence of temperature on the Young's equation was thus considered to be negligible. After friction measurement at elevated temperatures, which normally took around 13 s, the evaporation of water would reduce the mass of the droplet. At 80 °C, which is quite close to the boiling point of water, the evaporation was quite fast. As a result, a mass decrease (Δm) of 6.48% in the droplet was detected (Figure 6d). Considering the friction here also follows Amontons' law that the friction force is proportional to the normal force (here, the droplet weight, μg) with a constant friction coefficient of μ:

$$F = \mu mg,$$

a decrease of 6.48%, rather than ~27%, would be expected in the F_K. At lower temperatures, the evaporation of water was much slower, and the Δm was much lower, which meant an even smaller influence on the friction. This strongly suggests there could be other mechanisms contributing to the reduction in the friction force at high temperatures. The L_{a-r} during the kinetic friction was also monitored. At RT, an L_{a-r} of 1556.38 ± 12.13 µm was detected in the x direction, which decreased by 7.05% ($\Delta L_{a-r,x}$) to 1446.86 ± 3.06 µm as the temperature increased to ~80 °C (Figure 6e). The decrease in the L_{a-r} means the decrease in the contact area, which determined the contact point between the droplet and the surface and led to the decrease in the friction.

At room temperature, due to Newton's third law, the tested liquid–solid friction is expressed by the hydrodynamic resistance (F_H):

$$F = F_H,$$

which includes viscous forces in the droplet and the contact area of the droplet on surfaces [42]. As the contact area decreased at an elevated temperature, the F_H decreased ($F_H' < F_H$). Additionally, the ΔT in the droplet caused spatial variation in the surface tension (Figure 5d), adding a Marangoni force (F_M) [43,44], whose direction was the same as that of the flow in the droplet, to the droplet. The force balance can thus be reconsidered as follows:

$$F + F_M = F_H'.$$

That is, the Marangoni effect also contributes to the friction reduction. To summarize, the decrease in the mass and contact area and the Marangoni effect contribute together to the friction reduction (Figure 6f).

3.4. Liquid-Solid Friction on the GP-MA under Various Elongations

At a fixed elevated temperature, the influence of the ε on the friction force was then further investigated. At 40 °C, the $F_{S,x}$ sharply decreased by 25.10% from 63.72 ± 5.41 to 47.73 ± 3.99 µN, and the $F_{K,x}$ decreased by 30.67% from 40.36 ± 2.36 to 27.98 ± 0.84 µN, correspondingly, when the ε increased to 80% (Figure 7a). In contrast, the $F_{S,y}$ and $F_{K,y}$ increased from 63.26 ± 3.52 and 40.00 ± 3.09 µN to 70.92 ± 2.04 and 46.26 ± 3.60 µN, with a difference of 10.80% and 15.65%, respectively. As a result, the ΔF_K increased from 0.36 to 18.28 µN, suggesting an increase in the anisotropy with an increase in the ε. A similar phenomenon was also found in the GP-MA at 80 °C (Figure 7b). Generally, the friction forces, including $F_{S,x}$, $F_{S,y}$, $F_{K,x}$, and $F_{K,y}$, and the anisotropy, were all smaller than that at 40 °C and RT. This confirmed again that the increase in the temperature not only decreased the liquid–solid friction but also decreased the anisotropy.

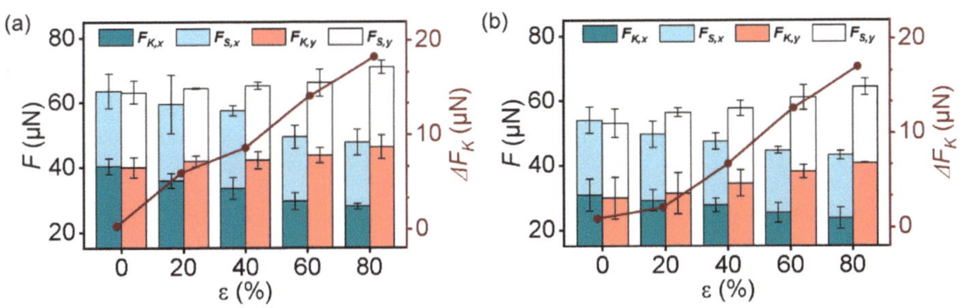

Figure 7. (**a**,**b**) The dependence of the friction forces along the x and y directions and the corresponding ΔF_K at (**a**) 40 °C and (**b**) 80 °C on ε. Each data point in (**a**,**b**) represents the mean value of at least five measurements. Standard deviations are indicated by error bars.

4. Conclusions

Here, we investigated the influence of the temperature on the friction of a water droplet on a GP-MA surface. The GP-MA was composed of GP micropillars supported by a thin layer of pure PDMS. The periods and roughness of the micropillars were regulated by changing the pre-stretching, ε, of the PDMS supporting layer. With the increase in the ε the CA and SA showed negligible differences along the orthogonal direction and could not reveal the anisotropy. With an increase in the ε (in direction x), the $F_{S,y}$ and $F_{K,y}$ increased while the $F_{S,x}$ and $F_{K,x}$ decreased. Meanwhile, the ε also increased the anisotropy (ΔF_K). Making use of the photothermal effect of graphene, the temperature of the GP-MA surface could be easily increased. The increase in the temperature decreased the $F_{S,x}$, $F_{S,y}$, $F_{K,x}$ and $F_{K,y}$, which originated from the reduced mass and contact area of the droplet and the introduced Marangoni flow in the droplet. Surprisingly, the anisotropy also decreased when the temperature increased, which was the result of the larger structural change in the x direction upon stretching. As the temperature range investigated here is quite common in our daily life, the study allows us to better understand the influence of temperature on wettability, which cannot be well revealed by contact angle measurements. In turn, the investigation here may also pave the way to invent new superwettability materials for high temperatures.

Author Contributions: Formal analysis, Z.L. and D.C.; Funding acquisition, L.X.; Investigation, Z.L. L.L. and Y.Z.; Methodology, L.L.; Project administration, L.X.; Software, K.X.; Supervision, L.X. Visualization, K.X. and X.Z.; Writing-original draft, Z.L.; Writing-review and editing, D.C. and L.X All authors have read and agreed to the published version of the manuscript.

Funding: This work was supported by the National Natural Science Foundation of China (51973165 52105296 and 62161160311) and the Fundamental Research Funds for the Central Universities (2042022kf1220).

Institutional Review Board Statement: Not applicable.

Data Availability Statement: The data presented in this study are available upon request from the corresponding author.

Acknowledgments: We acknowledge the nanofabrication assistance from the Center for Nanoscience and Nanotechnology at Wuhan University.

Conflicts of Interest: The authors declare no conflict of interest.

References

1. Chen, H.; Zhang, P.; Zhang, L.; Liu, H.; Jiang, Y.; Zhang, D.; Han, Z.; Jiang, L. Continuous Directional Water Transport on the Peristome Surface of *Nepenthes alata*. *Nature* **2016**, *532*, 85–89. [CrossRef]
2. Zheng, Y.; Gao, X.; Jiang, L. Directional Adhesion of Superhydrophobic Butterfly Wings. *Soft Matter* **2007**, *3*, 178–182. [CrossRef]
3. Feng, L.; Li, S.; Li, Y.; Li, H.; Zhang, L.; Zhai, J.; Song, Y.; Liu, B.; Jiang, L.; Zhu, D. Super-Hydrophobic Surfaces: From Natural to Artificial. *Adv. Mater.* **2002**, *14*, 1857–1860. [CrossRef]
4. Qian, Z.; Zhang, S.; Xue, L. Numerical Study of Mitigating Cloud Cavitation Shedding Using Biomimetic Protuberant Stripes. *J Fluids Eng.* **2022**, *144*, 091201. [CrossRef]
5. Xu, P.; Zhang, Y.; Li, L.; Lin, Z.; Zhu, B.; Chen, W.; Li, G.; Liu, H.; Xiao, K.; Xiong, Y.; et al. Adhesion Behaviors of Water Droplets on Bioinspired Superhydrophobic Surfaces. *Bioinspir. Biomim.* **2022**, *17*, 041003. [CrossRef] [PubMed]
6. Li, J.; Han, X.; Li, W.; Yang, L.; Li, X.; Wang, L. Nature-Inspired Reentrant Surfaces. *Prog. Mater. Sci.* **2023**, *133*, 101064. [CrossRef]
7. Yu, C.; Liu, M.; Zhang, C.; Yan, H.; Zhang, M.; Wu, Q.; Liu, M.; Jiang, L. Bio-Inspired Drag Reduction: From Nature Organisms to Artificial Functional Surfaces. *Giant* **2020**, *2*, 100017. [CrossRef]
8. Si, Y.; Dong, Z.; Jiang, L. Bioinspired Designs of Superhydrophobic and Superhydrophilic Materials. *ACS Cent. Sci.* **2018**, *4* 1102–1112. [CrossRef]
9. Parker, A.; Lawrence, C. Water Capture by a Desert Beetle. *Nature* **2001**, *414*, 33–34. [CrossRef]
10. Jiang, M.; Wang, Y.; Liu, F.; Du, H.; Li, Y.; Zhang, H.; To, S.; Wang, S.; Pan, C.; Yu, J.; et al. Inhibiting the Leidenfrost Effect above 1000 °C for Sustained Thermal Cooling. *Nature* **2022**, *601*, 568–572. [CrossRef]
11. Liu, C.; Lu, C.; Yuan, Z.; Lv, C.; Liu, Y. Steerable Drops on Heated Concentric Microgroove Arrays. *Nat. Commun.* **2022**, *13*, 3141 [CrossRef] [PubMed]

12. Zhao, W.; Jiang, Y.; Yu, W.; Yu, Z.; Liu, X. Wettability Controlled Surface for Energy Conversion. *Small* **2022**, *18*, 2202906. [CrossRef] [PubMed]
13. Li, J.; Hou, Y.; Liu, Y.; Hao, C.; Li, M.; Chaudhury, M.; Yao, S.; Wang, Z. Directional Transport of High-Temperature Janus Droplets Mediated by Structural Topography. *Nat. Phys.* **2016**, *12*, 606–612. [CrossRef]
14. Vakarelski, I.; Marston, J.; Chan, D.; Thoroddsen, S. Drag Reduction by Leidenfrost Vapor Layers. *Phys. Rev. Lett.* **2011**, *106*, 214501. [CrossRef]
15. Song, M.; Hu, D.; Zheng, X.; Wang, L.; Yu, Z.; An, W.; Na, R.; Li, C.; Li, N.; Lu, Z.; et al. Enhancing Droplet Deposition on Wired and Curved Superhydrophobic Leaves. *ACS Nano* **2019**, *13*, 7966–7974. [CrossRef]
16. Linke, H.; Alemán, B.; Melling, L.; Taormina, M.; Francis, M.; Dow-Hygelund, C.; Narayanan, V.; Taylor, R.; Stout, A. Self-Propelled Leidenfrost Droplets. *Phys. Rev. Lett.* **2006**, *96*, 154502. [CrossRef]
17. Lagubeau, G.; Le Merrer, M.; Clanet, C.; Quéré, D. Leidenfrost on a Ratchet. *Nat. Phys.* **2011**, *7*, 395–398. [CrossRef]
18. Li, Q.; Li, L.; Shi, K.; Yang, B.; Wang, X.; Shi, Z.; Tan, D.; Meng, F.; Liu, Q.; Hu, S.; et al. Reversible Structure Engineering of Bioinspired Anisotropic Surface for Droplet Recognition and Transportation. *Adv. Sci.* **2020**, *7*, 2001650. [CrossRef]
19. Zhang, J.; Li, L.; Xu, P.; Lei, Y.; Song, Q.; Liu, J.; Xiong, Y.; Yang, S.; Zhang, Y.; Xue, L. Anisotropic Wettability of Bioinspired Surface Characterized by Friction Force. *Biomimetics* **2022**, *7*, 108. [CrossRef]
20. Shi, K.; Li, Q.; Zhang, J.; Li, L.; Yang, B.; Hu, S.; Lei, Y.; Liu, Z.; Liu, S.; Xue, L. Quantitative Characterization of Surface Wettability by Friction Force. *Appl. Surf. Sci.* **2021**, *536*, 147788. [CrossRef]
21. Zhang, J.; Shi, K.; Xu, P.; Li, Q.; Xue, L. The dynamic regulation of friction force of a water droplet on goose bumps-inspired surfaces. *Surf. Tech.* **2021**, *50*, 66–73. [CrossRef]
22. Luo, X.; Lai, H.; Cheng, Z.; Liu, P.; Li, Y.; Yu, X.; Liu, Y. Slippery Shape Memory Polymer Arrays with Switchable Isotropy/Anisotropy and Its Application as a Reprogrammable Platform for Controllable Droplet Motion. *Chem. Eng. J.* **2021**, *403*, 126356. [CrossRef]
23. Wang, Y.; Lai, H.; Cheng, Z.; Zhang, H.; Liu, Y.; Jiang, L. Smart Superhydrophobic Shape Memory Adhesive Surface toward Selective Capture/Release of Microdroplets. *ACS Appl. Mater. Interfaces* **2019**, *11*, 10988–10997. [CrossRef] [PubMed]
24. Li, C.; Jiao, Y.; Zhang, Y.; Jiang, S.; Lv, X.; Wu, S.; Li, J.; Hu, Y.; Ye, J.; Liu, K.; et al. Noncontact All-In-Situ Reversible Reconfiguration of Femtosecond Laser-Induced Shape Memory Magnetic Microcones for Multifunctional Liquid Droplet Manipulation and Information Encryption. *Adv. Funct. Mater.* **2021**, *31*, 2100543. [CrossRef]
25. Hwang, H.; Papadopoulos, P.; Fujii, S.; Wooh, S. Driving Droplets on Liquid Repellent Surfaces via Light-Driven Marangoni Propulsion. *Adv. Funct. Mater.* **2022**, *32*, 2111311. [CrossRef]
26. Gao, C.; Wang, L.; Lin, Y.; Li, J.; Liu, Y.; Li, X.; Feng, S.; Zheng, Y. Droplets Manipulated on Photothermal Organogel Surfaces. *Adv. Funct. Mater.* **2018**, *28*, 1803072. [CrossRef]
27. Manabe, K.; Saito, K.; Nakano, M.; Ohzono, T.; Norikane, Y. Light-Driven Liquid Conveyors: Manipulating Liquid Mobility and Transporting Solids on Demand. *ACS Nano* **2022**, *16*, 16353–16362. [CrossRef]
28. Wang, F.; Liu, M.; Liu, C.; Zhao, Q.; Wang, T.; Wang, Z.; Du, X. Light-Induced Charged Slippery Surfaces. *Sci. Adv.* **2022**, *8*, eabp9369. [CrossRef]
29. Han, K.; Wang, Z.; Heng, L.; Jiang, L. Photothermal Slippery Surfaces towards Spatial Droplet Manipulation. *J. Mater. Chem. A* **2021**, *9*, 16974–16981. [CrossRef]
30. Zhao, D.; Sun, Y.; Liu, C.; Zhan, H.; Wang, M.; Liu, J.; Liu, Y. Temperature-Regulated Directional Bounce of Impacting Droplets on Gradient Grooves. *Surf. Coat. Technol.* **2018**, *356*, 132–137. [CrossRef]
31. Wang, X.; Tan, D.; Hu, S.; Li, Q.; Yang, B.; Shi, Z.; Das, R.; Xu, X.; Wu, Z.; Xue, L. Reversible Adhesion via Light-Regulated Conformations of Rubber Chains. *ACS Appl. Mater. Interfaces* **2019**, *11*, 46337–46343. [CrossRef] [PubMed]
32. Malard, L.; Pimenta, M.; Dresselhaus, G.; Dresselhaus, M. Raman Spectroscopy in Graphene. *Phys. Rep.* **2009**, *473*, 51–87. [CrossRef]
33. Khan, U.; O'Neill, A.; Lotya, M.; De, S.; Coleman, J. High-Concentration Solvent Exfoliation of Graphene. *Small* **2010**, *6*, 864–871. [CrossRef] [PubMed]
34. Wu, Z.; Ren, W.; Gao, L.; Liu, B.; Jiang, C.; Cheng, H. Synthesis of High-Quality Graphene with a Pre-Determined Number of Layers. *Carbon* **2009**, *47*, 493–499. [CrossRef]
35. Wang, Y.; Yang, R.; Shi, Z.; Zhang, L.; Shi, D.; Wang, E.; Zhang, G. Super-Elastic Graphene Ripples for Flexible Strain Sensors. *ACS Nano* **2011**, *5*, 3645–3650. [CrossRef]
36. Wang, X.; Shi, Z.; Meng, F.; Zhao, Y.; Wu, Z.; Lei, Y.; Xue, L. Interfacial Interaction-Induced Temperature-Dependent Mechanical Property of Graphene-PDMS Nanocomposite. *J. Mater. Sci.* **2020**, *55*, 1553–1561. [CrossRef]
37. Wang, X.; Yang, B.; Tan, D.; Li, Q.; Song, B.; Wu, Z.; del Campo, A.; Kappl, M.; Wang, Z.; Gorb, S.; et al. Bioinspired Footed Soft Robot with Unidirectional All-Terrain Mobility. *Mater. Today* **2020**, *35*, 42–49. [CrossRef]
38. Liu, K.; Vuckovac, M.; Latikka, M.; Huhtamäki, T.; Ras, R. Improving Surface-Wetting Characterization. *Science* **2019**, *363*, 1147–1148. [CrossRef]
39. Gao, N.; Geyer, F.; Pilat, D.; Wooh, S.; Vollmer, D.; Butt, H.; Berger, R. How Drops Start Sliding over Solid Surfaces. *Nat. Phys.* **2018**, *14*, 191–196. [CrossRef]
40. Liu, C.; Chang, Y.; Norris, T.; Zhong, Z. Graphene Photodetectors with Ultra-Broadband and High Responsivity at Room Temperature. *Nat. Nanotechnol.* **2014**, *9*, 273–278. [CrossRef]

41. Novoselov, K.; Fal'ko, V.; Colombo, L.; Gellert, P.; Schwab, M.; Kim, K. A Roadmap for Graphene. *Nature* **2012**, *490*, 192–200. [CrossRef] [PubMed]
42. Smith, J.; Dhiman, R.; Anand, S.; Reza-Garduno, E.; Cohen, R.; McKinley, G.; Varanasi, K. Droplet Mobility on Lubricant-Impregnated Surfaces. *Soft Matter* **2013**, *9*, 1772–1780. [CrossRef]
43. Brzoska, J.; Brochard-Wyart, F.; Rondelez, F. Motions of Droplets on Hydrophobic Model Surfaces Induced by Thermal Gradients. *Langmuir* **1993**, *9*, 2220–2224. [CrossRef]
44. Bjelobrk, N.; Girard, H.; Bengaluru Subramanyam, S.; Kwon, H.; Quéré, D.; Varanasi, K. Thermocapillary Motion on Lubricant-Impregnated Surfaces. *Phys. Rev. Fluids* **2016**, *1*, 063902. [CrossRef]

biomimetics

Article

Convergent Evolution of Adhesive Properties in Leaf Insect Eggs and Plant Seeds: Cross-Kingdom Bioinspiration

Thies H. Büscher * and Stanislav N. Gorb

Department of Functional Morphology and Biomechanics, Institute of Zoology, Kiel University, Am Botanischen Garten 9, 24118 Kiel, Germany
* Correspondence: tbuescher@zoologie.uni-kiel.de; Tel.: +49-431-880-4144

Abstract: Plants and animals are often used as a source for inspiration in biomimetic engineering. However, stronger engagement of biologists is often required in the field of biomimetics. The actual strength of using biological systems as a source of inspiration for human problem solving does not lie in a perfect copy of a single system but in the extraction of core principles from similarly functioning systems that have convergently solved the same problem in their evolution. Adhesive systems are an example of such convergent traits that independently evolved in different organisms. We herein compare two analogous adhesive systems, one from plants seeds and one from insect eggs, to test their properties and functional principles for differences and similarities in order to evaluate the input that can be potentially used for biomimetics. Although strikingly similar, the eggs of the leaf insect *Phyllium philippinicum* and the seeds of the ivy gourd *Coccinia grandis* make use of different surface structures for the generation of adhesion. Both employ a water-soluble glue that is spread on the surface via reinforcing fibrous surface structures, but the morphology of these structures is different. In addition to microscopic analysis of the two adhesive systems, we mechanically measured the actual adhesion generated by both systems to quantitatively compare their functional differences on various standardized substrates. We found that seeds can generate much stronger adhesion in some cases but overall provided less reliable adherence in comparison to eggs. Furthermore, eggs performed better regarding repetitive attachment. The similarities of these systems, and their differences resulting from their different purposes and different structural/chemical features, can be informative for engineers working on technical adhesive systems.

Keywords: glue; Phylliidae; Cucurbitaceae; fiber reinforcement; biomimetics; ivy gourd

Citation: Büscher, T.H.; Gorb, S.N. Convergent Evolution of Adhesive Properties in Leaf Insect Eggs and Plant Seeds: Cross-Kingdom Bioinspiration. *Biomimetics* **2022**, *7*, 173. https://doi.org/10.3390/biomimetics7040173

Academic Editor: Giuseppe Carbone

Received: 18 September 2022
Accepted: 20 October 2022
Published: 22 October 2022

1. Introduction

A core principle of biomimetics is to find inspiration for human problem solving in nature [1]. While several natural principles were successfully adapted in biomimetic studies in the past, sometimes a vague similarity to biological structures appears to be sufficient for some researchers to claim bioinspiration as a trademark to claim usefulness per se. However, such a top-down approach to back up technical innovations with supposed biological similarity does not necessarily use the full potential of biomimetic thinking [2]. Biological systems undoubtedly offer significant potential for inspiration for problem solving, as many functions in nature have evolved in response to specific environmental requirements and are subjected to continuous selection [3]. Several technical innovations are a result of investigation of examples from nature, for example, in the field of gripping devices in soft robotics [4–6]. However, the actual strength in investigating such phenomena lies in the understanding of the actual functional constraints these biological systems are adapted to and in disarticulation of the key functions. A great potential for finding inspiration in natural functional systems is especially present in systems that evolved convergently in different remotely related organisms. Within animals, one striking example of such a convergence is found in their adhesive systems [7]. As attachment, in general, is very important for many animals in different aspects of their life (nutrition, locomotion, dispersal, etc.) and various different taxa possess elaborate attachment systems [8,9]. These systems are widespread within the animal kingdom but occur in very different clades of animals and can have quite different morphology and functionality [7] The functions of the different kinds of attachment devices themselves in turn are often tuned to more general requirements, as they rely on the physical constraints of the interaction of the attachment organ and the substrate. These constraints are universal for all species, independently of their relatedness, and, if the environment is similar, result in a similar morphology [10–13]. Consequently, a similar morphology can occur convergently as a result of the similar conditions the systems are adapted to in phylogenetically distinct lineages.

Such a mechanism with a similar functionality in two organisms from phylogenetically distant lineages is represented by the adhesive mechanism of the eggs of the Philippine leaf insect *Phyllium philippinicum* Hennemann, Conle, Gottardo & Bresseel, 2009 (Phasmatodea, Phylliidae) and the seeds of the ivy gourd *Coccinia grandis* (L.) Voigt (Cucurbitaceae). Both systems include fibrillary surface structures on the surface and a glue component that is applied to the substrate. While the adhesive mechanism of leaf insect eggs has already been investigated [14,15], the adhesiveness of *C. grandis* seeds has, so far, no apparent notion in the literature. However, the seeds of this plant carry a similar adhesive system to that described for the eggs of walking leaf insects. Furthermore, they are so similar in appearance that they are easily confused with each other in the field because both species co-occur in similar environments. We became aware of this species and the similarity of both reproductive stages due to this confusion. During a field trip in Pasir Ris, Singapore (1°23′33.2″ N 103°55′33.6″ E), the supposed eggs of a Singaporean *Phyllium* sp. were found adhered to local plants (Figure 1B). However, closer inspection and incubation of these eggs revealed their true identity as C. *grandis* seeds (pers. comm. Wei-Song Lih).

Figure 1. Focal organisms. (**A–D**) *Coccinia grandis*: (**A**) Flowering plant (modified from [16], published under CCBY 4.0). (**B**) Seeds found in the field attached to the leaves of a different plant (provided by Lih Wei-Song). (**C**) Extracted dry seed before water contact. (**D**) Seed attached to a glass slide photographed through the slide. (**E–G**) *Phyllium philippinicum*: (**E**) Adult female (from [15] published under CCBY 4.0). (**F**) Dry egg before first water contact, lateral view. (**G**) Egg attached to a glass slide photographed through the slide. Scale bars: 1 mm.

As described for *P. philippinicum* eggs [14,15], these seeds display adhesive capability after activation with water (Figure 2). Likewise, they carry fibrillary adhesive structures on their surface, which expand after exposure to water and adapt to the geometry of the substrate's surface. Adhesion is also facilitated by a film of glue, which is distributed on the substrate. Details of this mechanism have been experimentally tested for leaf insect eggs, yielding a study on the influence of the roughness and surface chemistry [15] and the influence of different solvents on the activation of the exochorionic structures involved in it [14]. We aimed to investigate the mechanism of the seed adhesion of *C. grandis* in similar detail. This included an investigation of the morphology of the components involved in the adhesive mechanism and experimental characterization of its function under different substrate constraints. Both were used to compare the morphology and function of the two similar adhesive systems found in different kingdoms of life. The similarities and differences between the two species can be used to evaluate the common characteristics that are important for this kind of adhesive system and the modifications for specific tasks. Furthermore, it provides insights into the specific mechanisms in light of their ecological role, which might facilitate or prevent dispersal.

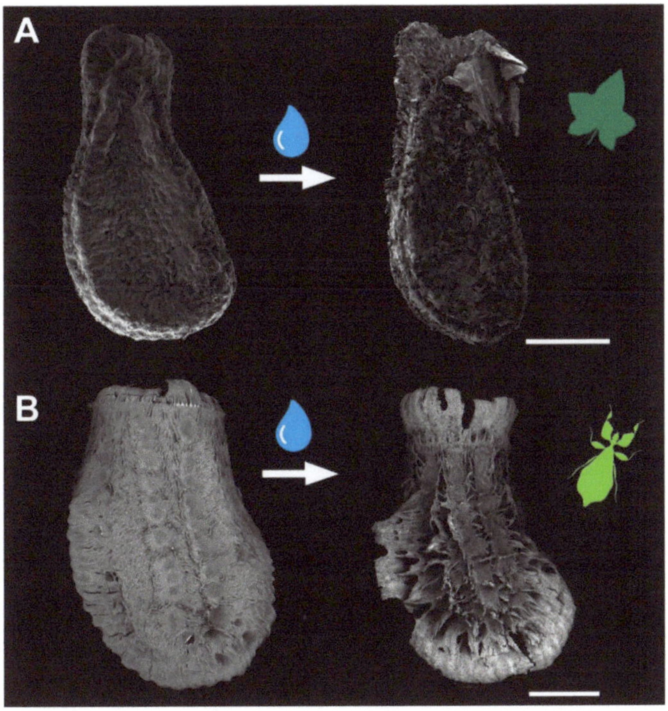

Figure 2. Water interactions of both adhesive systems. (**A**) Seed of *Coccinia grandis*. (**B**) Egg of *Phyllium philippinicum* (modified from [15] published under CCBY 4.0). The surface structures of both organisms are densely packed. Water contact induces the spreading of fibrillary adhesive structures, which carry glue for triggering adhesion in contact with the substrate. Scale bars: 1 mm.

Natural adhesive systems can be tuned to fulfill specific tasks, such as coping with the crystalline wax coverage of plants [17], with strong torrents in fast flowing water [18,19], or the challenging surface and motion of the host in the case of parasites [20]. Others cope with a variety of influences at once [7–9,21,22]. Comparative approaches, such as the one presented herein, can help to understand the main principles of these systems and provide insights into the essence of the common principle. Such knowledge is useful for biomechanics to isolate the key characteristics of natural systems and evaluate the adaptations for the actual tasks of the respective examples. Consequently, the comparative investigation of such two similar mechanisms yields an evaluation of the biomimetic potential of the underlying common principle. To test the similarity of two similar natural adhesive systems from two different kingdoms, namely plants and animals, we herein characterized the adhesive system of *C. grandis* seeds in light of the present knowledge of the adhesive system of *P. philippinicum* eggs and compared the two systems in terms of their morphology, the adhesive performance on varying surface roughness and surface chemistry, and the repeatability of adhesion, which has already been shown for the leaf insect eggs [15]. To provide a similar base of knowledge of *C. grandis* seeds, we conducted the same analysis of the morphology using light and scanning electron microscopy and mechanically tested the resulting adhesion with the same setup used for *P. philippinicum* eggs in previous studies.

We specifically investigated the following questions:

1. How does the morphology differ between *P. philippinicum* eggs and *C. grandis* seeds?
2. What influence does the substrate surface roughness have on the adhesion of *C. grandis* seeds?
3. What influence does the surface chemistry have on the adhesion of these seeds?
4. What are the similarities and differences between both examples in terms of their adhesive performance and the repeatability of adhesion?

The results are discussed in the background of their significance for biomimetics.

2. Materials and Methods

2.1. Specimens

This study explored two different focal objects: The seeds of the ivy gourd *Coccinia grandis* (L.) Voigt (Cucurbitaceae) and the eggs of the Philippine leaf insect *Phyllium philippinicum* Hennemann et al., 2009 (Phasmatodea, Phylliidae). Measurements of *P. philippinicum* eggs were used for comparison of the two mechanisms and were previously published in Büscher et al. [15]. Novel data for *C. grandis* were obtained using the same methodology used therein to warrant comparability. Eggs were obtained directly after oviposition from female insects from a captive breeding culture. *C. grandis* seeds were obtained from Danushka Hiruni (Kudaweda, Sri Lanka). They were harvested by mechanical extraction, dried, and stored in a dry environment until experimental use. Both the seeds and eggs were weighed with an AG204 Delta Range microbalance (Mettler Toledo, Greifensee, Switzerland; d = 0.1 mg).

2.2. Microscopic Visualization

Both the seeds and eggs were imaged prior to attachment using a microscope (M205, Leica Microsystems Ltd., Wetzlar, Germany). Furthermore, they were photographed while attached to microscopy glass slides from two directions (above and below the glass). Images were taken using the microscope camera Leica DFC420 (Leica Microsystems Ltd., Wetzlar, Germany). We recorded stacked multifocus images and merged them using the software Leica Application Suite (LAS) version 3.8.0 (Leica Microsystems Ltd., Wetzlar, Germany).

Further, the eggs and seeds were examined with the SEM Hitachi TM3000 (Hitachi High-technologies Corp., Tokyo, Japan) at an acceleration voltage of 10 kV to obtain overviews and with the SEM Hitachi S4800 (Hitachi High-technologies Corp., Tokyo, Japan) at an acceleration voltage of 5 kV to obtain the morphological details. Images were processed in Photoshop CS6 (Adobe Systems Inc., San Jose, CA, USA). The samples were either prepared in the untreated condition (before contact with water) or after contact with water and the corresponding attachment to the glass. Both untreated and detached samples were air-dried and sputtered with 10 nm gold-palladium.

2.3. Detachment Force Measurements

To compare the properties of both adhesive systems of eggs and seeds, the detachment forces of individual *C. grandis* seeds were measured with the same setup used in [15] for leaf insect eggs. The respective experimental samples were mounted on the standardized surfaces by individually placing them in droplets of distilled water (~100 µL) to activate the adhesive system. Afterwards, they were allowed to dry for 24–48 h and then individually connected to a force transducer (FORT1000, World Precision Instruments Inc., Sarasota, FL, USA) using bees wax to glue a horsehair onto the exposed side of the sample (Figure 3A). A BIOPAC Model MP100 and a BIOPAC TCI-102 system (BIOPAC Systems, Inc., Goleta, CA, USA) were used to record the detachment force–time curves of the samples from the substrates using the software Acqknowledge 3.7.0 (BIOPAC Systems Inc., Goleta, CA, USA). This was achieved by manually lowering the experimental substrates orthogonal to the sensor with a laboratory lifting platform at a speed of 2–3 cm/s. The maximum detachment force was determined by selecting the highest peak of the force–time curve. The detachment forces were measured in three set-ups:

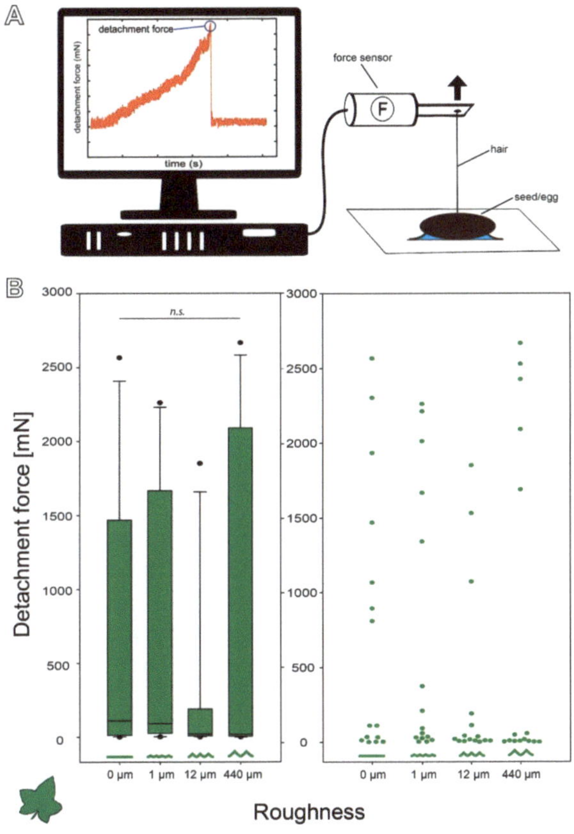

Figure 3. Detachment force measurements. (**A**) Schematic of the experimental setup with an example force–time curve of the detachment force measurement. (**B**) Detachment forces of *C. grandis* seeds on substrates with different roughness ($N = 15$ per substrate) represented by box plots (left) and jitter plots (right). Boxes indicate the 25th and 75th percentiles, the line represents the median, and the whiskers are the 10th and 90th percentiles. ***n.s.*** = no statistical difference $p = 0.55$; Kruskal–Wallis ANOVA on ranks).

1. Surface roughness

Four surfaces made of epoxy resin with different roughness were used as substrates for attachment of the samples (0, 1, 12, and 440 µm). For each substrate, 15 individual seeds and 32 individual eggs [15] were used.

2. Surface chemistry

Three surfaces with different surface free energy were used as substrates for the attachment of the samples. The surfaces had different water contact angles: 36.25 ± 1.15° (mean ± SD, n = 10) (hydrophilic), 83.38 ± 0.89° (hydrophobic), and 98.9 ± 0.47° (hydrophobic). For each substrate, 15 individual seeds and 20 individual eggs [15] were used.

3. Cyclic repetitions of attachment

The samples were subjected to repetitive individual pull-off measurements. After detachment, the same individual sample was reattached with a droplet of water. This procedure was repeated 6× for the eggs and 3× for the seeds. Furthermore, eggs were repetitively measured on the 0 µm epoxide substrate while the seeds were measured on all four different epoxide roughness test substrates. For each substrate, 15 individual seeds were used and 8 individual eggs [15] were used for the smooth substrate. If a sample did not adhere to the substrate at all, the detachment force of the individual sample was considered 0 mN, but the same individual was used again for subsequent measurements.

All measurements were carried out at a 20–23 °C temperature and 45.0–47.6% relative humidity. Except for the cyclic repetition experiments, neither the seeds nor eggs were used for more than one detachment force measurement.

2.4. Substrate Preparation

We used two different types of substrates: Epoxy resin with a range of surface roughness and glass with different wettability.

2.4.1. Glass

Microscope objective glass slides (Carl Roth GmbH & Co. KG, Karlsruhe, Germany) were cleaned with isopropylic alcohol and used untreated as the hydrophilic substrate. Clean glass sides were silanized following Voigt and Gorb [17] to reduce the surface free energy and used as a hydrophobic substrate. The wettability was quantified by measuring the water contact angle of the substrates (aqua Millipore, droplet size = 1 µL, sessile drop method; n = 10 per substrate) with an OCAH 200 (Dataphysics Instruments GmbH, Filderstadt, Germany). The contact angle of water was 36.25 ± 1.15° on untreated glass and 98.9 ± 0.47° on silanized glass.

2.4.2. Epoxy Resin

We used epoxy resin [23] and the two-step molding protocol of Salerno et al. [24] to obtain test substrates with different roughness. Glass with a 0 µm roughness, fine polishing papers (standardized roughness 1, 12 µm; Buehler, Lake Bluff, IL, USA), and industrial polishing paper with a 440 µm particle size were templates for the-two step molding. Negatives were created with polyvinylsiloxane-based dental wax (Colthéne/Whaledent AG, Altstatten, Switzerland) and filled with epoxy resin, which was cured at 70 °C for 24 h. The water contact angle of the epoxy resin was 83.38 ± 0.89° (mean ± SD, n = 10) [15].

2.5. Statistical Analysis

Statistical analyses were performed with SigmaPlot 12.0 (Systat Software Inc., San José, CA, USA). First, the data was tested for a normal distribution (Shapiro–Wilk test) and equal variance (Levene's test). Due to the non-normality or missing homoscedasticity in all comparisons, only non-parametric tests were chosen. The detachment forces of the seeds on varying substrate roughness and on substrates with different surface chemistry were compared with Kruskal–Wallis one-way analyses of variance (ANOVA) on ranks and Tukey's post hoc test. The novel data of the *C. grandis* seeds were compared with the previously reported data of *P. philippinicum* eggs [15] for these two scenarios for each substrate using the Mann–Whitney rank sum test. The repetitive measurements of both the seeds and the eggs were compared for each substrate with Friedman's repeated measures ANOVAs and Tukey's post hoc test.

3. Results

3.1. Morphology

Both the eggs and seeds are laterally flat and remarkably similar in their overall appearance despite the different affiliations of the two species within distinct kingdoms. Hence, since detailed descriptions of the morphology of both reproductive structures (eggs and seeds) can be found in [15] for *P. philippinicum* eggs and [25] for *C. grandis* seeds, we only focused on the functionally relevant features.

Both reproductive structures respond to water by expanding their fibrillar adhesive structures (**fas**). Prior to water contact, the *C. grandis* seed has a smooth surface due to the presence of a membrane (Figure 2A) under which the **fas** are tightly packed on the surface of the seed (Figure 4A,H). The **fas** of *C. grandis* seed are elongated, undivided filaments, which are mantled with a film of a hardened glue (Figure 4E,F). The length of the filaments is rather homogeneous and approximately 500 μm for the majority of them (see also [25]). Upon contact with water, the **fas** fan out and extend towards the substrate (Figure 4B–D). The tips of the **fas** make contact with the substrate and form a dense layer, adapting to the surface profile (Figure 4B). The glue is unevenly distributed along the length of the **fas** and accumulates on their tips, forming a continuous layer in combination with the **fas** (Figure 4B,C).

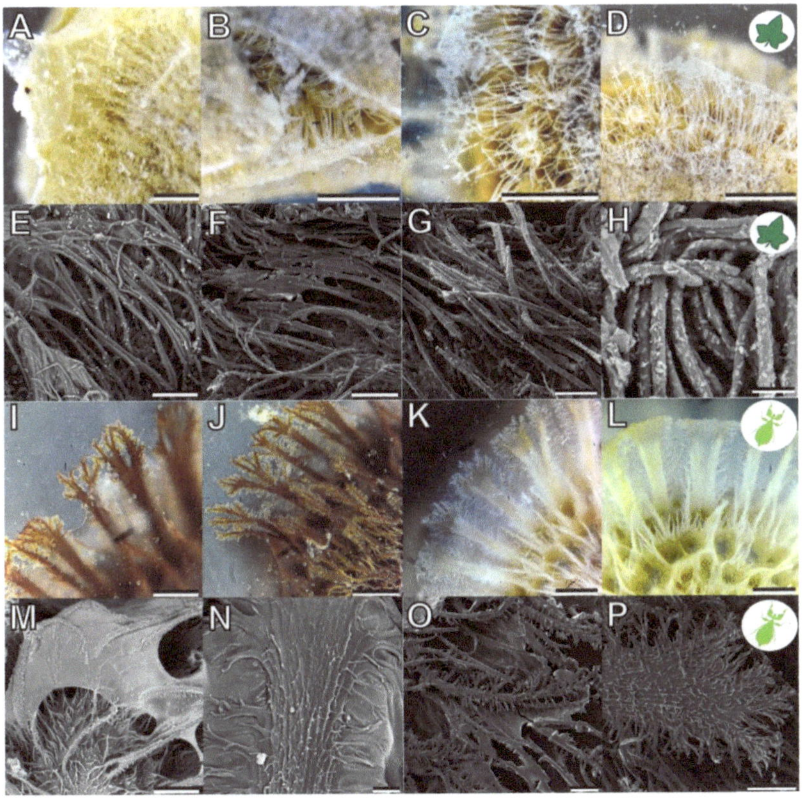

Figure 4. Morphology of the **fas** in the corresponding reproductive stages of the focal species. (**A–H**) Coccinia grandis seed. (**I–P**) Phyllium philippinicum egg. (**A–D**,**I–L**) Photographs of glue and **fas** interactions. (**E–H**,**M–P**) SEM of the glue and the **fas** morphology. Scale bars: (**A–D**,**I–L**) 300, (**E–G**,**N**,**O**) 50, (**H**) 20, (**M**) 150, and (**P**) 100 μm.

The **fas** of *P. philippinicum* also lie on the surface of the egg in the dry state, similar to the seeds (Figure 1F). However, both the distribution and the shape of the **fas** differ from the **fas** of *C. grandis* seeds. The so-called pinnae of these eggs are not homogenous in shape but consist of a central branch, which hierarchically splits into many finer terminal filaments (Figure 4I–L,N–P). Furthermore, they are oriented to the lateral rims of the egg and form two main rows and a collar at the operculum of the egg (Figure 1F,G). Smaller pinnae are present on the rest of the surface. The larger **fas** of these eggs are larger than the **fas** of the seeds but smaller ones are also present with fibrilllar structures in the heterogenous size range on the exochorion. Similar to the situation in seeds, a glue is present here, which mantles the **fas** in the dry state and spreads onto the substrate after water contact (Figure 4M–O). The **fas** respond to water by a similar expansion, forming a less continuous layer with the **fas** themselves in comparison to the one of seeds. The glue builds a closed film on the substrate in both organisms (Figure 4I–K). The tips of the pinnae carry less glue; instead, the glue is kept in the proximal space closer to the egg itself and is trapped between the **fas** and the egg (Figure 4I,J).

3.2. *Adhesion of C. grandis Seeds*

3.2.1. Influence of Substrate Roughness

The detachment forces of *C. grandis* seeds revealed a wide range of forces (Figure 3B). All four sets of measurements included particularly high detachment forces, but the overall distribution of the forces was strongly left skewed with much lower median detachment forces on all four substrates. However, the median detachment forces decreased with increasing substrate roughness. The median (±s.d.) detachment forces were 110.48 (±887.71) mN on 0 μm roughness, 91.23 (±883.72) mN on 1 μm, 20.37 (±598.02) mN on 12 μm, and 17.61 (±1086.81) mN on 440 μm. Nevertheless, several individual seeds detached at much higher pulling forces ranging up to 2600 mN on all four substrates (Figure 3B). This strong variation resulted in no significant differences between the detachment forces on the four substrates despite decreasing medians (Kruskal–Wallis one-way ANOVA on Ranks, H = 2.089, d.f. = 3, $p = 0.554$, $N = 15$ per roughness).

3.2.2. Influence of Surface Chemistry

The *C. grandis* seeds attached strongly to the hydrophilic substrate (Figure 5A). While the medium detachment force of the seeds on the substrate with a water contact angle of 36° wzs 1651.78 (±1083.55) mN, it was significantly lower (Kruskal–Wallis one-way ANOVA on Ranks, H = 10.992, d.f. = 2, $p = 0.005$, $N = 15$ per substrate; Tukey's test $p < 0.05$) on the substrates with a contact angle of 83° (110.48 (±887.71) mN) and 99° (37.16 (±668.30) mN). The forces did not differ statistically between the substrates with contact angles of 83° and 99° (Tukey's test $p > 0.05$).

3.2.3. Cyclic Repetition

The sequence of the detachment repetitions of *C. grandis* seeds on the four substrates with different roughness is shown in Figure 6B–E. The detachment forces significantly decreased on all four substrates from the first to the third cycle. While all individual seeds adhered in the first cycle, different amounts of seeds did not adhere in the second and third cycle depending on the substrate they were measured on. While the initial detachment force differed depending on the substrate (see Section 3.2.1), the subsequent cycles all showed reduced detachment forces (Figure 6B–E). On the smooth substrate, the median detachment force significantly decreased from 110.48 (±887.71) mN in the first cycle to 1.86 (±304.59) mN in the third cycle (Friedman repeated-measures ANOVA on ranks, $\chi^2 = 8.98$, d.f. = 2, $p = 0.011$; Tukey's test $p < 0.05$). While 100% of the seeds adhered in the first cycle, only 47% adhered in the second cycle and 60% in the third cycle (Figure 6B). A substrate roughness of 1 μm resulted in the median detachment forces significantly decreasing from 91.23 (±883.72) mN in the first cycle to 5.50 (±220.87) mN in the third cycle (repeated-measures ANOVA, F = 3.51, d.f. = 2, $p = 0.044$; Tukey's test $p < 0.05$). Compared to the other substrates, a higher fraction of seeds adhered in the later cycles: 100% in the first cycle, 67% in the second cycle, and 60% in the third cycle (Figure 6C). Measurements on the 12 μm rough substrate revealed a significant decrease in the detachment force from 20.37 (±598.02) mN in the first cycle to 0.00 (±62.24) mN in the third cycle (Friedman repeated-measures ANOVA on ranks, $\chi^2 = 10.308$, d.f. = 2, $p = 0.006$; Tukey's test $p < 0.05$). The 12 μm roughness had the strongest impact on the attachment ratio: 100% of the seeds adhered in the first cycle, only 33% adhered in the second cycle, and 20% in the third cycle (Figure 6D). On the roughest substrate, the eggs showed a significant decrease in the median detachment forces from 17.61 (±1086.67) mN in the first cycle to 0.63 (±184.67) mN in the third cycle (Friedman repeated-measures ANOVA on ranks, $\chi^2 = 9.927$, d.f. = 2, $p = 0.007$; Tukey's test $p < 0.05$). In the subsequent repetitions, they adhered more reliably, again with 100% of the seeds adhering in the first cycle, 60% in the second cycle, and 53% in the third cycle (Figure 6E).

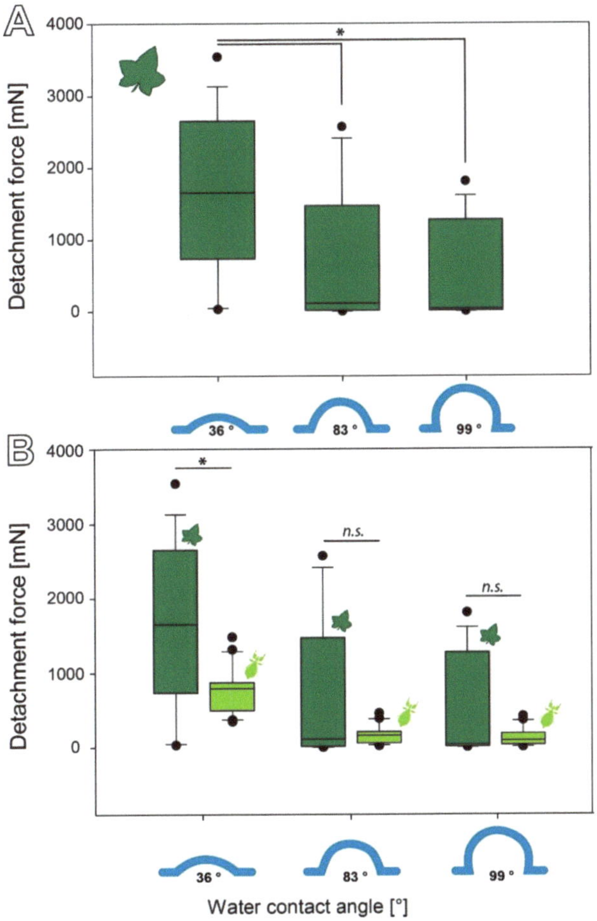

Figure 5. Influence of the surface wettability on the detachment forces. (**A**) Detachment force of Coccinia grandis seeds on substrates with different surface chemistry ($N = 15$ per substrate) * $p \leq 0.05$ (Kruskal–Wallis one-way ANOVA on Ranks). (**B**) Comparison of the detachment forces of seeds and eggs ($N_{seeds} = 15$; $N_{eggs} = 20$). ***n.s.*** = no statistical difference; * $p \geq 0.05$; Mann–Whitney rank sum test). Boxes indicate the 25th and 75th percentiles, the line represents the median, and the whiskers are the 10th and 90th percentiles.

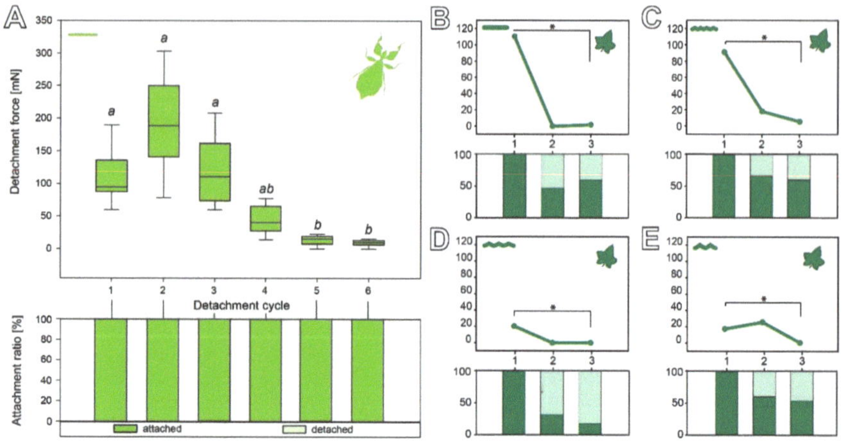

Figure 6. Sequential detachment force measurements. (**A**) *P. philippinicum* eggs. Line plot (above) and the corresponding count of attached and detached eggs (attachment ratio, below) for sequential repetitions ($N = 8$). Boxes indicate the 25th and 75th percentiles, the line represents the median and the whiskers are the 10th and 90th percentiles. Lowercase letters indicate statistical similarity boxes with the same letters are not statistically different (Friedman repeated-measures ANOVA on ranks, Tukey's post hoc test, $p < 0.05$). (**B–E**) Detachment force of *C. grandis* seeds during repetitive detachments on different substrates ($N = 15$ per substrate). Line plots (above) and the corresponding attachment ratio (below) for sequential measurements; dots represent the median. (**B**) 0 μm. (**C**) 1 μm (**D**) 12 μm. (**E**) 440 μm. * $p \leq 0.05$ (Friedman repeated-measures ANOVA on ranks or repeated measures ANOVA, respectively; Tukey's post hoc test, $p < 0.05$).

3.3. Comparison of the Attachment Capability between Eggs and Seeds

The detachment forces on substrates with different roughness show a similar pattern of substrate dependence for both the eggs and seeds (Figure 7A). Both reproductive stages revealed no statistically significant difference in regard to the roughness within the respective species (see Section 3.2, [15]); however, comparison of the medians of each experimental group yields a different behavior between the two adhesive systems. While the median detachment force of the *C. grandis* seeds strictly decreases with an increasing roughness the median detachment forces of the *P. philippinicum* eggs are higher on the 1 and 12 μm rough substrates than on the smooth and rougher ones (Figure 7A). Nevertheless, due to the strong variation of the seeds, there was no statistically significant difference between the eggs and seeds on any of the four substrates. There was no significant difference between both reproductive stages on 0 (Mann–Whitney rank sum test, $U = 179.00$, $T = 421.00$ $N_{eggs} = 32$, $N_{seeds} = 15$, $p = 0.167$), 1 (Mann–Whitney rank sum test, $U = 214.00$, $T = 334.00$ $N_{eggs} = 32$, $N_{seeds} = 15$, $p = 0.561$), 12 (Mann–Whitney rank sum test, $U = 179.00$, $T = 299.00$ $N_{eggs} = 32$, $N_{seeds} = 15$, $p = 0.167$), and 440 μm (Mann–Whitney rank sum test, $U = 219.00$ $T = 334.00$, $N_{eggs} = 32$, $N_{seeds} = 15$, $p = 0.914$).

The surface chemistry affected the detachment forces in both species. Both showed significantly decreasing detachment forces with an increasing water contact angle of the substrate (Figure 5, [15]). Hydrophilic substrates (water contact angle of 36°) caused the highest detachment forces in both cases, but the seeds adhered significantly stronger to this substrate than the eggs (Mann–Whitney rank sum test, $U = 81.00$, $T = 339.00$, $N_{eggs} = 20$ $N_{seeds} = 15$, $p = 0.022$). The hydrophobicity of the substrate resulted in lower detachment forces but no significant difference between the eggs and seeds for an 83° (Mann–Whitney rank sum test, $U = 142.00$, $T = 278.00$, $N_{eggs} = 20$, $N_{seeds} = 15$, $p = 0.803$) and 99° water contact angle (Mann–Whitney rank sum test, $U = 140.00$, $T = 260.00$, $N_{eggs} = 20$, $N_{seeds} = 15$ $p = 0.751$).

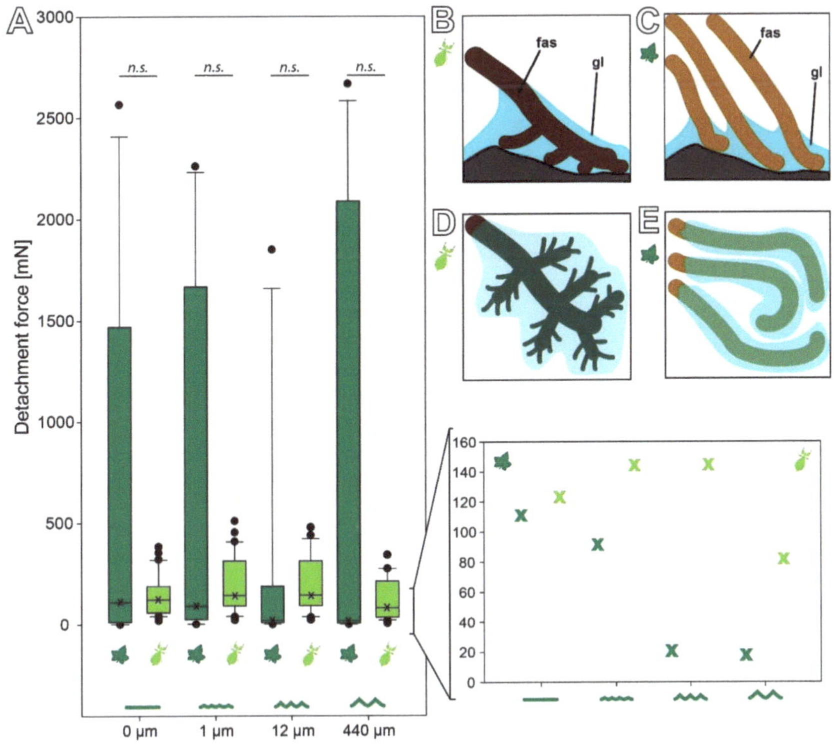

Figure 7. Comparison of the detachment forces of seeds and eggs on substrates with different substrate roughness. (**A**) Detachment forces are represented by box plots. Boxes indicate the 25th and 75th percentiles, the line represents the median, and the whiskers are the 10th and 90th percentiles. ***n.s.*** = no statistical difference $p > 0.05$; Mann–Whitney rank sum test). Eggs ($N = 32$) are shown in light green, seeds ($N = 15$) in dark green. The enlargement shows the median of the respective detachment forces for clearer representation. X = median. (**B–E**) Schemes of the contact formation with rough substrates (**B**,**C**) and the glue distribution (**D**,**E**) of the expansions of eggs (**B**,**D**) and seeds (**C**,**E**). **fas** = fibrillary adhesive structure; gl = glue.

During repetitive detachment events, eggs and seeds performed differently on two main aspects (Figure 6). The eggs of *P. philippinicum* retain repeatable attachment capability over some cycles. Although the detachment forces were significantly lower in the fifth and sixth cycle compared to the first three cycles (Friedman repeated-measures ANOVA on ranks, $\chi^2 = 35.358$, d.f. = 5, $p \leq 0.001$; Tukey's test, $p < 0.05$), all of the eggs adhered rather strongly and none failed to make sufficient contact during the attachment process. The seeds, in contrast, showed a fast decay of the detachment force starting from the second cycle and a high failure rate during the attachment repetitions, especially on the substrate with a 12 µm roughness (Figure 6B–E).

4. Discussion

The overall appearance and the general adhesive mechanism are similar in the reproductive stages of the two species examined here. However, the details of the specific adhesive performance differ between the two species and result in different advantages and disadvantages for both. While the offspring of the leaf insect *P. philippinicum* relies on the presence of suitable foodplants and should be protected during embryonic development [26,27], *C. grandis* seeds need to come into contact with soil for germination and to put down roots [26]. Consequently, the actual role of the adhesiveness of the two species differs, which is reflected in the differences in both the morphology and the resulting functionality.

4.1. Comparison between Seed and Egg Adhesive Systems

4.1.1. Morphology

In addition to the similar overall appearance, *C. grandis* seeds and *P. philippinicum* eggs share several morphological characteristics. These include (1) the presence of fibrillary adhesive structures (**fas**) on the outer surface, (2) the presence of glue, and (3) the response to water of both components (Figure 2). Naturally, as one object is a plant and the other is an animal, these shared features are not homologous to each other. The specific shape of the **fas** differs in detail (Figure 4), which leads to some differences in their functionality. Obviously, the **fas** of both species are formed by completely different structures of different chemical and developmental origins.

The eggs of phasmids are distinct from those of most other insect orders. The main specialty is the strong, hardened outer shell [28]. The egg capsule consists of two layers: the endochorion and the exochorion, which are both multi-layered [29,30]. The exochorion consists of different layers, which are structurally and chemically different, and is a product of the follicle cells [31]. Most noteworthy are the thick layer of calcium carbonate and the layer of calcium oxalate, with both being particularly tough and unique among insects [30,32]. The apomorphic toughening of the outer chorion enabled modifications for a plethora of different functions on the surface, for example, the adhesive system of the pinnae of *P. philippinicum*. The pinnae are formed on the outer surface of the egg capsule as secondary (follicular) secretions [31]. They are of variable length and width and hierarchically split several times.

The seeds of *C. grandis*, in contrast, carry more uniform **fas**. These are slender and of a rather uniform length (~500 µm). The outer layer (testa) of this seed generally consists of four layers, of which the outermost is an epidermal layer [25]. The epidermis of the closely related *Cocconia abyssinica* (Lam.) Cogn. is, according to Holstein [25], disintegrated and the cell walls of the epidermal cells form 500-µm-long **fas**. Other sources interpret the fibrils of the outer surface of *Coccinia* seeds of other species to be a fibrillose testa [33,34], or a disintegrated exotesta in *C. grandis* in particular [35]. It is likely that the **fas** of all *Coccinia* seeds with such a fibrillary surface are formed by the epidermis. In comparison, the **fas** of *C. grandis* seeds are more uniform and slender while the **fas** of *P. philippinicum* eggs are of a variable length and thickness and show hierarchically branching.

There is, to our knowledge, no notion of the glue of *C. grandis* seeds (Figure 4B–D) in the literature. However, the seeds are encapsuled in a hyaline juicy envelope within the fruit, which seems to originate from carpellary tissue [36], which might partially remain on the seed surface and develop adhesive properties on the **fas**. Other plant seeds produce mucilage envelopes on their surface [37,38], which can be either pectin [39], hemicellulose, or cellulose dominated [40] and facilitate adhesion of the seeds. The glue of *C. grandis* could also originate from either of these two mechanisms. The glue of the *P. philippinicum* eggs, in contrast, is apparently produced by females as a tertiary secretion (extraovarian) [31], mantles the surface of the egg [14,15], and is kept by the **fas** (Figure 4I–P). This glue is probably proteinaceous and includes at least two functional groups, one hydrophilic and oriented towards the substrate and the other hydrophobic and associated with the surface of the egg [14].

The fas themselves, without the glue, do not show substantial adhesive capability. The feet of different animals, both vertebrates and invertebrates, are equipped with fibrillary adhesive hairs as well [7]. These are adhesive because of the compliance of the flexible tips of the adhesive hairs that approach the substrate enough to enable van der Waals interactions with the substrate [7]. The **fas** of both seeds and eggs would, in principle, be able to do so as well, but the **fas** are likely not as flexible and compliant with the substrate

4.1.2. Adhesive Performance

The main difference between the adhesive performance of the *P. philippinicum* eggs and the C. grandis seeds is actually not the adhesive strength. Although the detachment forces measured reached higher maximum values for the plant seeds (Figure 5B or Figure 7A), the median forces are rather similar, or in some cases even higher for the leaf insect eggs. This is a result of the reliability and efficiency of the adhesion of eggs. The attachment system of the eggs performs consistently in a similar range of forces on different substrates while the detachment forces of seeds revealed strong deviations and a strongly left skewed distribution: most measurements actually resulted in very low forces, but only few measurements of very high forces were obtained. This is likely a result of two aspects of these mechanisms: the properties of the glue and the shape of the **fas**.

The hierarchical splitting and unequal distribution of the pinnae of the eggs of *P. philippinicum* results in a more reliable adaptation to the substrate geometry (Figure 7B). To maximize the attachment force, any adhesive system needs to maximize the actual contact area [7,41]. In comparison to the straight **fas** of the seeds, the pinnae of eggs seem to adapt more efficiently to rough substrates. The same applies to the repetition of the attachment events. Especially on the 12 μm roughness, the detachment forces of the seeds were low. Furthermore, the attachment ratios of the seeds of *C. grandis* were strongly reduced on this roughness (Figure 6D). On the one hand, the diameter of the **fas** of the seeds is approximately in the range of this roughness (Figure 4H), conflicting with proper contact formation. On the other hand, the distribution of the glue makes a difference for both reproductive stages. The glue of the leaf insect eggs forms a dense film, mantling the pinnae and large fractions of the egg surface (Figure 4), while the glue of the plant seeds is mainly distributed on the tips of the fibrils. This potentially results in the higher depth of the surface adaptation and thicker films of glue for the eggs compared to the plant seeds (Figure 7B). Consequently, the eggs make more reliable contact, especially to rough substrates.

On substrates with different surface chemistry but similar topography, both seeds and eggs showed similarly decreasing detachment forces with an increasing water contact angle. As the surface topography was the same for all three substrates, differences between the two species are the result of the glue properties. The *C. grandis* seeds adhered significantly stronger to hydrophilic substrates. While most plant seed glues are polysaccharides [38,39,42], the glue of the eggs of this particular species of leaf insects has been hypothesized to be a glycoprotein [14,15]. The majority of egg glues in insects are proteinaceous [43–49] and the amphiphily of the glue properties could well be achieved with glycoproteins, such as in other insect glues [50]. Therefore, a glycoprotein remains a plausible explanation for this particular adhesive system, but the chemical structure of the glue still warrants further investigation. Due to the distance in the phylogenetic relation between *C. grandis* and *P. philippinicum*, it is most likely that these convergently evolved glues in the two species originate from different chemical groups. Comparing the glue of *C. grandis* seeds to other plant seed glues, the forces are similar to the ones reported for cellulose-based mucilage envelopes [38]. However, the basis on which the glue is applied to the substrate is largely different. While the cellulose fibrils and pectines in the known seed glues are anchored in an undisintegrated cell wall, the epidermal cell wall of the *C. grandis* seeds is significantly modified. Nevertheless, the net adhesive forces of other seeds are often much stronger and often more reliable [38], although they do not possess similar specialized **fas** for adaptation to the substrate. The linearity of the seed **fas** causes a less homogenous distribution of the glue film on the substrate compared to the eggs' pinnae. This interferes with reliable contact formation with the substrate. Therefore, the variance of the detachment forces of the seeds can be higher. If, by chance, a good contact is formed, the detachment forces can be quite high; however, it is also quite likely, due to the unspecialized fibrils, that the contact is unpredictable and can be quite unreliable. The **fas** of *C. grandis* are present from the very beginning as disintegration of the cell wall and spread out after contact with water, whereas the fibers of mucilaginous seeds appear due to their uncoiling from the cell wall after the first hydration [38]. Functionally, the contac formation of both types of seed appendages is similar and the fibrils of both kinds car adapt to the surface profile, but the origin of the fibrils differs. As a result, the size of the cell wall originating **fas** is larger in comparison to the cellulose fibrils of the majority o mucilaginous seeds and might provide less efficient contact. In contrast to cellulose fibrils the glue generating the adhesion in the seeds examined herein liquefies again with water contact and this can be repeated several times. However, the detachment force decreases over repeating cycles, as the glue is washed off and partly remains on the substrates with every cycle. This effect is stronger in the *C. grandis* seeds as the hierarchically splitting pinnae of the *P. philippinicum* eggs strongly keep the glue on the surface of the egg.

4.1.3. Ecological Differences

Plants and insects are naturally rather different in terms of their demands on their environments. Often, insects are a threat that plants tend to avoid. Interestingly, leaf insects in particular, visually imitate the leaves of plants to avoid their own predators (spiders other insects, mammals, birds, and lizards [27,51]). This kind of camouflage evolved quite early in phasmids in general [52–57] and leaf insects in particular [58,59]. One result of this type of camouflage is a strong sexual dimorphism in Phylliidae [60] due to the fact that females are sedentary and imitate leaves in the canopy and males need to be mobile to find their mates to reproduce [61]. As a result, the eggs of all phylliids are dropped from the place where the females hide. This results in three aspects of concern, which might require attachment to some kinds of substrates: (1) Eggs dropped to the forest floor might be subjected to flightless parasitic wasps (e.g., Amiseginae), which are often specialized for particular phasmid species [51,62,63]. Attaching the eggs in higher levels of the for- est is a widespread strategy to avoid these parasitoids, which evolved independently in many phasmid lineages [64,65]. (2) Localization of the offspring close to the foodplant could guarantee suitable food for the offspring [14,15,66]. (3) Attachment can be used for dispersal [26], as has been shown for many seeds as well [40]. All three scenarios require strong, reliable attachment, as shown for the adhesive system of *P. philippinicum*. Based on the shape of the eggs and their appendages, several other phylliid species likely possess an adhesive capability as well [67], but some seem to have different glue properties and seem to attach better on hydrophobic than hydrophilic substrates (pers. obs). Furthermore other eggs of several unrelated phasmids also carry glue [26], and some seem to represent similar non-permanent water-responsive adhesive mechanisms [68]. The evolution of such egg surface structures and adhesive systems on eggs is likely a similar complex evolu- tionary scenario, comparable to other aspects of phasmatodean evolution [69,70], such as the tarsal adhesive systems. These also result from complex environmental conditions and are shaped by interactions with various substrates [7,71–75]. The preferred food plants that are documented for this insect species are *Mangifera indica* L. (Anacardiaceae), *Nephelium lappaceum* L. (Sapindaceae), and *Psidium guajava* L. (Myrtaceae) and the surface characteristics that are potential adhesive sites are discussed in [15]. Likely, the rough hydrophilic bark is beneficial for attachment. In contrast to *C. grandis*, the seeds of these plants carry no adhesive capabilities, as far as it has been documented.

For the *C. grandis* seeds, in contrast, it is essential to reach the ground for germination. Therefore, a strong reliable adhesive system is disadvantageous for reproduction. While the eggs can adhere several times, which can be useful for site optimization and to ensure suitable conditions for embryonic development, the *C. grandis* seeds adhere once with note- worthy strength. Presumably, they will be washed off their substrates with the first rain contact and then reach the soil for further germination. The mucilage glue of other plant species is common and studied most in plant species in arid environments or disturbed habitats [38]. Glue-based anchoring in plants has some different advantages compared to that in insects, but some advantages are congruent. Seed glue can be advantageous to sustain proper microenvironments, for example, by retaining humidity or anchoring in a suitable regime [37,76–79]. This particularly includes anchoring to the ground for germination [80,81]. For plants, which in contrast to insects do not have a moving reproductive stage, dispersal plays a big role and is often facilitated by glues. Long-distance dispersal can, for example, be mediated by migratory animals if the seeds are glued to the feathers of birds or the fur of mammals [79,82–85]. However, for *C. grandis*, there are no firsthand observations of actual seed dispersal. Nevertheless, mammals and birds are reported to be attracted by fruits and potentially disperse the seeds. These include fruit bats [86,87] and birds [88]. Other larger potential dispersers [89] include humans and elephants [90,91], which feed on the fruits and potentially disperse the seeds via endozoochory. Transport within animals plays a role for some plant seeds [92]. However, for *C. grandis*, the passage of the digestive tract of an animal is not required for successful germination [25] [pers. observation]. Whether the seeds would survive the passage through the digestive system of animals is not known so far [25]. In turn endozoochory is rather unlikely for phasmids in general, and even less likely for leaf insects in particular. It might be possible for the eggs to survive digestion by birds if the gravid female is consumed [93], but feeding experiments of individual eggs to birds exclude the chance of survival of bird digestion for most phasmid eggs [93,94]. In contrast to seeds, insect eggs are, if adhesive, rather designed for specialized tasks and usually adapted to specific attachment sites [17,95–100].

4.2. Biomimetic Implications

Insects are considered suitable sources for bioinspiration [101] and the same applies for plants [102–104]. Both insect egg and plant seed glues studied herein are considered useful templates for water-based glues [14,38,105]. Due to their degradability and potential biocompatibility, they or their derivatives can be potentially directly used for biomedical applications [43]. The differences in the two systems can propagate bioinspiration in two directions. While the adhesive system of the eggs works more reliably for durable long-term adhesion, the seeds adhere stronger but are less reliable. However, the common features of both systems can also provide general considerations for the design of fiber-reinforced glue-based adhesives. Both systems make use of fibrillary structures for glue application and simultaneous structural reinforcement. Fiber reinforcement, in general, can increase the mechanical stability [106,107] and reduce the likelihood of failure of the glue itself (cohesive failure) [108]. Hierarchical splitting of the reinforcing and glue-applying structures can increase the adaption to the substrate roughness [109,110], such as in the leaf insect egg. Unbranching fibers, in contrast, can be used for the short-term initial adhesion but are less useful for reliable long-term adhesion. Hierarchical splitting of these surface structures can increase the contact reliability and reduce the required amount of glue [109,111], whereas the introduction of any fibrillary reinforcements at least increases the stability. Some insect egg proteins are cured by glycosylation [112]. As the system investigated herein works in an enzyme-free environment, it is unlikely that the glue is activated by enzymes but cured by water uptake, e.g., glycation.

These results might inspire technological applications that reduce the required amount of glue, reducing the material cost or yielding more sustainable adhesive systems. As several other species of Phylliidae and other Phasmatodea carry very different exochorionic structures on their eggs, which are likely involved in adhesion as well, future studies can also further explore the advantages and disadvantages of different modifications of this system in a comparative experimental setting. Natural adhesive systems offer various similar fibrillary adhesive structures that can be informative for biomimetics. These also include temporary adhesive systems such as the hairy adhesive systems of invertebrates or the largely dry fibrous adhesive systems of geckos [7].

5. Conclusions

The adhesive systems of the leaf insect *P. philippinicum* and the ivy gourd *C. grandis* consist of similar main components: fibrillary adhesive structures and glue. Both adhesive systems convergently yield strong adhesive forces but perform with different reliability, which correlates with the autecological demands of both reproductive stages. While the fas (pinnae) of *P. philippinicum* hierarchically split, the **fas** of *C. grandis* (disintegrated epidermal cells) are more uniform straight fibrils. Both systems facilitate adaption to different surface roughness and perform particularly well on hydrophilic substrates after activation by water contact. While insect eggs attach more reliably in avoidance of parasitoids and foodplant association, the seeds of the ivy gourd are dependent on contact with the soil for germination. Therefore, the strong initial adhesion is usually not repetitive in the ivy gourd. The eggs of the walking leaf, in contrast, are capable of repetitive reattachment over several cycles. Both adhesive systems convergently make use of reinforcing fibers in the glue system and adjust to the surface profile; however, the straight fibers of seeds apparently perform less reliably and are more suitable for initial, temporary attachment while the hierarchically splitting adhesive structures of eggs make more reliable contact and apparently store more glue on the surface. In addition to the choice of a particular morphology for biomimetic applications, the specific requirements can be tuned with different glues in the adhesive system. Nevertheless, both types of glue exemplified by *C. grandis* and *P. philippinicum* are potential candidates for water-soluble biocompatible glues. This study exemplifies the benefits of studying similar mechanisms for a comparison of different perspectives of different but convergently evolved systems for biomimetics The specific requirements result in different modifications of similar mechanisms for the respective tasks and enable an assessment of the underlying constraints. Examination of further similar adhesive mechanisms on the eggs of other walking leaf species, other phasmid species in general, and the seeds of further plant species with similar fibrillary adhesive systems might yield more insights into the different modifications, which can represent an informative toolbox for biomimetics.

Supplementary Materials: The following supporting information can be downloaded at: https://www.mdpi.com/article/10.3390/biomimetics7040173/s1, Excel Sheet S1: force measurements.

Author Contributions: Conceptualization, S.N.G. and T.H.B.; methodology, S.N.G. and T.H.B.; software, T.H.B.; validation, S.N.G. and T.H.B.; formal analysis, T.H.B.; investigation, T.H.B.; resources, S.N.G.; data curation, T.H.B.; writing—original draft preparation, T.H.B.; writing—review and editing, S.N.G.; visualization, T.H.B.; supervision, S.N.G.; project administration, T.H.B.; funding acquisition, S.N.G. All authors have read and agreed to the published version of the manuscript.

Funding: This research was funded by German Science Foundation, grant number DFG GO 995/43-1

Institutional Review Board Statement: Not applicable.

Data Availability Statement: The raw data of the pull-off experiments of both the *Phyllium philippinicum* eggs and the *Coccinia grandis* seeds from substrates with different surface roughness and surface chemistry are made available via the Supplementary Materials.

Acknowledgments: We are very thankful to Wei-Song Lih (Singapore) for sharing the primary observation of the adhesive seeds. Susanne Petersen (Botanical Garden of Kiel University, Germany) for helped during seed identification and Kerstin Behrends (Germany) during force measurements We thank Linus Reck and Alexander Kovalev (Functional Morphology and Biomechanics, Kiel University, Germany) for their kind help with microscopy methods.

Conflicts of Interest: The authors declare no conflict of interest. The funders had no role in the design of the study; in the collection, analyses, or interpretation of data; in the writing of the manuscript, or in the decision to publish the results.

References

1. Baur, F.; Gorb, S.N. How the Bee Releases Its Leg Attachment Devices. In *Biona Report*; Wisser, A., Nachtigall, W., Eds.; Gustav Fischer: Stuttgart, Germany, 2000; Volume 15, pp. 295–297.
2. Ng, L.; Elgar, M.A.; Stuart-Fox, D. From bioinspired to bioinformed: Benefits of greater engagement from biologists. *Front. Ecol. Evol.* **2021**, *9*, e11234. [CrossRef]
3. Bell, G. *Selection*; Oxford University Press: Oxford, UK, 2007; 699p.
4. Langowski, J.K.A.; Sharma, P.; Shoushtari, A.L. In the soft grip of nature. *Sci. Robot.* **2020**, *5*, abd9120. [CrossRef] [PubMed]

5. Winand, J.; Büscher, T.H.; Gorb, S.N. Learning From Nature: A Review on Biological Gripping Principles and Their Application to Robotics. In *Soft Robotics*; Monkman, G., Ed.; Bentham Science Publishers: Sharjah, United Arab Emirates, 2022; pp. 21–59.
6. Hughes, J.; Culha, U.; Giardina, F.; Guenther, F.; Rosendo, A.; Iida, F. Soft manipulators and grippers: A review. *Front. Robot. AI* **2016**, *3*, 20. [CrossRef]
7. Büscher, T.H.; Gorb, S.N. Physical constraints lead to parallel evolution of micro- and nanostructures of animal adhesive pads: A review. *Beilstein J. Nanotechnol.* **2021**, *12*, 725–743. [CrossRef] [PubMed]
8. Beutel, R.G.; Gorb, S.N. Evolution of attachment structures and phylogeny of Hexapoda (Arthropoda). *Zoology* **2001**, *104*, 68.
9. Gorb, S.N. *Attachment Devices of Insect Cuticle*; Springer: Dordrecht, The Netherlands, 2001; 305p.
10. Büscher, T.H.; Buckley, T.R.; Grohmann, C.; Gorb, S.N.; Bradler, S. The evolution of tarsal adhesive microstructures in stick and leaf insects (Phasmatodea). *Front. Ecol. Evol.* **2018**, *6*, 69. [CrossRef]
11. Büscher, T.H.; Gorb, S.N. Subdivision of the neotropical Prisopodinae Brunner von Wattenwyl, 1893 based on features of tarsal attachment pads (Insecta, Phasmatodea). *ZooKeys* **2017**, *645*, 1–11. [CrossRef] [PubMed]
12. Büscher, T.H.; Grohmann, C.; Bradler, S.; Gorb, S.N. Tarsal attachment pads in Phasmatodea (Hexapoda: Insecta). *Zoologica* **2019**, *164*, 1–94.
13. Büscher, T.H.; Kryuchkov, M.; Katanaev, V.L.; Gorb, S.N. Versatility of Turing patterns potentiates rapid evolution in tarsal attachment microstructures of stick and leaf insects (Phasmatodea). *J. R. Soc. Interface* **2018**, *15*, 20180281. [CrossRef] [PubMed]
14. Büscher, T.; Lohar, R.; Kaul, M.-C.; Gorb, S.N. Multifunctional adhesives on the eggs of the leaf insect *Phyllium philippinicum* (Phasmatodea: Phylliidae): Solvent influence and biomimetic implications. *Biomimetics* **2020**, *5*, 66. [CrossRef] [PubMed]
15. Büscher, T.H.; Quigley, E.; Gorb, S.N. Adhesion performance in the eggs of the Philippine leaf insect *Phyllium philippinicum* (Phasmatodea: Phylliidae). *Insects* **2020**, *11*, 400. [CrossRef] [PubMed]
16. Tau'olunga. Coccinia Grandis Fruit. Available online: https://commons.wikimedia.org/wiki/Coccinia_grandis#/media/File:Coccinia_grandis_fruit.jpg (accessed on 15 May 2022).
17. Voigt, D.; Gorb, S.N. Egg attachment of the asparagus beetle *Crioceris asparagi* to the crystalline waxy surface of *Asparagus officinalis*. *Proc. R. Soc. B* **2010**, *277*, 895–903. [CrossRef] [PubMed]
18. Kang, V.; Johnston, R.; van de Kamp, T.; Faragó, T.; Federle, W. Morphology of powerful suction organs from blephariceridlarvae living in raging torrents. *BMC Zool.* **2019**, *4*, 187. [CrossRef]
19. Kang, V.; White, R.T.; Chen, S.; Federle, W. Extreme suction attachment performance from specialised insects living in mountain streams (Diptera: Blephariceridae). *eLife* **2021**, *10*, 63250. [CrossRef]
20. Büscher, T.H.; Petersen, D.S.; Bijma, N.N.; Bäumler, F.; Pirk, C.W.W.; Büsse, S.; Heepe, L.; Gorb, S.N. The exceptional attachment ability of the ectoparasitic bee louse *Braula coeca* (Diptera, Braulidae) on the honeybee. *Physiol. Entomol.* **2021**, *19*, 170. [CrossRef]
21. Beutel, R.G.; Gorb, S.N. A revised interpretation of the evolution of attachment structures in Hexapoda with special emphasis on Mantophasmatodea. *Arthropod Syst. Phylogeny* **2006**, *64*, 3–25.
22. Petersen, D.S.; Kreuter, N.; Heepe, L.; Büsse, S.; Wellbrock, A.H.; Witte, K.; Gorb, S.N. Holding tight to feathers–structural specializations and attachment properties of the avian ectoparasite Crataerina pallida (Diptera, Hippoboscidae). *J. Exp. Biol.* **2019**, *221*, jeb179242.
23. Spurr, A.R. A low-viscosity epoxy resin embedding medium for electron microscopy. *J. Ultrastruct. Res.* **1969**, *26*, 31–43. [CrossRef]
24. Salerno, G.; Rebora, M.; Gorb, E.V.; Gorb, S.N. Attachment ability of the polyphagous bug *Nezara viridula* (Heteroptera: Pentatomidae) to different host plant surfaces. *Sci. Rep.* **2018**, *8*, 10975. [CrossRef]
25. Holstein, N. Monograph of *Coccinia* (Cucurbitaceae). *PhytoKeys* **2015**, *54*, 1–166. [CrossRef]
26. O'Hanlon, J.C.; Jones, B.R.; Bulbert, M.W. The dynamic eggs of the Phasmatodea and their apparent convergence with plants. *Naturwissenschaften* **2020**, *107*, 34. [CrossRef] [PubMed]
27. Brock, P.D.; Büscher, T.H. *Stick and Leaf Insects of the World*; NAP Editions: Verrieres le Buisson, France, 2022.
28. Hinton, H.E. *Biology of Insect Eggs*; Pergamon Press: Oxford, UK, 1981; Volume 1.
29. Moscona, A. Studies of the egg of *Bacillus libanicus* (Orthoptera, Phasmidae). I. The egg envelopes. *Q. J. Microsc. Sci.* **1950**, *91*, 183–193. [PubMed]
30. Van de Kamp, T.; Greven, H. Struktur des spezialisierten und unspezialisierten Chorion des Eis der Stabschr ecke *Malacomorpha cyllarum* (Phasmatodea). *Entomol. Gen.* **2008**, *31*, 64–74. [CrossRef]
31. Mazzini, M.; Carcupino, M.; Fausto, A.M. Egg chorion architecture in stick insects (Phasmatodea). *Int. J. Insect Morphol. Embryol.* **1993**, *22*, 391–415. [CrossRef]
32. Moscona, A. Studies of the egg of *Bacillus libanicus* (Orthoptera, Phasmidae). *Q. J. Microsc. Sci.* **1950**, *91*, 195–203.
33. Jeffrey, C. *Cucurbitaceae*; University of Chicago Press: Chicago, IL, USA, 2000.
34. Keraudren-Aymonin, M. *Curcubitaceae: Flore du Cameroun*, 6th ed.; Muséum National D'histoire Naturelle: Paris, France, 1967.
35. De Wilde, W.J.; Duyfjes, B.E.; Phonsena, P.; van Der Ham, R.W. Miscellaneous South East Asian Cucurbit News IV. *Thai For. Bull. Bot.* **2014**, *39*, 1–22.
36. Chakravorti, A.K. Cytology of *Coccinia indica* W. & A. with reference to the behaviour of its sex-chromosomes. *Proc. Indian Acad. Sci. B* **1948**, *27*, 74–86.
37. Kreitschitz, A. Mucilage formation in selected taxa of the genus *Artemisia* L. (Asteraceae, Anthemideae). *Seed Sci. Res.* **2012**, *22*, 177–189. [CrossRef]

38. Kreitschitz, A.; Kovalev, A.; Gorb, S.N. Plant seed mucilage as a glue: Adhesive properties of hydrated and dried-in-contact seed mucilage of five plant species. *Int. J. Mol. Sci.* **2021**, *22*, 1443. [CrossRef]
39. Kreitschitz, A.; Kovalev, A.; Gorb, S.N. Slipping vs. sticking: Water-dependent adhesive and frictional properties of *Linum usitatissimum* L. seed mucilaginous envelope and its biological significance. *Acta Biomater.* **2015**, *17*, 152–159. [CrossRef]
40. Kreitschitz, A.; Kovalev, A.; Gorb, S.N. "Sticky invasion"—The physical properties of *Plantago lanceolata* L. seed mucilage. *Beilstein J. Nanotechnol.* **2016**, *7*, 1918–1927. [CrossRef] [PubMed]
41. Johnson, K.L.; Kendall, K.; Roberts, A.D. Surface energy and the contact of elastic solids. *Proc. R. Soc. Lond. A* **1971**, *324*, 301–313.
42. Yang, X.; Baskin, J.M.; Baskin, C.C.; Huang, Z. More than just a coating: Ecological importance, taxonomic occurrence and phylogenetic relationships of seed coat mucilage. *PPEES* **2012**, *14*, 434–442. [CrossRef]
43. Li, D.; Huson, M.G.; Graham, L.D. Proteinaceous adhesive secretions from insects, and in particular the egg attachment glue of *Opodiphthera* sp. moths. *Arch. Insect Biochem. Physiol.* **2008**, *69*, 85–105. [CrossRef]
44. Beament, J.W.L.; Lal, R. Penetration through the egg-shell of *Pieris brassicae* (L.). *Bull. Entomol. Res.* **1957**, *48*, 109–125. [CrossRef]
45. Riley, R.C.; Forgash, A.J. *Drosophila melanogaster* eggshell adhesive. *J. Insect Physiol.* **1967**, *13*, 509–517. [CrossRef]
46. Burkhart, C.N.; Stankiewicz, B.A.; Pchalek, I.; Kruge, M.A.; Burkhart, C.G. Molecular composition of the louse sheath. *J. Parasitol.* **1999**, *85*, 559. [CrossRef]
47. Jin, Y.; Chen, Y.; Jiang, Y.; Xu, M. Proteome analysis of the silkworm (*Bombyx mori.* L) colleterial gland during different development stages. *Arch. Insect Biochem. Physiol.* **2006**, *61*, 42–50. [CrossRef]
48. Betz, O. Adhesive Exocrine Glands in Insects: Morphology, Ultrastructure, and Adhesive Secretion. In *Biological Adhesive Systems—From Nature to Technical and Medical Application*; von Byern, J., Grunwald, I., Eds.; Springer: Wien, Germany; New York, NY, USA, 2010; pp. 111–152.
49. Burgess, I.F. Do nit removal formulations and other treatments loosen head louse eggs and nits from hair? *Med. Vet. Entomol.* **2010**, *24*, 55–61. [CrossRef]
50. Graham, L.D. Biological Adhesives from Nature. In *Encyclopedia of Biomaterials and Biomedical Engineering*; Wnek, G.E., Bowlin, G.L., Eds.; Informa Healthcare: New York, NY, USA, 2008; pp. 236–253.
51. Bedford, G.O. Biology and ecology of the Phasmatodea. *Ann. Rev. Entomol.* **1978**, *23*, 125–149. [CrossRef]
52. Tihelka, E.; Cai, C.; Giacomelli, M.; Pisani, D.; Donoghue, P.C.J. Integrated phylogenomic and fossil evidence of stick and leaf insects (Phasmatodea) reveal a Permian-Triassic co-origination with insectivores. *R. Soc. Open Sci.* **2020**, *7*, 201689. [CrossRef] [PubMed]
53. Bank, S.; Buckley, T.R.; Büscher, T.H.; Bresseel, J.; Constant, J.; de Haan, M.; Dittmar, D.; Dräger, H.; Kahar, R.S.; Kang, A.; et al. Reconstructing the nonadaptive radiation of an ancient lineage of ground-dwelling stick insects (Phasmatodea: Heteropterygidae). *Syst. Entomol.* **2021**, *46*, 487–507. [CrossRef]
54. Bradler, S.; Buckley, T.R. Stick insect on unsafe ground: Does a fossil from the early Eocene of France really link Mesozoic taxa with the extant crown group of Phasmatodea? *Syst. Entomol.* **2011**, *36*, 218–222. [CrossRef]
55. Engel, M.S.; Wang, B.; Alqarni, A.S. A thorny, 'anareolate' stick-insect (Phasmatidae s.l.) in Upper Cretaceous amber from Myanmar, with remarks on diversification times among Phasmatodea. *Cretac. Res.* **2016**, *63*, 45–53. [CrossRef]
56. Ren, D.; Shih, C.K.; Gao, T.; Yao, Y. *Rhythms of Insect Evolution: Evidence from the Jurassic and Cretaceous in Northern China*; Wiley Blackwell: Hoboken, NJ, USA, 2019.
57. Yang, H.; Shi, C.; Engel, M.S.; Zhao, Z.; Ren, D.; Gao, T. Early specializations for mimicry and defense in a Jurassic stick insect. *Natl. Sci. Rev.* **2021**, *8*, nwaa056. [CrossRef] [PubMed]
58. Wedmann, S.; Bradler, S.; Rust, J. The first fossil leaf insect: 47 million years of specialized cryptic morphology and behavior. *Proc. Natl. Acad. Sci. USA* **2007**, *104*, 565–569. [CrossRef]
59. Bank, S.; Cumming, R.T.; Li, Y.; Henze, K.; Le Tirant, S.; Bradler, S. A tree of leaves: Phylogeny and historical biogeography of the leaf insects (Phasmatodea: Phylliidae). *Commun. Biol.* **2021**, *4*, 932. [CrossRef]
60. Cumming, R.T.; Le Tirant, S.; Teemsma, S.N.; Hennemann, F.H.; Willemse, L.; Büscher, T.H. Lost lovers linked at long last: Elusive female *Nanophyllium* mystery solved after a century of being placed in a different genus (Phasmatodea, Phylliidae). *ZooKeys* **2020**, *969*, 43–84. [CrossRef]
61. Boisseau, R.P.; Büscher, T.H.; Klawitter, L.J.; Gorb, S.N.; Emlen, D.J.; Tobalske, B.W. Multi-modal locomotor costs favor smaller males in a sexually dimorphic leaf-mimicking insect. *BMC Ecol. Evol.* **2022**, *22*, 39. [CrossRef]
62. Kimsey, L.; Dewhurst, C.; Nyaure, S. New species of egg parasites from the Oil Palm Stick Insect (*Eurycantha insularis*) in Papua New Guinea (Hymenoptera, Chrysididae, Phasmatodea, Phasmatidae). *J. Hymenopt. Res.* **2013**, *30*, 19–28. [CrossRef]
63. Baker, E. An online taxonomic database of the stick insect (Phasmida) egg-parasitising subfamilies Amiseginae and Loboscelidiinae (Hymenoptera: Chrysididae). *Biodivers. Data J.* **2016**, *4*, e7441. [CrossRef]
64. Goldberg, J.; Bresseel, J.; Constant, J.; Kneubühler, B.; Leubner, F.; Michalik, P.; Bradler, S. Extreme convergence in egg-laying strategy across insect orders. *Sci. Rep.* **2015**, *5*, 7825. [CrossRef] [PubMed]
65. Robertson, J.A.; Bradler, S.; Whiting, M.F. Evolution of oviposition techniques in stick and leaf insects (Phasmatodea). *Front. Ecol. Evol.* **2018**, *6*, 125. [CrossRef]
66. Hennemann, F.H.; Conle, O.V.; Gottardo, M.; Bresseel, J. On certain species of the genus *Phyllium* Illiger, 1798, with proposals for an intra-generic systematization and the descriptions of five new species from the Philippines and Palawan (Phasmatodea: Phylliidae: Phylliinae: Phylliini). *Zootaxa* **2009**, *2322*, 1–83. [CrossRef]

67. Cumming, R.T.; Le Tirant, S.; Büscher, T.H. Resolving a century-old case of generic mistaken identity: Polyphyly of *Chitoniscus* sensu lato resolved with the description of the endemic New Caledonia *Trolicaphyllium* gen. nov. (Phasmatodea, Phylliidae). *ZooKeys* **2021**, *1055*, 1–41. [CrossRef] [PubMed]
68. Crispino, E.B.; Ghirotto, V.M.; Engelking, P.W. Contributions to the knowledge of *Ceroys* (Miroceroys) Piza, 1936 (Phasmatodea: Heteronemiidae): Two new mossy stick insects from the Atlantic Forest of Brazil. *Zootaxa* **2022**, *5134*, 34–60. [CrossRef] [PubMed]
69. Bank, S.; Bradler, S. A second view on the evolution of flight in stick and leaf insects (Phasmatodea). *BMC Ecol. Evol.* **2022**, *22*, 62. [CrossRef] [PubMed]
70. Whiting, M.F.; Bradler, S.; Maxwell, T. Loss and recovery of wings in stick insects. *Nature* **2003**, *421*, 264–267. [CrossRef] [PubMed]
71. Büscher, T.H.; Gorb, S.N. Complementary effect of attachment devices in stick insects (Phasmatodea). *J. Exp. Biol.* **2019**, *222*, jeb209833. [CrossRef]
72. Büscher, T.H.; Becker, M.; Gorb, S.N. Attachment performance of stick insects (Phasmatodea) on convex substrates. *J. Exp. Biol.* **2020**, *223*, jeb226514. [CrossRef] [PubMed]
73. Labonte, D.; Federle, W. Functionally different pads on the same foot allow control of attachment: Stick insects have load-sensitive "heel" pads for friction and shear-sensitive "toe" pads for adhesion. *PLoS ONE* **2013**, *8*, e81943. [CrossRef]
74. Labonte, D.; Struecker, M.-Y.; Birn-Jeffery, A.; Federle, W. Shear-sensitive adhesion enables size-independent adhesive performance in stick insects. *Proc. R. Soc. B* **2019**, *286*, 20191327. [CrossRef] [PubMed]
75. Pattrick, J.G.; Labonte, D.; Federle, W. Scaling of claw sharpness: Mechanical constraints reduce attachment performance in larger insects. *J. Exp. Biol.* **2018**, *221*, jeb188391. [CrossRef] [PubMed]
76. Grubert, M. Studies on the distribution of myxospermy among seeds and fruits of angiospermae and its ecological importance. *Acta Biol. Venez.* **1974**, *8*, 315–551.
77. Kreitschitz, A.; Valles, J. Achene morphology and slime structure in some taxa of *Artemisia* L. and *Neopallasia* L. (Asteraceae). *Flora* **2007**, *202*, 570–580. [CrossRef]
78. Western, T.L. The sticky tale of seed coat mucilages: Production, genetics, and role in seed germination and dispersal. *Seed Sci. Res.* **2012**, *22*, 1–25. [CrossRef]
79. Huang, Z.; Huc, Z.; Gutterman, Y. Structure and function of mucilaginous achenes of *Artemisia monosperma* inhabiting the Negev desert of Israel. *Israel J. Plant Sci.* **2000**, *48*, 255–266. [CrossRef]
80. García-Fayos, P.; Bochet, E.; Cerdà, A. Seed removal susceptibility through soil erosion shapes vegetation composition. *Plant Soil* **2010**, *334*, 289–297. [CrossRef]
81. Engelbrecht, M.; García-Fayos, P. Mucilage secretion by seeds doubles the chance to escape removal by ants. *Plant Ecol.* **2012**, *213*, 1167–1175. [CrossRef]
82. Yu, L.; Yakubov, G.E.; Zeng, W.; Xing, X.; Stenson, J.; Bulone, V.; Stokes, J.R. Multi-layer mucilage of *Plantago ovata* seeds: Rheological differences arise from variations in arabinoxylan side chains. *Carbohydr. Polym.* **2017**, *165*, 132–141. [CrossRef] [PubMed]
83. Kozlowski, T.T. *Seed Biology: Importance, Development, and Germination*; Elsevier Science: Oxford, UK, 1972.
84. Carlquist, S. The biota of long-distance dispersal. V. Plant dispersal to pacific islands. *Bull. Torrey Bot. Club* **1967**, *94*, 129–162. [CrossRef]
85. Vazacová, K.; Münzbergová, Z. Simulation of seed digestion by birds: How does it reflect the real passage through a pigeon's gut? *Folia Geobot.* **2013**, *48*, 257–269. [CrossRef]
86. Elangovan, V.; Marimuthu, G.; Kunz, T.H. Temporal patterns of resource use by the short-nosed fruit bat, *Cynopterus sphinx* (Megachiroptera: Pteropodidae). *J. Mammal.* **2001**, *82*, 161–165. [CrossRef]
87. Ruby, J.; Nathan, P.T.; Balasingh, J.; Kunz, T.H. Chemical composition of fruits and leaves eaten by short-nosed fruit bat, *Cynopterus sphinx*. *J. Chem. Ecol.* **2000**, *26*, 2825–2841. [CrossRef]
88. Bhatt, D.; Kumar, A. Foraging ecology of red-vented bulbul Pycnonotus cafer in Haridwar, India. *Forktail* **2001**, *16*, 109–110.
89. Voigt, J.O. *Hortus Suburbanus Calcuttensis: A Catalogue of the Plants Which Have Been Cultivated in the Hon. East India Company's Botanical Garden, Calcutta, and in the Serampore Botanical Garden, Genera*; Bishop's College Press: Calcutta, India, 1845.
90. Zimmermann, A. *Die Cucurbitaceen 1*; Gustav Fischer: Jena, Germany, 1922.
91. Mubalama, L. Population and distribution of elephants (*Loxodonta africana africana*) in the central sector of the Virunga National Park, eastern DRC. *Pachyderm* **2000**, *28*, 44–55.
92. Kreitschitz, A.; Haase, E.; Gorb, S.N. The role of mucilage envelope in the endozoochory of selected plant taxa. *Sci. Nat.* **2020**, *108*, 2. [CrossRef]
93. Suetsugu, K.; Funaki, S.; Takahashi, A.; Ito, K.; Yokoyama, T. Potential role of bird predation in the dispersal of otherwise flightless stick insects. *Ecology* **2018**, *99*, 1504–1506. [CrossRef]
94. Shelomi, M. Phasmid eggs do not survive digestion by quails and chickens. *J. Orthoptera Res.* **2011**, *20*, 159–162. [CrossRef]
95. Salerno, G.; Rebora, M.; Piersanti, S.; Büscher, T.H.; Gorb, E.V.; Gorb, S.N. Oviposition site selection and attachment ability of *Propylea quatuordecimpunctata* and *Harmonia axyridis* from the egg to the adult stage. *Physiol. Entomol.* **2021**, *217*, 20–37. [CrossRef]
96. Al Bitar, L.; Gorb, S.N.; Zebitz, C.P.W.; Voigt, D. Egg adhesion of the codling moth *Cydia pomonella* L. (Lepidoptera, Tortricidae) to various substrates: I. Leaf surfaces of different apple cultivars. *Arthropod-Plant Interact.* **2012**, *603*, 471–488. [CrossRef]
97. Al Bitar, L.; Gorb, S.N.; Zebitz, C.P.W.; Voigt, D. Egg adhesion of the codling moth *Cydia pomonella* L. (Lepidoptera, Tortricidae) to various substrates: II. Fruit surfaces of different apple cultivars. *Arthropod-Plant Interact.* **2014**, *8*, 57–77. [CrossRef]

98. Cogley, T.P.; Anderson, J.R.; Weintraub, J. Ultrastructure and function of the attachment organ of warble fly eggs (Diptera: Oestridae: Hypodermatinae). *Int. J. Insect Morphol. Embryol.* **1981**, *10*, 7–18. [CrossRef]
99. Gaino, E.; Rebora, M. Synthesis and function of the fibrous layers covering the eggs of *Siphlonurus lacustris* (Ephemeroptera, Siphlonuridae). *Acta Zool.* **2001**, *82*, 41–48. [CrossRef]
100. Cogley, T.P.; Cogley, M.C. Morphology of the eggs of the human bot fly, *Dermatobia hominis* (L. Jr.) (Diptera: Cuterebridae) and their adherence to the transport carrier. *Int. J. Insect Morphol. Embryol.* **1989**, *18*, 239–248. [CrossRef]
101. Gorb, S.N.; Gorb, E.V. Aquatic Insects as a Source for Biomimetics. In *Aquatic Insects: Behavior and Ecology*; Del-Claro, K., Guillermo, R., Eds.; Springer: Cham, Switzerland, 2019.
102. Speck, O.; Speck, T. Functional morphology of plants—A key to biomimetic applications. *New Phytol.* **2021**, *231*, 950–956. [CrossRef] [PubMed]
103. Gorb, S.; Speck, T. Biological and biomimetic materials and surface. *Beilstein J. Nanotechnol.* **2017**, *8*, 403–407. [CrossRef]
104. Masselter, T.; Scharf, U.; Speck, T. Plants and animals as concept generators for the development of biomimetic cable entry systems. *J. Bionic Eng.* **2008**, *5*, 167–173. [CrossRef]
105. Schaber, C.F.; Kreitschitz, A.; Gorb, S.N. Friction-active surfaces based on free-standing anchored cellulose nanofibrils. *ACS Appl. Mater. Interfaces* **2018**, *10*, 37566–37574. [CrossRef]
106. Jain, D.; Sahni, V.; Dhinojwala, A. Synthetic adhesive attachment discs inspired by spider's pyriform silk architecture. *J. Polym. Sci. Part B Polym. Phys.* **2014**, *52*, 553–560. [CrossRef]
107. Wang, L.; Culha, U.; Iida, F. A dragline-forming mobile robot inspired by spiders. *Bioinspir. Biomim.* **2014**, *9*, 16006. [CrossRef] [PubMed]
108. Wolff, J.O. Structural Effects of Glue Application in Spiders—What Can We Learn from Silk Anchors? In *Bio-Inspired Structured Adhesives—Biological Prototypes, Fabrication, Tribological Properties, Contact Mechanics and Novel Concepts*; Heepe, L., Xue, L., Gorb, S.N., Eds.; Springer International Publishing: Basel, Switzerland, 2017; pp. 63–80.
109. Gorb, S.N. Biological attachment devices: Exploring nature's diversity for biomimetics. *Philos. Trans. R. Soc. Lond. A* **2008**, *366*, 1557–1574. [CrossRef] [PubMed]
110. Scherge, M.; Gorb, S.N. *Biological Micro- and Nanotribology*; Springer: Berlin/Heidelberg, Germany, 2001.
111. Hennemann, O.-D. Kleben von Kunststoffen. Anwendung, Ausbildung, Trend. *Kunststoffe* **2000**, *90*, 184–188.
112. Lei, Y.; Guo, K.; Zhang, Y.; Zhang, X.; Qin, L.; Wang, X.; Zhu, H.; Guo, Y.; Yang, W.; Li, B.; et al. Adhesive property and mechanism of silkworm egg glue protein. *Acta Biomater.* **2021**, *134*, 499–512. [CrossRef] [PubMed]

biomimetics

MDPI

Article

The Involvement of Cell-Type-Specific Glycans in *Hydra* Temporary Adhesion Revealed by a Lectin Screen

Sofia Seabra [1,2], Theresa Zenleser [2], Alexandra L. Grosbusch [2], Bert Hobmayer [2] and Birgit Lengerer [2,*]

[1] Institute Superior Técnico, University of Lisboa, Av. Rovisco Pais, 1049-001 Lisboa, Portugal
[2] Institute of Zoology and Center of Molecular Biosciences Innsbruck, University of Innsbruck, Technikerstr. 25, 6020 Innsbruck, Austria
* Correspondence: birgit.lengerer@uibk.ac.at

Abstract: *Hydra* is a freshwater solitary polyp, capable of temporary adhesion to underwater surfaces. The reversible attachment is based on an adhesive material that is secreted from its basal disc cells and left behind on the substrate as a footprint. Despite *Hydra* constituting a standard model system in stem cell biology and tissue regeneration, few studies have addressed its bioadhesion. This project aimed to characterize the glycan composition of the *Hydra* adhesive, using a set of 23 commercially available lectins to label *Hydra* cells and footprints. The results indicated the presence of N-acetylglucosamine, N-acetylgalactosamine, fucose, and mannose in the adhesive material. The labeling revealed a meshwork-like substructure in the footprints, implying that the adhesive is mainly formed by fibers. Furthermore, lectins might serve as a marker for *Hydra* cells and structures, e.g., many labeled as glycan-rich nematocytes. Additionally, some unexpected patterns were uncovered, such as structures associated with radial muscle fibers and endodermal gland cells in the hypostome of developing buds.

Keywords: underwater bioadhesion; reversible attachment; adhesives; glue; carbohydrates

Citation: Seabra, S.; Zenleser, T.; Grosbusch, A.L.; Hobmayer, B.; Lengerer, B. The Involvement of Cell-Type-Specific Glycans in *Hydra* Temporary Adhesion Revealed by a Lectin Screen. *Biomimetics* **2022**, *7*, 166. https://doi.org/10.3390/biomimetics7040166

Academic Editors: Stanislav N. Gorb and Thies Henning Büscher

Received: 19 August 2022
Accepted: 13 October 2022
Published: 15 October 2022

Publisher's Note: MDPI stays neutral with regard to jurisdictional claims in published maps and institutional affiliations.

1. Introduction

The ability of organisms to attach to surfaces is entitled biological adhesion and is present in various organisms. Bioadhesion processes are diverse and complex and play a crucial role in organism survival and basic functions [1]. In glue-based adhesive systems, the attachment is mediated by the secretion of an adhesive material and can either be temporary or permanent [2]. The biochemical composition of the secreted adhesive material varies among organisms and is difficult to characterize [3]. It is generally stated that in temporarily adhering animals, such as *Hydra*, the glue is mainly constituted of proteins and carbohydrates [2,4]. Studies on aquatic temporary adhesives predominately focus on the identification of proteins, but, especially in temporary adhesion, carbohydrates are abundant in the secreted material [2]. Adhesion-related glycans have mostly been detected through histological stains such as Alcain blue and lectin-binding assays [2]. Using lectin-based methods, glycans have consistently been detected in the adhesive of non-permanently adhering animals such as sea urchins [5,6], sea stars [7,8], flatworms [9–11], and limpets [12]. Moreover, aquatic adhesive proteins are often highly glycosylated [5–7,13,14]. This post-translational modification significantly changes proteins characteristics and has to be taken into account in any biomimetic approach.

The cnidarian *Hydra* is a solitary polyp, inhabiting shallow freshwater bodies. It attaches itself to underwater surfaces through the secretion of an adhesive material and can repeatedly voluntarily detach and reattach [15]. *Hydra* is a classic and simple model system for pattern formation, regeneration, and stem cell biology research [16–20]. Structurally, *Hydra* has a single apical-basal axis with radial symmetry, and two layers of epithelial cells (the endoderm and the ectoderm) separated by an extracellular matrix (the mesoglea)

(Figure 1). The *Hydra* body is composed by the head, the gastric region, and the foot, where the animals attaches itself with its basal disc (Figure 1A,B). The basal disc cells produce and secrete the adhesive material, and four morphological distinct secretory granule types (HSGI to IV) have been described. The cells are characterized by an irregular rectangular like shape, water vacuoles, and numerous secretory granules accumulating at the aboral end, the area of attachment [15]. In contrast to animals with a duo-gland adhesive system, *Hydra* lacks dedicated de-adhesive gland cells that secrete a substance to weaken the bond between the animals and the substrate [4]. In *Hydra*, the detachment process likely occurs due to muscle contractions [15]. It was proposed that the individual basal disc cells retract from the surface, with the movement starting at the outer rim of the basal disc and moving towards the center. Upon detachment, an underwater transparent footprint composed by the secreted adhesive material, is left on the substrate. The footprint is formed by the secretion and blending of the contents of the adhesive granules [15]. Expression analysis in combination with mass spectrometry of the secreted footprints revealed 21 footprint-specific proteins [21]. These proteins presumably ensure adhesion and cohesion and contain domains that mediate protein–protein and protein–carbohydrate interactions. Remarkably, eight of these proteins are annotated with glycan-binding domains, such as galactose and chitin binding domains [21]. Periodic acid Schiff (PAS) staining revealed the presence of glycans within some of the basal disc secretory granules, but the glycan composition of the *Hydra* adhesive is unkown [15,22].

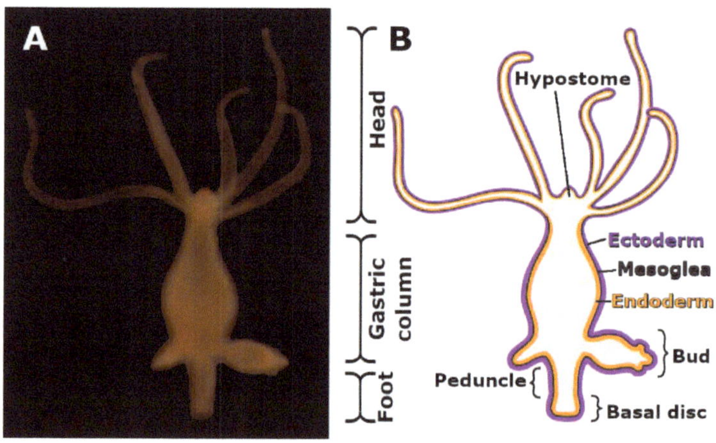

Figure 1. (**A**) Picture and (**B**) morphological scheme of an adult asexually reproducing *Hydra* polyp

Here, we characterize the glycan composition of the *Hydra* adhesive material, using a set of 23 commercially available lectins. We applied the lectins to label *Hydra* tissue including whole-mount animals and macerated basal disc cells, and the secreted footprints. Overall, eight lectins detected the footprints left behind on the substrate, indicating the presence of N-acetylglucosamine, N-acetylgalactosamine, fucose, and mannose in the adhesive material. The secreted adhesive appeared fibrillar and formed a dense meshwork with an accumulation of material at the cell borders of the formerly attached basal disc cells. Furthermore, our results indicated a high abundance of glycans within mature nematocytes and revealed some unexpected pattern, for example glycans associated with radial muscle fibers or within the hypostome of developing buds.

2. Methods

2.1. Hydra Culture

All experiments were carried out with individuals of *Hydra magnipapillata* strain 105, which were bred and kept in mass cultures at the Institute of Zoology, University of

Innsbruck. *Hydra* cultures were kept in growth chambers at 18 °C in *Hydra* culture medium and fed five times per week with *Artemia nauplii*. Before any experiment, the animals were starved for 24 h.

2.2. Whole-Mount Lectin Labeling

For the 23 used lectins, the full names, abbreviations, and sugar specificities are listed in Supplementary Table S1. Whole-mount animals were relaxed in 2% urethane in culture medium for 3–5 min. They were subsequently fixed using three different conditions: 4% paraformaldehyde (PFA) in phosphate buffered saline (PBS) overnight at 4 °C, 4% PFA in PBS for 1 h at room temperature (RT) and Lavdowsky fixative (ethanol:formamid:acetic acid:distilled water—50:10:4:40) for 4 h at RT. Samples were washed several times in Tris-buffered saline (TBS, pH 8.0) supplemented with 5 mM $CaCl_2$ and 0.1% Triton X (TBS-T). Unspecific background staining was blocked by pre-incubation in TBS-T containing 3% (w/v) bovine serum albumin (BSA) for 2 h at RT. Biotinylated lectins (Vector) were diluted in BSA-T to a final concentration of 10 (for WGA) or 25 μg/mL (all other lectins) and applied to the samples for 2 h at RT. After three washes of 10 min each in TBS-T (with 0.05% Triton X), the animals were incubated in Dylight488-conjugated-streptavidin (Vector), Phalloidin-Atto 565, and DAPI diluted (1:500, 1:1000, and 1:10,000, respectively) in BSA-T for 1 h at RT. After three washes in TBS-T (0.05% Triton X), the samples were mounted in Vectashield. During the waiting periods, the samples were placed on a shaker (40 rot/min). Control reactions were performed by substituting the lectins with TBS-T-BSA. The samples were analyzed with a Leica DM5000 microscope or with a Leica SP5 II confocal scanning microscope. As the intensity of the labeling varied among different lectins (see Table 1), the images of the most strongly stained specimens (+++) and of weakly stained (+) specimens had to be taken at different exposure times to sufficiently visualize them without over- or underexposure. The negative control images were taken with the same, longer exposure time as the weakly stained specimen. With the Leica SP5 II confocal scanning microscope, z-stacks were acquired and maximum z-projected.

2.3. Footprint Lectin Labeling

Footprints were collected by placing animals on microscope glass slides, submerged in *Hydra* medium and allowing them to attach overnight. The next day, the animals were gently detached using a glass pipet and the slides were washed with distilled water. The slides were then fixed in 4% PFA in PBS for 1 h at RT. The labeling was performed as for the whole-mount lectin labeling, but without adding phalloidin and DAPI. Footprint double stainings were performed using biotinylated lectins (Vector) (25 μg/mL) and Starlight conjugated Streptavidin (Vector) (1:500 diluted) and fluorescein conjugated WGA (Vector) (10 μg/mL).

2.4. Hydra and Corresponding Footprint Labeling after Voluntary and Forced Detachment

Hydras were placed on microscope glass slides, submerged in *Hydra* medium and allowed to attach under observation. Whenever a *Hydra* detached voluntarily, the animal and the glass slides were instantly fixed in PFA for 1 h at RT. Attachment times for voluntarily detached animals ranged from 2 to 33 min (n = 15). Additionally, *Hydras* were forcibly detached by the investigator after 35 to 95 min attachment time (n = 15) and animals and glass slides were fixed the same way. Samples were washed several times in TBS-T and blocked in BSA-T for 1 h at RT. The samples were then incubated in fluorescein conjugated WGA (Vector) (10 μg/mL) in BSA-T for 2 h at RT, washed several times and mounted in Vectashield. The labeling was analyzed with a Leica DM5000 microscope.

2.5. Single Basal Disc Cells Lectin Labeling

Fifteen budless polyps were cut at 20% of the body column, separating the foot from the anterior part. The basal disc cells were obtained by maceration of a single *Hydra* foot, by adding 1 drop of maceration solution (acetic acid:glycerol:distilled water—1:1:14) per animal [23]. Upon 1 h of incubation at RT, cell separation was driven by mechanical forces by gently tapping on the tube and gently pipetting up and down. For fixation, 1 drop of 8% formaldehyde was added per drop of maceration solution. A total of 50 μL of the resulting solution was added to a pre-treated slide (coated with gelatin), spread into a rectangulare shape and left to dry for 1 h at RT. After this, lectin staining was performed following the protocol used for the footprints (starting at the first washing step). The labeling was analyzed with a Leica SP5 II confocal scanning microscope.

3. Results

3.1. Lectin Labeling of Hydra Tissues and Footprint Secretions

We performed lectin labeling of whole-mount animals and macerated basal disc cells and footprints using 23 commercially available lectins. As the fixative and fixation duration can influence the outcome of the labeling, we performed the whole-mount labeling under three standard conditions: animals fixed with PFA over night at 4 °C, fixed with PFA for one hour at room temperature, and fixed with Lavdowsky for four hours at room temperature. For maceration experiments, the best results (regarding cell morphology and staining intensity) were obtained with a fixation with formaldehyde. Footprints were labeled without fixation and after one hour of PFA fixation at room temperature, without any apparent difference in the results (Figure S1). All labeling results are summarized in Table 1, indicating the intensity of the staining of the different cells and structures. Details on the sugar moieties recognized by the lectins are listed in Table 1.

Out of the 23 tested lectins, 17 led to a distinct labeling in whole-mount animals (Table 1) and six (Elderberry bark lectin, Jacalin, *Maackia amurensis* lectin II, Peanut agglutinin, *Sonaum tuberosum* lectin, *Sophora Japonica* agglutinin) did not react with any *Hydra* tissue (Figure S2). However, *Sonaum tuberosum* lectin (STL) led to blurry staining surrounding erupted nematocytes' threads, potentially reacting with the capsules' contents (Figure S3). Due to our focus on *Hydra* adhesive secretions, we grouped the results into the categories: "lectins detecting the secreted footprints" (eight lectins) and "lectin labeling of universal and positional distinct *Hydra* cell types and associated structures" (nine lectins) From the nine lectins that did not label the footprints, four lectins (*Erythrina cristagalli* lectin, *Pisum sativum* agglutinin, *Griffonia* (*Bandeiraea*) *simplicifolia* lectin I, *Dolichos bilforus* agglutinin) detected the basal disc. Based on the labeling of the basal disc, the glycans recognized by these four lectins might play a role in adhesion, but as they were not detected in the footprints, they are likely not a major component of the adhesive material Detailed descriptions of these results can be found in the Supplementary Material section "whole-mount labeling of lectins detecting the basal disc but not the footprints" (Figure S4)

Table 1. Overview of lectin labeling in *Hydra* whole-mounts, basal disc cells, and footprints. +++, strong labeling, ++ intermediate labeling, + weak labeling, X not performed. a—patchy staining, associated with muscle fibers; b—cell membranes; c—subcellular structures; d—gland cells in the hypostome of developing buds; e—base of cnidocil; f—erupted nematocytes' tubules; g—dot-like; h—secretory granules; i—variant intensity among individual footprints.

			Gastric Region			Tentacles				Foot		
							Nematocytes					
Lectin	Acronym	Fixative	Endoderm	Overall Ectoderm Surface	Nematocytes, Nematoblasts or Vacuoles	Tentacle Surface	Capsules	Opercolum	Cnidocil	Basal Disc Surface	Basal Disc Cells	Footprints
		o.n PFA		+++		+++ g	++	++	++	++		
Wheat germ agglutinin	WGA	1 h PFA		+++		+++ g	++	++	++	++	++ h	+++
		4 h Lavdwosky			+++	+++ g	++	++	++	++		
		o.n PFA		+++		++ g	+++	++		++		
Succinylated wheat germ agglutinin	sWGA	1 h PFA		+++		++ g	+++	++		++	++ h	+++
		4 h Lavdwosky			+++	++ g	+++	++		++		
		o.n PFA				++ g	++	++				
Datura Stramonium lectin	DSL	1 h PFA		+		++ g	++	++			++	+ i
		4 h Lavdwosky		++ g		++ g	++	++				
		o.n PFA					+++					
Lycopersicon esculentum (tomato) lectin	LEL	1 h PFA		+++ g		+++ g	+++			++	+	+++
		4 h Lavdwosky	+++ b				+++					
		o.n PFA		++ g		++ g	+++	+++				
Soybean agglutinin	SBA	1 h PFA		++ g		++ g	+++	+++			+++	+++
		4 h Lavdwosky		++ g		++ g	+++	+++				
		o.n PFA		+						+		
Ricinus communis agglutinin	RCA	1 h PFA		+						+	+++ h	+ i
		4 h Lavdwosky		+						++		
		o.n PFA		+++								
Ulex europaeus agglutinin I	UEA I	1 h PFA		+++							+++ h	+ i
		4 h Lavdwosky		+++			++ f		++			
		o.n PFA		+		+++	+	++	+++	++		
Concanavaline A	Con A	1 h PFA		+		+++	+	++	+++	++	+++	++
		4 h Lavdwosky		+		+++	+	++	+++	++		
		o.n PFA				+++ g						
Erythrina cristagalli lectin	ECL	1 h PFA		++ g		++ g				+++ g	x	
		4 h Lavdwosky		++ g	+++	++ g				+++ g		
		o.n PFA		+		++			+++	++		
Pisum sativum agglutinin	PSA	1 h PFA		++		++			++		x	
		4 h Lavdwosky			+++ f	++						
		o.n PFA		+			+++	+++		+++		
Griffonia (Bandeiraea) *simplicifolia* lectin I	GSL I	1 h PFA		+			+++	+++		+++	x	
		4 h Lavdwosky			+		++	+++		+++		
		o.n PFA		++			+++	+++				
Dolichos bilforus agglutinin	DBA	1 h PFA			++		+++	+++		++	x	
		4 h Lavdwosky			++		+++	+++		+		
		o.n PFA	+++ a	++								
Phaseolus vulgaris erythro agglutinin	PHA-E	1 h PFA	+++ a	++							x	
		4 h Lavdwosky	+++ b	+++								
		o.n PFA	+++ a,b	++								
Phaseolus vulgaris leuco agglutinin	PHA-L	1 h PFA	+++ a	++							x	
		4 h Lavdwosky	+++ b	+++ b								
		o.n PFA			++ c							
Griffonia (*Bandeiraea*) *simplicifolia* lectin II	GSL II	1 h PFA			++ c						x	
		4 h Lavdwosky	++ d		++							
		o.n PFA							+++ e			
Lens culinaris agglutinin	LCA	1 h PFA							+++ e		x	
		4 h Lavdwosky			++	++	++ f					
		o.n PFA		+		+++ g	++	++	+++			
Vicia villosa agglutinin	VVL	1 h PFA		+		+++ g	++	++	+++		x	
		4 h Lavdwosky		++		+++ g	++	++				

3.2. Lectins Detecting the Secreted Footprints

Eight lectins labeled the secreted footprints, indicating the presence of the corresponding glycans in the secreted adhesive (Table 1). In the whole-mount *Hydra* labeling, these eight lectins detected diverse structures from the overall animal surface to the nematocytes (sting cells) in the tentacles (Table 1). For example, *Wheat germ* agglutinin (WGA) (Figure 2A–D) and *Succinylated wheat germ* agglutinin (sWGA) (Figure 2E–G) detected the overall ectoderm surface (Figure 2A,B,E), the basal disc surface (Figure 2C,D), dot-like structures in the ectoderm of the tentacles, and the nematocyte capsules and operculum (Figure 2F). After Lavdowsky fixation, both lectins additionally detected the developing nematoblasts and nematocytes present in the gastric region (Figure 2G). Remarkably, the intensity and the pattern of the basal disc surface staining varied among individual *Hydras* not relying on the fixation method (Figure 2C,D). *Datura Stramonium* lectin (DSL), *Lycopersicon esculentum* (tomato) lectin (LEL), and soybean agglutinin (SBA) detected dot-like structures and the nematocysts in the tentacles (Figure S5). DSL and SBA also reacted with the operculum, which was not stained with LEL. The basal disc surface was not detected, with the exception of an intermediate strong labeling with LEL after one hour of PFA fixation. *Ricinus communis* agglutinin (RCA) and *Ulex europaeus* agglutinin I (UEA I) labeled the overall ectoderm surface and RCA additionally labeled the basal disc surface (Figure S5A–H). Concanavaline A (Con A) reacted strongly with the tentacle surface and all structures of the nematocytes, weakly with the overall ectoderm surface and intermediately with the basal disc surface (Figure S5I–L).

Footprint structure

The eight lectins detecting the footprints indicated the presence of several glycans in the secreted adhesive material (Table 1). The footprint structure was the same in all lectin labelings, but the staining intensity varied. WGA led to one of the strongest signals (Figure 3) resembling previous descriptions of the *Hydra* footprint [15]. The secreted material accumulated at the basal disc cell borders, resulting in an imprint of the formerly attached basal disc cells (Figure 3A). At higher magnification, a hole in this net was occasionally present likely because the cell at this location had not secreted its adhesive content (Figure 3B). The footprints consisted of a fibrous material resulting in an inhomogeneous, meshwork-like structure (Figure 3B). In approximately half of the observed footprints, a small-to-middle-sized area in the middle of the footprints was devoid of any secreted material (Figure 3C). On one footprint, discarded cells were observed at this location (Figure 3D). To determine if the same structures were labeled with different lectins, we performed double lectin labeling of fluorescein-conjugated WGA and the seven other lectins (Figure 3E). The difference in the staining intensity made some comparisons difficult, but the labeling overlapped WGA in all cases (Figure 3E). Four of the eight lectins (WGA, sWGA, DSL, and LEL) are known to bind to N-acetylglucosamine (GlcNAc) residues. WGA, sWGA, DSL, and LEL share similar binding preferences and bind to multimers of GlcNAc, chitobiose, and terminal GlcNAc. With the exception of DSL, these four lectins led to the strongest footprint labeling. With DSL, the intensity varied among individual footprints from weak to strong (Table 1). SBA and RCA detect N-acetylgalactosamine (GalNAc) and galactose residues. SBA led to a strong labeling, while for RCA the intensity varied in between footprints. A similar variability was observed with UEA I, a lectin reacting with α-linked fucose. Con A caused an intermediate footprint labeling and is known to bind α-linked mannose. Based on these results, we presume that the *Hydra* adhesive footprint contains GlcNAc, GalNAc, fucose and mannose to varying degrees.

Throughout our labeling experiments, we observed many footprints that were folded in at the edges (Figure 3A) and/or smudgy in some areas. Additionally, the basal disc surface pattern appeared variant between individual *Hydras*. We assumed that this was an artefact caused by our method to collect the footprints and *Hydras* by forcibly detaching them from the surface with a glass pipet. To test this, we let *Hydras* attach to glass slides and observed how long they stayed attached before voluntarily detaching on their own. The moment a *Hydra* detached, the footprint and corresponding *Hydra* were immediately fixed in PFA and lectin labeling with fluorescein-conjugated WGA was performed ($n = 15$). Furthermore, half of the *Hydras* were forcibly detached with a glass pipet and fixed and stained the same way ($n = 16$). The only difference we could observe between voluntarily self-detached and forcibly detached specimens was the tissue integrity of the basal disc (Figure 4). While, in voluntarily detached *Hydras*, the basal disc was always intact (15 out of 15) (Figure 4A), in forcibly detached animals the basal disc was damaged in half of the *Hydras* (8 out of 16) (Figure 4C,E). We again noted a variation in the staining intensity and appearance of the basal disc surface, but could not correlate those variations to the mode of detachment or the time the *Hydra* had stayed attached before fixation. On most basal discs, the cell borders were strongly stained, causing the characteristic net-like pattern (Figure 4A). This labeling likely stems from the adhesive material accumulating there during attachment. However, we also frequently observed a staining of the cell surfaces, likely representing freshly secreted adhesive (Figure 4E). Occasionally, footprints appeared thick and blurry, but surprisingly this was independent of their attachment time (the example in Figure 4B only attached for 11 min). Additionally, if footprints were smudgy, folded in, or incomplete (Figure 4D,F) was independent of attachment time and mode of detachment.

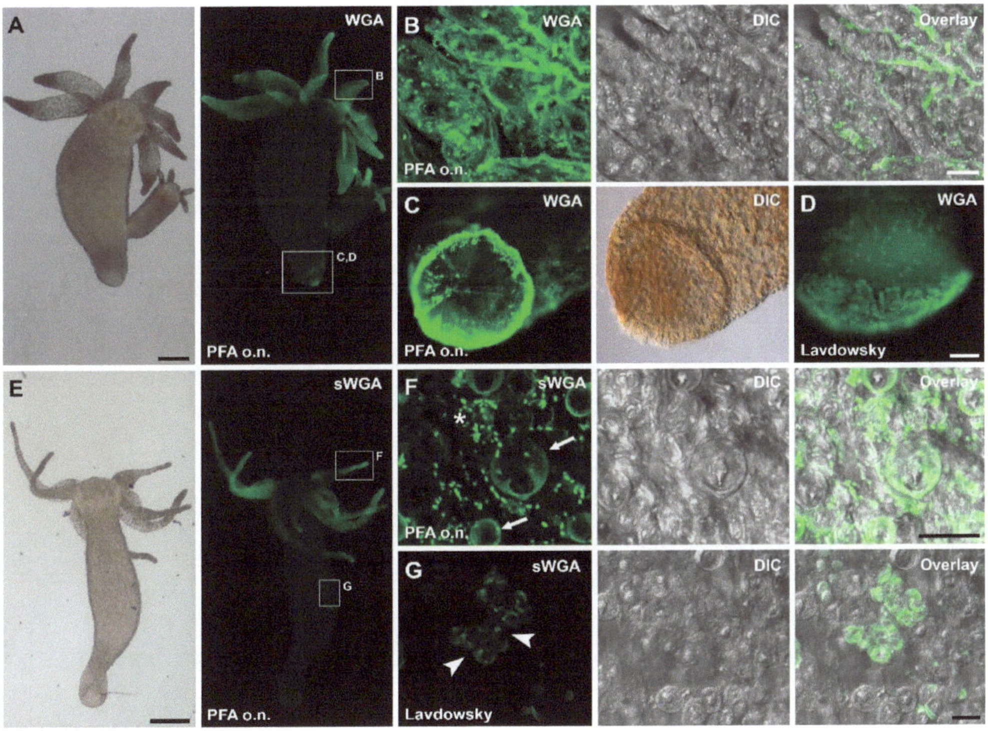

Figure 2. Lectin labeling of *Hydra* whole-mounts with (**A–D**) WGA and (**E–G**) sWGA. (**A**) WGA and (**E**) sWGA labeling of whole-mount individuals. (**B**,**F**) Overall ectoderm surface, ellipsoid structures, nematocytes capsules, and operculum were labeled with both lectins. Basal disc surface staining for (**C**) overnight PFA and (**D**) Lavdowsky fixatives. Note that the intensity and the pattern of the disc surface staining varied among individual *Hydras*, without relying on the fixation method. (**G**) After Lavdowsky fixation, developing nematoblasts and nematocytes were labeled in the body column for both lectins (sWGA is shown in the figure). Arrows highlight nematocytes and arrowheads point towards developing nematoblasts. The asterisk indicates ellipsoid structures. The fixation method used is indicated in the images. Scale bars: (**A**,**E**) 500 μm; (**C**,**D**) 100 μm; (**B**,**F**,**G**) 20 μm.

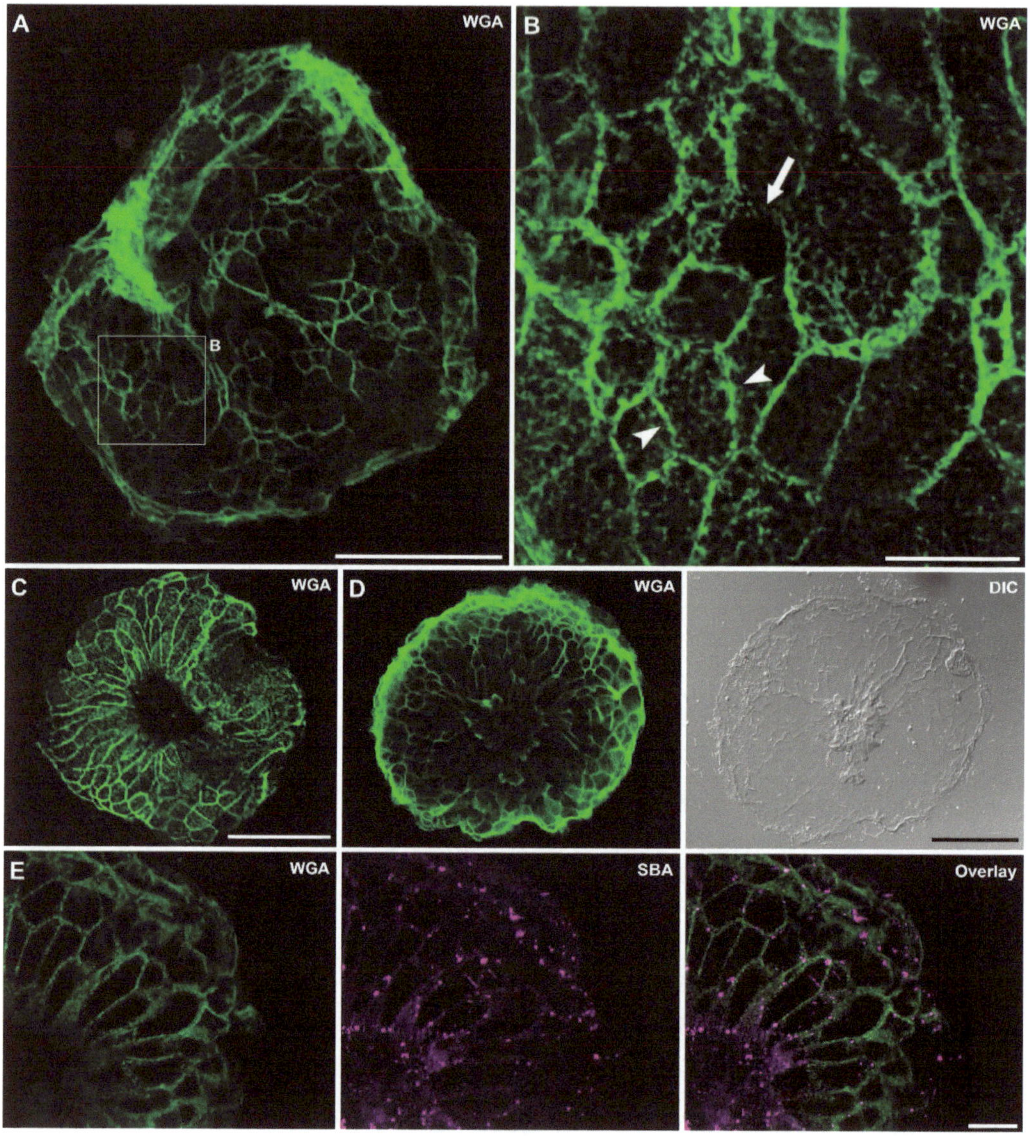

Figure 3. Lectin labeling of *Hydra* footprints with WGA (**A–D**) and SBA/WGA double labeling (**E**) (**A**) Footprint overview. (**B**) Footprint structure at a higher magnification; note the meshwork-like appearance. Arrowheads highlight the imprint of the basal disc cell borders and the arrow indicates a hole in the footprint. (**C**) Footprint with an empty area in the middle and (**D**) discarded cells at the same position. **(E)** Footprint detail of SBA and WGA double staining, note that the staining overlaps Scale bars: (**A**,**C**,**D**) 100 µm; (**B**,**E**) 20 µm.

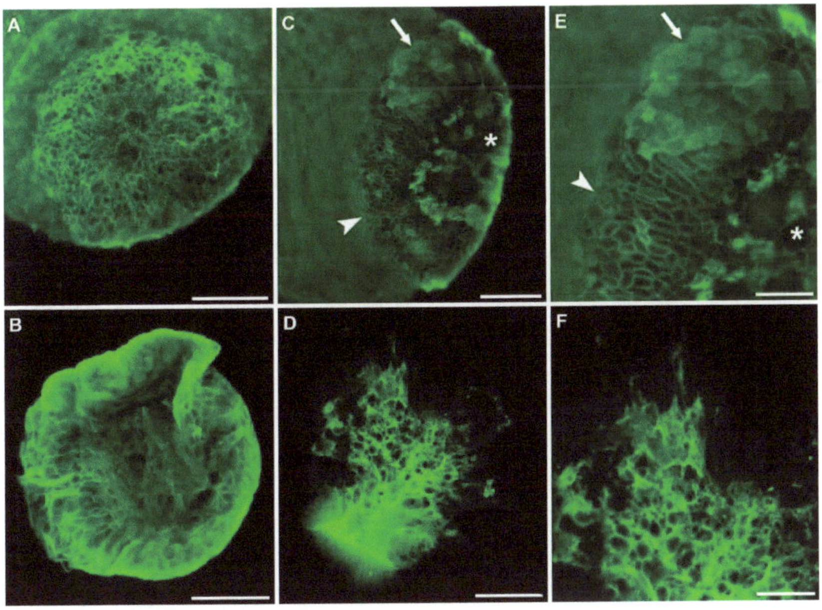

Figure 4. Representative lectin labeling of *Hydra* basal disc and the corresponding footprint with WGA. (**A**) Basal disc of a voluntarily detached *Hydra* and (**B**) corresponding footprint after 11 min attachment. (**C**,**E**) Basal disc of a forcibly detached animal and (**D**,**F**) corresponding footprint after 78 min attachment. Arrowheads indicate the adhesive material accumulated at the basal disc cell borders and arrows point to the adhesive material on the surface of the cells. Asterisks indicate damaged tissue. Scale bars: (**A–D**) 100 μm; (**E**,**F**) 50 μm.

Localization of footprint-specific glycans within the basal disc cells

As we were unable to visualize basal disc granules in the whole-mount samples, we performed lectin labeling of the eight footprint-specific lectins on separated (macerated) basal disc cells (Figure 5). Four secretory granule types can be distinguished in these cells: the large HSGI and HSGII, with HSGI likely representing immature HSGII, and the smaller and numerous HSGIII and HSGIV (Figure 5A) [15]. Basal disc cells are characterized by the four types of secretory granules, which are denser at the aboral end of the cells, by irregular water vacuoles, and by oriented actin filament bundles (myonemes) (Figure 5A) [15]. As expected, the four lectins WGA, sWGA, UEA I, and RCA (Figure 5B–E) reacted with numerous granules, which accumulated at the aboral end of the cells (Figure 5D,E). Based on their size and localization, they likely correspond to type III and/or IV granules. The larger HSGI and HSGII were not labeled. Surprisingly, no granular staining was detected for LEL, SBA, DSL, or Con A (Figure 5F–I). LEL and SBA reacted with vacuoles and small intracellular structures, which could not be identified (Figure 5F,G). DSL labeling of basal disc cells resulted in no or occasionally a weak labeling of vacuoles (Figure 5H). Con A reacted strongly with the cytoplasm throughout the cell, leaving only the nucleus unlabeled (Figure 5I). Overall, these results showed that the footprint material detected by WGA, sWGA, UEA I, and RCA was produced and secreted from the HSGIII and/or HSGIV granules. For the other four lectins, the origin of the detected glycans in the footprints remained unclear.

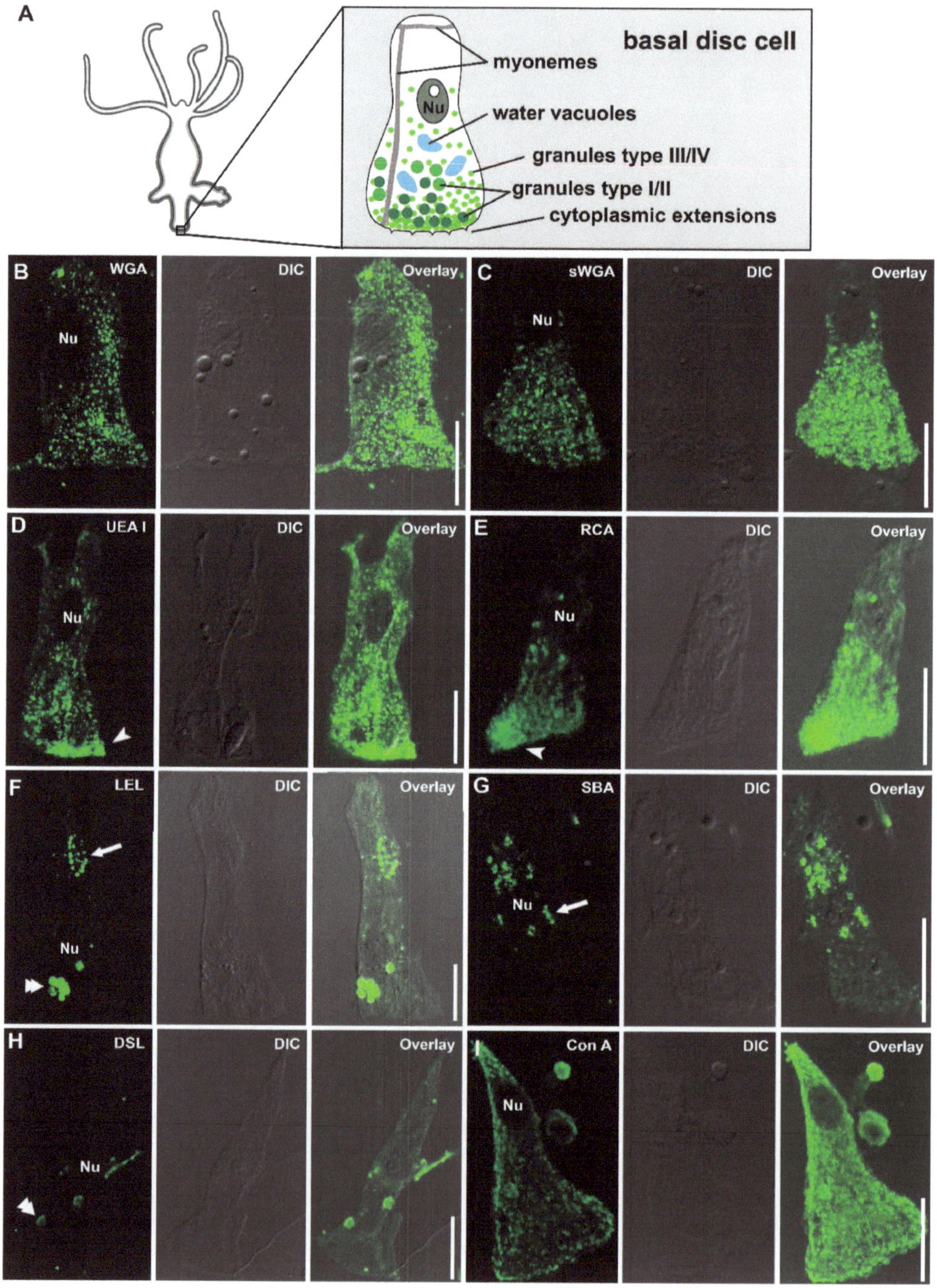

Figure 5. (**A**) Morphological scheme and (**B–I**) lectin labeling of macerated *Hydra* basal disc cells. Lectin labeling with (**B**) WGA, (**C**) sWGA, (**D**) UEA I, (**E**) RCA, (**F**) LEL, (**G**) SBA, (**H**) DSL, and (**I**) Con A. Arrow heads indicate denser concentrations of granules, double arrows point to labeled vacuoles and arrows to the dot-like structures. Scale bars: 20 μm.

3.3. Lectin Labeling of Universal and Positional Distinct Hydra Cell Types and Associated Structures

In addition to the glycans potentially involved in adhesion, the lectin screen revealed some common glycan patterns in whole-mount *Hydras*. The most prevalent stained structures were the overall ectoderm surface (14), nematocytes (12), and dot-like structures in the ectoderm (7) (Table 1). Exemplary images of frequent staining results are shown on the example of *Lens culinaris* agglutinin (LCA) and *Vicia villosa* agglutinin (VVL) in Figure 6. LCA strongly reacted with substructures of the nematocytes, with fixative-depending pattern. After PFA fixation, LCA only reacted with the base of the cnidocil (Figure 6A,B), whereas after Lavdowsky fixation, the whole capsules of the nematocytes were labeled (Figure 6C). Additionally, nematoblasts in the gastric region, as well as the surface of the tentacles, were intermediately stained and erupted tubules of nematocytes were labeled. VVL weakly labeled dot-like structures in the ectoderm of the gastric region and the tentacles (Figure 6D). These structures likely correspond to the subapical secretory granules of the ectodermal cells. Additionally, various parts of the nematocytes were labeled (Figure 6E,F). The nematocyte staining varied among fixatives, with the cnidocil only being stained after PFA fixation (Figure 6E), and not after Lavdowsky fixation (Figure 6F).

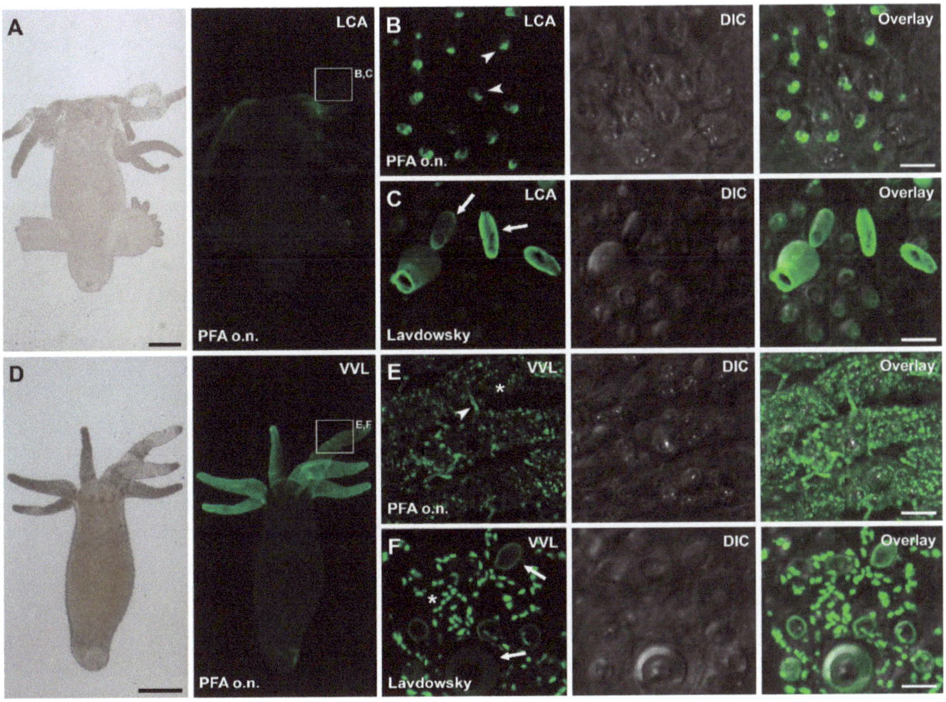

Figure 6. Lectin labeling of *Hydra* whole-mounts with (**A–C**) LCA and (**D–F**) VVL. (**A–C**) LCA labeling of a (**A**) whole-mount individual. Detailed view of stained structures in the tentacles, showing (**B**) cnidocil base staining and (**C**) nematocyst staining, fixed with PFA overnight or Lavdwosky for 4 h, respectively. (**D–F**) VVL labeling of a (**D**) whole-mount individual. Detailed view of stained structures in the tentacles, showing (**E**) cnidocil, nematocytes and dot-like structures and (**F**) nematocytes and dot-like structures, fixed with PFA overnight or Lavdwosky for 4 h, respectively. The fixation method used is indicated in the images. Arrowheads highlight the cnidocil base and the cnidocil, arrows indicate the nematocytes capsules and asterisks the dot-like structures. Scale bars: (**A,D**) 500 μm; (**B,C,E,F**) 10 μm.

In addition to these frequent labeling patterns, some unexpected labeling results occurred in whole-mount *Hydra*. For example, PHA-L and PHA-E labeled structures associated with radial muscle fibers and GSL II detected endodermal gland cells in the hypostome of developing buds (Figure S7). Detailed descriptions of these non-adhesion related results are presented in the Supplementary Materials section: "lectin labeling of universal and positional distinct *Hydra* cell types and associated structures".

4. Discussion

4.1. Glycan Distribution in Whole-Mount Hydra

Ultrasensitive mass spectrometry has revealed that the overall *Hydra* glycome consists of heavily fucosylated N- and O-glycans [24]. As these experiments have been performed with whole *Hydra* polyps, the localization of the detected glycans has not been determined We used a set of 23 commercially available lectins to determine glycan distribution within whole-mount *Hydras* and to distinguish glycans present in its secreted adhesive material. The lectins were selected to cover a wide range of common glycan moieties. *Hydra* is covered by a thick, layered glycocalyx composed of a high concentration of sulfated glycosaminoglycans (GAG) [25]. The glycocalyx is only well-preserved when cryo-based fixation methods (high-pressure freezing and freeze-substitution) are used [25,26]. With standard chemical fixation, such as the fixatives that were used (PFA and Lavdowsky), the glycocalyx shrinks and outer layers are lost. Nonetheless, 14 out of the tested 23 lectins labeled the overall ectoderm surface in varying degrees. Due to the technical limitations of the chemical fixation, this list is likely not exhaustive. In addition to the overall surface staining, seven lectins detected dot-like structures in the ectoderm. Based on their position and size, they likely corresponded to subapical secretory granules of the ectodermal epithelial cells, which are secreting the glyocalyx's components [25]. Mostly, the labeling appeared stronger at the tentacles compared to the gastric region. If this was an artefact due to the poor conservation of the glycocalyx or if there is a positional difference in the glycocalyx's composition remains unknown.

4.2. Hydra Footprints Are Built up by a Fibrillar Material

In other temporarily adhering animals, such as sea stars, the amount of secreted adhesive material varies depending on the surface composition [27] and the strength and duration of attachment [28]. In *Hydra*, no correlation between the thickness of the footprints and attachment time could be determined. Furthermore, sea star footprints are formed by a thin homogenous film covering the surface and a thick meshwork on top of it [8,27]. In contrast to this, our labeling revealed that *Hydra* footprints were formed by a fibrillar dense meshwork. Basal disc cell borders could be distinguished by an accumulation of adhesive material, resembling previous descriptions [15]. Additionally, we occasionally observed holes in the net, probably resulting from the non-secretion of the basal disc cell at this location. In half of all footprints, the middle area was devoid of any secreted material. This could potentially indicate that *Hydra* is able to create a vacuum under its basal disc to increase attachment strength. Alternatively, this could also result from the aboral pore located in the middle of the basal disc [29] and/or old epidermal cells, which are supposed to be discarded at this location. In one case, we observed discarded cells in the middle of the footprints, supporting the latter explanation.

The process of voluntary detachment from a secreted adhesive is still unclear. In the marine flatworm *Macrostomum lignano*, it has been proposed that a negatively charged molecule is secreted and outcompetes the binding of the positively charged adhesive to the glycocalyx [13]. In sea stars and sea urchins, an enzymatic detachment through proteinases has been postulated [30,31]. In contrast to these animals, *Hydra* lacks a dedicated de-adhesive gland. In *Hydra* the position and orientation of myonemes in its basal disc might allow for a mechanical detachment. Moreover, video analyzes of detaching individuals support the theory that muscular contractions in the basal disc are involved in the detachment process [15]. In our study, many footprints appeared to be smeared and folded in at the rim, regardless of whether the polyps were detached by force or detached voluntarily on their own. This observation is in line with the theory of a mechanical detachment.

4.3. Glycans Detected in the Secreted Footprint

We identified eight lectins that reacted with the secreted *Hydra* footprint. Surprisingly, only four (WGA, sWGA, UEA I, and RCA) equally detected granules in the basal disc cells, the subcellular structures in which the adhesive is stored until secretion [15]. All four lectins labeled small numerous granules that likely correspond to type III or IV. Both granule types have been described to be PAS-positive, highlighting that they are rich in glycans [15]. That the other four lectins (DSL, LEL, SBA, and ConA) did not react with any secretory granules could be an artefact caused by a limited accessibility of the glycans in the densely packed granules, as has previously been observed in sea star adhesive granules [7].

The fact that glycans are prevalent in temporary adhesives indicate an essential role in the adhesion process, but their function is still speculative. It has been proposed that cohesive strength is achieved through glycan–protein interactions, involving glycoproteins and proteins with glycan-binding functional domains [13]. Glycosylation could also enhance protein-binding ability and make proteins more resistant to degradation [6]. There is a high variability of glycans in the adhesive material found in between species [6,10,11,32,33]. Even in animals of the same phylum, the adhesive glycan composition is variant. For example, in the sea urchin *Paracentrotus lividus*, five lectins (GSL II, WGA, STL, LEL, and SBA) label the adhesive material [5], but in three other sea urchin species these lectins lead to different results [6]. In the sea star *Asterias rubens*, four lectins detect the footprints (DBA, WGA, RCA, and Con A), while in *Asterina gibbosa* this was only true for one lectin (Con A) [7,8]. This variability might be caused by an adaptation of the species to their respective environment, but further research is required to unravel the cause and functional relevance of this inconsistency in the glycan composition.

The glycans detected in the secreted adhesive are often part of the glycosylated proteins [5–7,12,13,34]. However, their function during attachment is mostly unknown. In *M. lignano*, the function of a glycosylated adhesive protein has been determined [13]. The adhesive protein is associated with GalNac residues and can be detected by the lectin PNA [9,13]. The glycoprotein binds to the surface during attachment and, upon functional knock-down, the animals are unable to attach themselves [13]. In *Hydra*, eight lectins reacted with the footprints left on the substrate, indicating that the *Hydra* adhesive contains GlcNAc, GalNAc, fucose, and mannose to varying degrees. It is unknown if these glycans are part of glycosylated proteins. Nonetheless, the presence of the enzyme glycosyl hydrolase AbfB [21] and three subunits of a Dolichyl-diphospho-oligosaccharide-protein glycosyltransferase [20] in the basal disc cells indicate that at least some glycans might be part of glycosylated proteins.

4.4. Lectins as Markers for Hydra Nematocytes

We found that 17 out of 23 lectins labeled whole-mount *Hydra* in a distinct pattern. Notably, 12 lectins reacted with fully differentiated nematocytes in the tentacles (Table 1). In *Hydra*, four different types of nematocytes can be distinguished: the holotrichous and the atrichous isorhizas (spineless), the desmonemes (small and with a tightly coiled tubule) and the stenoteles (large and with a prominent stiletto apparatus at the base of their tubules) [35]. In the tentacles, the mature nematocytes are incorporated into large battery cells, containing all the different types [36]. We observed no difference in the lectin labeling for the four types, except that sometimes the labeling intensity varied slightly. These results indicated that the glycan composition was similar among all four nematocyte types.

The nematocyte capsules (nematocysts) consist of an extracellular matrix-like composition of proteins and GAG, and protein–carbohydrate interactions mediate their capsule assembly [35]. Additional to structural proteins, such as minicollagens, C-type lectin NOWA, and spinalin [37], nematocysts are rich in chondroitin, which is a sulfated GAG, composed of a chain of alternating sugars (GalNAc and glucuronic acid) [38]. The chondroitin is present in form of proteoglycans [38] and GAG biosynthesis inhibition, using a β-D-Xyloside treatment, results in the complete absence of mature nematocysts in the tenta-cles [39]. This indicates that the GAG plays a crucial role in the capsule assembly and might serve as a scaffold for the structural proteins [39]. Our results confirm that nematocytes contain a high amount of glycans and indicate the presence of GluNAc, GalNAc, and man-nose residues in the capsules, the operculum, and the cnidocil. Anti-chondroitin antibodies mainly react with differentiating nematoblasts, whereas, in mature nematocytes, only the operculum is stained [39]. The capsules' walls harden during maturation, which might limit the antibodies' access [39]. Accordingly, we could not observe any labeling of the nematocysts' tubules in intact nematocytes, but several lectins labeled erupted nematocysts tubules. Furthermore, the lectin STL caused a blurry labeling surrounding erupted tubules which might indicate that the content of the capsules also contained glycans. However erupted nematocysts were not observed in all samples; therefore, our results might not be exhaustive. Additionally, the fixation method influenced the lectin labeling outcome Mature nematocytes in the tentacles were labeled after both fixations, but developing nematoblasts were only stained after Lavdowsky fixation.

4.5. Biomimetic Approaches and Their Limitations

Adhesives that perform under wet conditions or even underwater would have broad applications in the engineering and medical fields. Natural, aquatic adhesives might serve as a source for bio-inspired synthetic counterparts [40]. Thus far, biomimetic approaches mainly focused on adhesives produced by permanent adhering animals, like mussels [41] In recent years, the adhesives produced by temporarily adhering animals have gained increasing attention. In contrast to permanent adhesion, temporarily adhering animals can repeatedly detach and reattach [4]. The involved adhesive proteins are not conserved among phyla, but share reoccurring characteristics, such as a biased amino acid distribution repetitive regions, and prevalent protein domains [2]. For example, the cohesive proteins of sea stars, sea urchins, limpets, and flatworms contain calcium-binding epidermal growth factor (EGF)-like domains, galactose-binding lectin domains, discoidin domains (also known as F5/8 type C domains), von Willebrand Factor type D domains, and trypsin inhibitor-like cysteine rich domains [12,13,30,42]. Two fragments of the sea star cohesive protein that comprise these domains have been recombinantly produced in bacteria [43,44] These recombinant proteins not only self-assemble and adsorb on various surfaces, they also show no cytotoxic effects on cell cultures [43]. These results are highly promising and show the potential of recombinantly produced adhesive proteins for biomedical applications Nevertheless, the approach has its limitations, as recombinant production via bacteria is restricted to single proteins and fails to reproduce any post-translational modifications of the proteins. The natural sea star adhesive consists of a set of 16 proteins [28], of which many are glycosylated [7]. The recombinant proteins, therefore, only represent a fraction of the natural adhesive. To replicate the adhesive strength achieved in the natural system the protein interactions and the role of the prevalent glycans need to be investigated However, tools to test gene and protein function in sea stars are not available. In *Hydra* the needed molecular tools, such as gene knock-down and knock-out, are well established Previous findings show that the *Hydra* adhesive contains proteins with glycan-binding domains [21]. Here, we identified the glycans N-acetylglucosamine, N-acetylgalactosamine fucose, and mannose in the adhesive, which might be relevant to the proposed glycan-protein interactions. Our findings now lay the basis for further functional investigations on glycan and protein function.

5. Conclusions

Bio-inspired adhesives present themselves as a high potential substitute to the cur-rently used synthetic adhesives. The unraveling of the molecular composition of bioadhe-sives is crucial to provide models for bio-inspired technologies. *Hydra* constitutes a standard model in stem cell biology and tissue regeneration, but few studies have addressed its un derwater attachment ability. This project aimed to identify the glycans present in the *Hydra* secreted adhesive material, complementing previous transcriptomic and proteomic work. Our results indicate the presence of N-acetylglucosamine, N-acetylgalactosamine, fucose, and mannose in the secreted adhesive material. Furthermore, we observed a meshwork-like substructure in the footprints that implies that the adhesive is mainly formed by fibers. Additionally, we showed that commercially available lectins can be used as markers for several *Hydra* cell types and structures, such as nematocytes, endodermal gland cells, and cell membranes.

Supplementary Materials: The following supporting information can be downloaded at: https://www.mdpi.com/article/10.3390/biomimetics7040166/s1, Figure S1: WGA labeling of *Hydra* footprints using fresh (unfixed) footprints and footprints fixed for 1 h with PFA; Table S1: Overview of lectin binding specificity according to the manufacturer Vector laboratories; Figure S2: Lectin labeling of *Hydra* whole-mounts with EBL, Jacalin, Mal II, SJA, STL, PNA omitting a lectin and using only the Streptavidin-Dylight488 conjugate (negative control); Figure S3: STL labeling of *Hydra* erupted nematocytes' tubules; Figure S4: Lectin labeling of *Hydra* whole-mounts with ECL and PSA; Figure S5: Lectin labeling of *Hydra* whole-mounts with DSL, LEL and SBA; Figure S6: Lectin labeling of *Hydra* whole-mounts with RCA, UEA I and Con A; Figure S7: Lectin labeling of *Hydra* whole-mounts with PHA-L and GSLII.

Author Contributions: Conceptualization, B.L.; Methodology, S.S. and T.Z.; Formal Analysis, B.L. and B.H.; Investigation, S.S., T.Z., A.L.G., B.H. and B.L.; Writing—Original Draft Preparation, S.S. and B.L.; Writing—Review and Editing, S.S., A.L.G., B.H. and B.L.; Visualization, S.S. and A.L.G.; Supervision, B.L. and B.H.; Funding Acquisition, B.L. All authors have read and agreed to the published version of the manuscript.

Funding: B.L. is funded by an ESPRIT grant of the Austrian Science Fund (FWF): [ESP 15].

Institutional Review Board Statement: Not applicable.

Data Availability Statement: Not applicable.

Acknowledgments: The authors would like to thank Nathalie Kolb for *Hydra* culture maintenance.

Conflicts of Interest: The authors declare that they have no competing financial interests or personal relationships that could have influenced the work reported in this paper.

References

1. Ditsche, P.; Summers, A.P. Aquatic versus terrestrial attachment: Water makes a difference. *Beilstein J. Nanotechnol.* **2014**, *5*, 2424–2439. [CrossRef]
2. Davey, P.A.; Power, A.M.; Santos, R.; Bertemes, P.; Ladurner, P.; Palmowski, P.; Clarke, J.; Flammang, P.; Lengerer, B.; Hennebert, E.; et al. Omics-based molecular analyses of adhesion by aquatic invertebrates. *Biol. Rev.* **2021**, *96*, 1051–1075. [CrossRef]
3. Hennebert, E.; Maldonado, B.; Ladurner, P.; Flammang, P.; Santos, R. Experimental strategies for the identification and characterization of adhesive proteins in animals: A review. *Interface Focus* **2015**, *5*, 20140064. [CrossRef]
4. Lengerer, B.; Ladurner, P. Properties of temporary adhesion systems of marine and freshwater organisms. *J. Exp. Biol.* **2018**, *221*, jeb182717. [CrossRef]
5. Simão, M.; Moço, M.; Marques, L.; Santos, R. Characterization of the glycans involved in sea urchin Paracentrotus lividus reversible adhesion. *Mar. Biol.* **2020**, *167*, 125. [CrossRef]
6. Gaspar, L.; Flammang, P.; José, R.; Luis, R.; Ramalhosa, P.; Monteiro, J.; Nogueira, N.; Canning-Clode, J.; Santos, R. Interspecific Analysis of Sea Urchin Adhesive Composition Emphasizes Variability of Glycans Conjugated With Putative Adhesive Proteins. *Front. Mar. Sci.* **2021**, *8*, 1–16. [CrossRef]
7. Hennebert, E.; Wattiez, R.; Flammang, P. Characterisation of the carbohydrate fraction of the temporary adhesive secreted by the tube feet of the sea star Asterias rubens. *Mar. Biotechnol.* **2011**, *13*, 484–495. [CrossRef]
8. Lengerer, B.; Bonneel, M.; Lefevre, M.; Hennebert, E.; Leclère, P.; Gosselin, E.; Ladurner, P.; Flammang, P. The structural and chemical basis of temporary adhesion in the sea star Asterina gibbosa. *Beilstein J. Nanotechnol.* **2018**, *9*, 2071–2086. [CrossRef]
9. Lengerer, B.; Hennebert, E.; Flammang, P.; Salvenmoser, W.; Ladurner, P. Adhesive organ regeneration in Macrostomum lignano. *BMC Dev. Biol.* **2016**, *16*, 20. [CrossRef]
10. Bertemes, P.; Pjeta, R.; Wunderer, J.; Grosbusch, A.L.; Lengerer, B.; Gruner, K.; Knapp, M.; Mertens, B.; Andresen, N.; Hess, M.W.; et al. (Un)expected Similarity of the Temporary Adhesive Systems of Marine, Brackish, and Freshwater Flatworms. *Int. J. Mol. Sci.* **2021**, *22*, 12228. [CrossRef]

11. Bertemes, P.; Grosbusch, A.L.; Geschwindt, A.; Kauffmann, B.; Salvenmoser, W.; Mertens, B.; Pjeta, R.; Egger, B.; Ladurner, P. Sticking Together an Updated Model for Temporary Adhesion. *Mar. Drugs* **2022**, *20*, 359. [CrossRef]
12. Kang, V.; Lengerer, B.; Wattiez, R.; Flammang, P. Molecular insights into the powerful mucus-based adhesion of limpets (*Patella vulgata* L.). *Open Biol.* **2020**, *10*, 200019. [CrossRef]
13. Wunderer, J.; Lengerer, B.; Pjeta, R.; Bertemes, P.; Kremser, L.; Lindner, H.; Ederth, T.; Hess, M.W.; Stock, D.; Salvenmoser, W.; et al. A mechanism for temporary bioadhesion. *Proc. Natl. Acad. Sci. USA* **2019**, *116*, 4297–4306. [CrossRef]
14. Dreanno, C.; Matsumura, K.; Dohmae, N.; Takio, K.; Hirota, H.; Kirby, R.R.; Clare, A.S. An alpha2-macroglobulin-like protein is the cue to gregarious settlement of the barnacle Balanus amphitrite. *Proc. Natl. Acad. Sci. USA* **2006**, *103*, 14396–14401. [CrossRef]
15. Rodrigues, M.; Leclère, P.; Flammang, P.; Hess, M.W.; Salvenmoser, W.; Hobmayer, B.; Ladurner, P. The cellular basis of bioadhesion of the freshwater polyp Hydra. *BMC Zool.* **2016**, *1*, 3. [CrossRef]
16. Hobmayer, B.; Jenewein, M.; Eder, D.; Eder, M.K.; Glasauer, S.; Gufler, S.; Hartl, M.; Salvenmoser, W. Stemness in Hydra—A current perspective. *Int. J. Dev. Biol.* **2012**, *56*, 509–517. [CrossRef]
17. Bosch, T.C.; Anton-Erxleben, F.; Hemmrich, G.; Khalturin, K. The Hydra polyp: Nothing but an active stem cell community. *Dev. Growth Differ.* **2010**, *52*, 15–25. [CrossRef]
18. Holstein, T.W. The role of cnidarian developmental biology in unraveling axis formation and Wnt signaling. *Dev. Biol.* **2022**, *487*, 74–98. [CrossRef]
19. Vogg, M.C.; Galliot, B.; Tsiairis, C.D. Model systems for regeneration: Hydra. *Development* **2019**, *146*, dev177212. [CrossRef]
20. Siebert, S.; Farrell, J.A.; Cazet, J.F.; Abeykoon, Y.; Primack, A.S.; Schnitzler, C.E.; Juliano, C.E. Stem cell differentiation trajectories in. *Science* **2019**, *365*, 1–8. [CrossRef]
21. Rodrigues, M.; Ostermann, T.; Kremeser, L.; Lindner, H.; Beisel, C.; Berezikov, E.; Hobmayer, B.; Ladurner, P. Profiling of adhesive-related genes in the freshwater cnidarian Hydra magnipapillata by transcriptomics and proteomics. *Biofouling* **2016**, *32*, 1115–1129. [CrossRef]
22. Philpott, D.E.; Chaet, A.B.; Burnett, A.L. A study of the secretory granules of the basal disk of hydra. *J. Ultrastruct. Res.* **1966**, *14*, 74–84. [CrossRef]
23. David, C.N. Quantitative Method for Maceration of Hydra Tissue. *Wilhelm Roux Arch. Fur Entwickl. Der. Org.* **1973**, *171*, 259–268. [CrossRef]
24. Sahadevan, S.; Antonopoulos, A.; Haslam, S.M.; Dell, A.; Ramaswamy, S.; Babu, P. Unique, polyfucosylated glycan-receptor interactions are essential for regeneration of *Hydra magnipapillata*. *ACS Chem. Biol.* **2014**, *9*, 147–155. [CrossRef]
25. Bottger, A.; Doxey, A.C.; Hess, M.W.; Pfaller, K.; Salvenmoser, W.; Deutzmann, R.; Geissner, A.; Pauly, B.; Altstatter, J.; Munder, S.; et al. Horizontal gene transfer contributed to the evolution of extracellular surface structures: The freshwater polyp Hydra is covered by a complex fibrous cuticle containing glycosaminoglycans and proteins of the PPOD and SWT (sweet tooth) families. *PLoS ONE* **2012**, *7*, e52278. [CrossRef]
26. Holstein, T.W.; Hess, M.W.; Salvenmoser, W. Preparation techniques for transmission electron microscopy of Hydra. *Methods Cell Biol.* **2010**, *96*, 285–306. [CrossRef]
27. Hennebert, E.; Viville, P.; Lazzaroni, R.; Flammang, P. Micro- and nanostructure of the adhesive material secreted by the tube feet of the sea star *Asterias rubens*. *J. Struct. Biol.* **2008**, *164*, 108–118. [CrossRef]
28. Algrain, M.; Hennebert, E.; Bertemes, P.; Wattiez, R.; Flammang, P.; Lengerer, B. In the footsteps of sea stars: Deciphering the catalogue of proteins involved in underwater temporary adhesion. *Open Biol.* **2022**, *12*, 220103. [CrossRef]
29. Shimizu, H.; Takaku, Y.; Zhang, X.M.; Fujisawa, T. The aboral pore of hydra: Evidence that the digestive tract of hydra is a tube not a sac. *Dev. Genes Evol.* **2007**, *217*, 563–568. [CrossRef]
30. Pjeta, R.; Lindner, H.; Kremser, L.; Salvenmoser, W.; Sobral, D.; Ladurner, P.; Santos, R. Integrative Transcriptome and Proteome Analysis of the Tube Foot and Adhesive Secretions of the Sea Urchin. *Int. J. Mol. Sci.* **2020**, *21*, 946. [CrossRef]
31. Hennebert, E.; Leroy, B.; Wattiez, R.; Ladurner, P. An integrated transcriptomic and proteomic analysis of sea star epidermal secretions identifies proteins involved in defense and adhesion. *J. Proteom.* **2015**, *128*, 83–91. [CrossRef]
32. Pjeta, R.; Wunderer, J.; Bertemes, P.; Hofer, T.; Salvenmoser, W.; Lengerer, B.; Coassin, S.; Erhart, G.; Beisel, C.; Sobral, D.; et al. Temporary adhesion of the proseriate flatworm *Minona ileanae*. *Philos. Trans. R. Soc. B Biol. Sci.* **2019**, *374*, 20190194. [CrossRef] [PubMed]
33. Zeng, F.; Wunderer, J.; Salvenmoser, W.; Hess, M.W.; Ladurner, P.; Rothbächer, U. Papillae revisited and the nature of the adhesive secreting collocytes. *Dev. Biol.* **2019**, *448*, 183–198. [CrossRef] [PubMed]
34. Clarke, J.L.; Davey, P.A.; Aldred, N. Sea anemones (*Exaiptasia pallida*) use a secreted adhesive and complex pedal disc morphology for surface attachment. *BMC Zool.* **2020**, *5*, 5. [CrossRef]
35. Ozbek, S. The cnidarian nematocyst: A miniature extracellular matrix within a secretory vesicle. *Protoplasma* **2011**, *248*, 635–640. [CrossRef] [PubMed]
36. Beckmann, A.; Ozbek, S. The nematocyst: A molecular map of the cnidarian stinging organelle. *Int. J. Dev. Biol.* **2012**, *56*, 577–582. [CrossRef]
37. Balasubramanian, P.G.; Beckmann, A.; Warnken, U.; Schnolzer, M.; Schuler, A.; Bornberg-Bauer, E.; Holstein, T.W.; Ozbek, S. Proteome of *Hydra* nematocyst. *J. Biol. Chem.* **2012**, *287*, 9672–9681. [CrossRef] [PubMed]

38. Yamada, S.; Morimoto, H.; Fujisawa, T.; Sugahara, K. Glycosaminoglycans in *Hydra magnipapillata* (Hydrozoa, Cnidaria): Demonstration of chondroitin in the developing nematocyst, the sting organelle, and structural characterization of glycosaminoglycans. *Glycobiology* **2007**, *17*, 886–894. [CrossRef]
39. Adamczyk, P.; Zenkert, C.; Balasubramanian, P.G.; Yamada, S.; Murakoshi, S.; Sugahara, K.; Hwang, J.S.; Gojobori, T.; Holstein, T.W.; Ozbek, S. A non-sulfated chondroitin stabilizes membrane tubulation in cnidarian organelles. *J. Biol. Chem.* **2010**, *285*, 25613–25623. [CrossRef]
40. Almeida, M.; Reis, R.L.; Silva, T.H. Marine invertebrates are a source of bioadhesives with biomimetic interest. *Mater. Sci. Eng. C* **2019**, *108*, 110467. [CrossRef]
41. Ahn, B.K. Perspectives on Mussel-Inspired Wet Adhesion. *J. Am. Chem. Soc.* **2017**, *139*, 10166–10171. [CrossRef]
42. Hennebert, E.; Wattiez, R.; Demeuldre, M.; Ladurner, P.; Hwang, D.S.; Waite, J.H.; Flammang, P. Sea star tenacity mediated by a protein that fragments, then aggregates. *Proc. Natl. Acad. Sci. USA* **2014**, *111*, 6317–6322. [CrossRef]
43. Lefevre, M.; Flammang, P.; Aranko, A.S.; Linder, M.B.; Scheibel, T.; Humenik, M.; Leclercq, M.; Surin, M.; Tafforeau, L.; Wattiez, R.; et al. Sea star-inspired recombinant adhesive proteins self-assemble and adsorb on surfaces in aqueous environments to form cytocompatible coatings. *Acta Biomater.* **2020**, *112*, 62–74. [CrossRef] [PubMed]
44. Lefevre, M.; Ederth, T.; Masai, T.; Wattiez, R.; Leclere, P.; Flammang, P.; Hennebert, E. Disentangling the Roles of Functional Domains in the Aggregation and Adsorption of the Multimodular Sea Star Adhesive Protein Sfp1. *Mar. Biotechnol.* **2021**, *23*, 724–735. [CrossRef]

biomimetics

MDPI

Review

Studying Stickiness: Methods, Trade-Offs, and Perspectives in Measuring Reversible Biological Adhesion and Friction

Luc M. van den Boogaart [1,2], Julian K. A. Langowski [1,*] and Guillermo J. Amador [1,*]

1 Experimental Zoology Group, Department of Animal Sciences, Wageningen University & Research, De Elst 1, 6708 WD Wageningen, The Netherlands

2 Department of BioMechanical Engineering, Faculty of Mechanical, Maritime and Materials Engineering, Delft University of Technology, Mekelweg 2, 2628 CD Delft, The Netherlands

* Correspondence: julian.langowski@wur.nl (J.K.A.L.); guillermo.amador@wur.nl (G.J.A.)

Abstract: Controlled, reversible attachment is widely spread throughout the animal kingdom: from ticks to tree frogs, whose weights span from 2 mg to 200 g, and from geckos to mosquitoes, who stick under vastly different situations, such as quickly climbing trees and stealthily landing on human hosts. A fascinating and complex interplay of adhesive and frictional forces forms the foundation of attachment of these highly diverse systems to various substrates. In this review, we present an overview of the techniques used to quantify the adhesion and friction of terrestrial animals, with the aim of informing future studies on the fundamentals of bioadhesion, and motivating the development and adoption of new or alternative measurement techniques. We classify existing methods with respect to the forces they measure, including magnitude and source, i.e., generated by the whole body, single limbs, or by sub-structures. Additionally, we compare their versatility, specifically what parameters can be measured, controlled, and varied. This approach reveals critical trade-offs of bioadhesion measurement techniques. Beyond stimulating future studies on evolutionary and physicochemical aspects of bioadhesion, understanding the fundamentals of biological attachment is key to the development of biomimetic technologies, from soft robotic grippers to gentle surgical tools.

Keywords: biological adhesion; friction; contact mechanics; biomimetics; force sensor; bioinspiration

Citation: van den Boogaart, L.M.; Langowski, J.K.A.; Amador, G.J. Studying Stickiness: Methods, Trade-Offs, and Perspectives in Measuring Reversible Biological Adhesion and Friction. *Biomimetics* **2022**, 7, 134. https://doi.org/10.3390/biomimetics7030134

Academic Editors: Stanislav N. Gorb and Thies Henning Büscher

Received: 15 August 2022
Accepted: 11 September 2022
Published: 15 September 2022

1. Introduction

Controlled reversible attachment is a key adaptation across diverse terrestrial animal groups that exhibit various locomotory modes and encounter complex three-dimensional environments. Sticking to vertical or overhanging substrates requires a combination of strong adhesion (i.e., attachment force perpendicular to a substrate) and strong friction (i.e., attachment force parallel to a substrate) [1]. Among spiders, insects, tree frogs, and geckos, various versatile attachment strategies have evolved. The adhesive pads on the limbs of geckos and spiders rely on what is commonly referred to as 'dry' adhesion, thought to be dominated by weak intermolecular forces [2–4], while those of insects and tree frogs are believed to rely also on what is referred to as 'wet' adhesion—liquid-mediated interactions, such as capillary and viscous forces [5–7]. In addition to the adhesive pads on their limbs, animals may utilize other body parts to control or aid their attachment, such as generating friction through other tarsal segments in insects [8] or through the belly in tree frogs [9,10], or using claws to mechanically interlock with asperities on substrates [11,12]. These mechanisms have been studied in animals that vary in size across several orders of magnitude—from insects and spiders of a couple of milligrams to geckos and tree frogs of several hundreds of grams in mass [13].

Some animals can rapidly establish and reverse attachment, with stride frequencies of up to 10 steps per second for geckos or even 100 steps per second for mites [8]. To achieve such rapid reversibility, animals presumably control the strength of their attachment via shear-sensitive adhesive pads and control peeling by varying the angle between their limb and the substrate [8,14,15]. Furthermore, there is increasing evidence that shearing and peeling also contribute to self-cleaning during locomotion [16–19].

The fundamental understanding of rapid and reversible attachment of biological systems can inform many biomimetic applications that benefit humans in daily life. Reversible adhesion finds applications in sticky tapes, robotic grippers [20–22], and climbing robots [23–25]. The development of surgical tools may be inspired by the strategies and mechanisms used by animals, specifically for the manipulation of delicate and slippery tissues inside the human body [26–28]. Other applications can be found in agriculture and architecture, such as the development of grippers for autonomous harvesting robots [29], protecting crops from animal pests [30,31], improving pollination of flowers [32,33], protecting buildings from termites [34], or safeguarding people from disease vectors such as mosquitoes and ticks [35–37].

Accurate measurements of adhesion and friction forces are crucial for unravelling the fundamental mechanisms of biological adhesion, or bioadhesion. In order to understand and transfer the underlying principles of bioadhesion into biomimetic applications, physicochemical models of attachment need to be developed and validated against experimentally measured attachment forces, or derived parameters such as normal or shear stresses. As adhesive forces correlate strongly with contact area [38], normalizing adhesion forces to average adhesive stresses using contact areas provides a scale-independent representation of adhesive capacity [3]. However, measuring these parameters accurately poses a number of challenges.

To measure maximum adhesion and friction performance, one needs to detach the animal from a substrate through external forcing. These external forces can be applied globally, as a field, like gravitational or centrifugal forces, or locally by pulling on parts of the animal, for example through a tether. Such forces can be applied to the entire animal, one of its organs, or its sub-structures. In force measurements of live animals, behavior needs to be considered. When an animal moves freely it might employ behavioral strategies that are different than when it is perturbed, constrained, or sedated. Isolating individual limbs (or sub-structures) can help to control for animal behavior; however, extrapolating measurements on a single limb to the whole animal may lead to errors due to assumptions and oversimplifications. For example, in some animals, it has been found on the limb-level that larger adhesive pads generate stronger adhesion per unit area [13,39–41], which, however, may be explained through behavioral adaptations on the whole-organism-level (i.e., active shearing of the pad for adhesion control; [41]).

Given the many parameters that can influence adhesion and friction, such as temperature, humidity, and substrate properties, as well as the hierarchy of biological attachment devices (Figure 1), many factors need to be considered in the design of a bioadhesion study. In this review, we give an overview of the methods used for measuring contact forces in animal attachment studies, and discuss their trade-offs. This review limits itself to methods used in studies on terrestrial animals because they have direct implications for applications that humans encounter in their daily, (mostly) terrestrial life. However, many of the methods presented here are also used in studies on aquatic bioadhesive systems.

We conclude this review with a novel perspective on force measurement methods focusing on force magnitudes and how they are generated by and/or applied to the animal. To this extent, we will review the most-used force measurement methods considering whole animals (Figure 1A), isolated limbs (Figure 1B), and their sub-structures (Figure 1C), and whether the animals experience global or local forcing. Additionally, we address relevant parameters that can be measured, controlled, and varied in the different methods. This overview provides guidance for scientists that are new to the field of bioadhesion, and presents key challenges in measurement methodology that need to be overcome to advance the field. To assist those new to the field, we also provide a glossary of technical terms at the end.

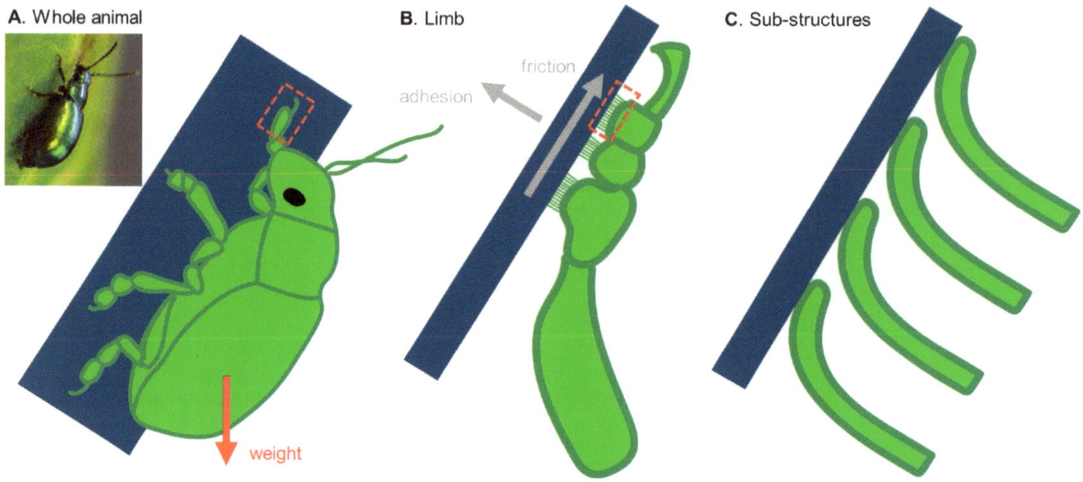

Figure 1. Levels at which bioadhesion can be studied in an animal. Schematics of a beetle sticking to a sloped substrate, showing (**A**) the whole animal, (**B**) its limb, and (**C**) sub-structures (fibres, setae or spatulae, depending on species). Inset of (**A**) shows a green dock beetle *Gastrophysa viridula* on a dock leaf *Rumex* spp.

2. Force Measurement Methods

In the past four decades, numerous measurements methods have been used in bioadhesion research. Table 1 outlines these methods, including the animals they have been used on, variables they measure and control for, and the ranges of force magnitudes they are capable of measuring. In the ensuing text of this section, we describe the methods in detail and elaborate on how they have been implemented in previous bioadhesion studies.

Table 1. Summary of bioadhesion measurement methods.

Level [1]	Forcing [2]	Method	Configuration	Subject Class	Dependent Variables	Independent Variables	Measurable Range	Study
Wh	Gl	3D force platforms	Single platform	Geckos	Reaction force	Walking direction	-	[42,43]
Wh	Gl			Tree frogs	Reaction force	Walking direction		[44]
Li	Gl			Insects	Reaction force	-		[45,46]
Wh	Gl		Force Measurement Array (FMA)	Geckos	Reaction force	Surface roughness		[43]
Wh	Gl			Tree frogs	Reaction force	Surface roughness, platform angle		[15,47,48]
Wh	Gl	Photo-elastic gelatin	-	Insects	Reaction force	-	-	[49]
Wh	Gl	Frustrated total internal reflection (FTIR)	-	Insects	Contact area	Load		[50]
Wh	Gl		-	Tree frogs	Contact area	Substrate curvature		[47,51]
Wh	Gl		Rotation platform	Tree frogs	Contact area	Surface roughness		[9,10]
Li	Gl	Optic tactile	-	Geckos	Normal stress	Load angle		[52]

Table 1. *Cont.*

Level [1]	Forcing [2]	Method	Configuration	Subject Class	Dependent Variables	Independent Variables	Measurable Range	Study
Wh	Gl	Rotation platforms	-	Arachnids	Adhesion %	Surface roughness	{0.7 mN, –} SF = {0.1, 7.0}	[37]
Wh	Gl			Insects	Adhesion %	Surface type, roughness, and structure		[53,54]
Wh	Gl			Tree frogs	Adhesion and shear force	Surface roughness		[10,18,55]
Wh	Gl	Force centrifuges	Adhesion	Insects	Adhesion force	Angular velocity, subject orientation	{500 µN, 500 mN}	[40,41,56–60]
Wh	Gl		Friction	Insects	Dynamic friction force	Surface chemistry and roughness, angular velocity		[37,60–64]
Wh	Lo	Tethered studies	Adhesion	Geckos	Adhesion force	Load	{200 µN, 10 mN}	[65]
Wh	Lo		Friction	Insects	Static friction force	Surface chemistry and roughness		[30–32,54,57,66–70]
Li	Lo	1D (uniaxial) force transducers	Adhesion	Insects	Adhesive force	Preload, retraction speed	{80 µN, 100 mN}	[19,71,72]
Li	Lo		Friction	Geckos	Friction force	Surface curvature and roughness, retraction speed		[73]
Li	Lo	2D (biaxial) force transducers	-	Geckos	Friction force	Surface chemistry, preload		[2,3]
Li	Lo			Insects	Friction force	Surface roughness, humidity, preload, sliding speed, retraction speed		[41,74,75]
Li	Lo			Tree frogs	Friction force	Surface roughness, preload		[18,47,76]
Li	Lo	Multiaxial force transducers	3-axis	Geckos	Friction force	Drag direction		[77]
Li	Lo		6-axis	Geckos	Friction force	Substrate roughness		[78]
Su	Lo	Atomic force miscroscopy (AFM)	-	Geckos	Adhesion force	Surface roughnes and chemistry, humidity, preload	{200 pN, 1 µN}	[3,79–81]
Su	Lo			Insects	Adhesion force	Surface roughness, humidity		[36]
Li	Lo			Insects	Adhesion force	Buffer presence		[82]

[1] Wh = Whole animal, Li = Limb, Su = Sub-structure; [2] Gl = Global forcing, Lo = Local forcing.

2.1. Global External Forcing

Force platforms (Figure 2A) are the most commonly used method to measure the contact forces of climbing animals. Conventional three-dimensional (3D) force platforms allow for the measurement of the magnitude and direction of ground reaction forces during locomotion and attachment. These measurements can be used to characterize gait patterns of studied animals, determining attachment forces through calculating stabilizing moments during locomotion [42]. The simplest setups consist of a single force platform for recording reaction forces [42,44].

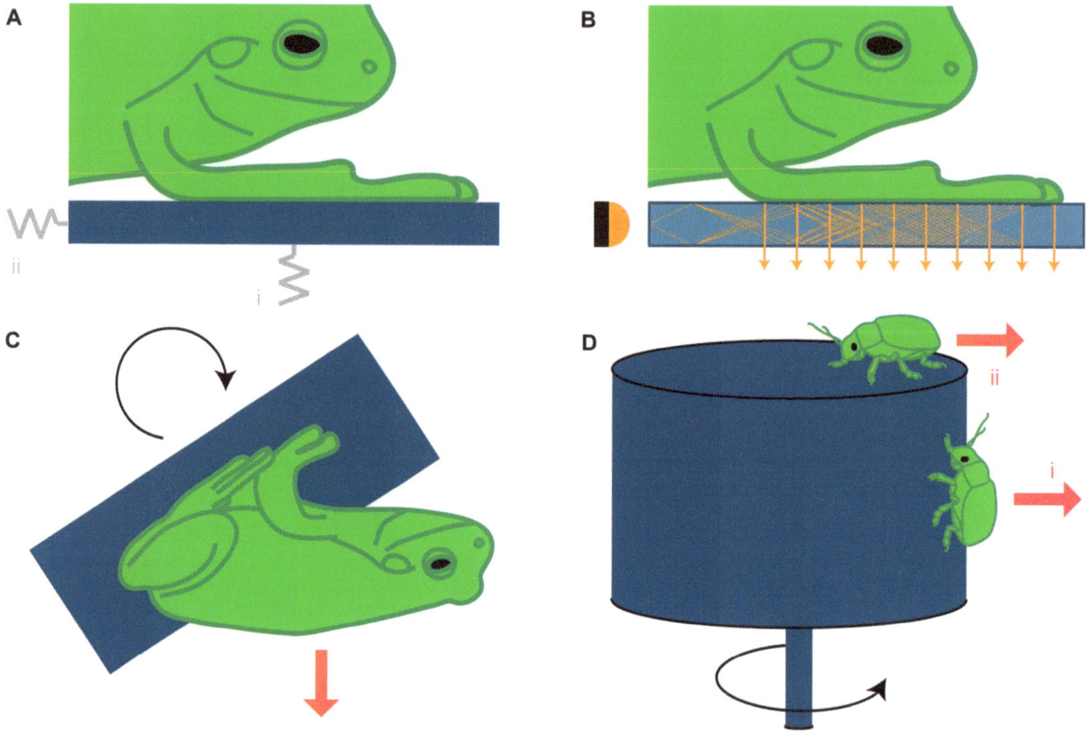

Figure 2. Measuring attachment forces with global forcing. (**A**) Force platform interacting with the limb of an unconstrained animal. The springs (in grey) represent capabilities of measuring adhesion (i; normal to substrate) and friction (ii; parallel to substrate). (**B**) Optic sensor based on frustrated total internal reflection (FTIR) to measure the contact area of adhesive pads. The yellow lines represent light reflected inside the transparent substrate, while the yellow arrows represent light that escapes the substrate when it is reflected by the adhesive pads in contact. (**C**) Rotation platform where the animal is gradually rotated around a horizontal axis until the component of gravitational force (red arrow) normal to the substrate exceeds the animal's adhesive capability. (**D**) Centrifuge system where the rotational velocity gradually increases until the centrifugal force (red arrows) exceeds the animal's (i) adhesive or (ii) frictional capabilities.

The main limitation of using a single force platform for the entire animal is the inability to distinguish force contributions from individual legs. To compensate for this limitation, later studies present experimental setups with an increased spatial resolution by using multiple platforms in force measurement arrays (FMAs). FMAs have been predominantly used to investigate the gait patterns of lizards [43] and tree frogs [15,47,48]. Reinhardt et al. [45] and Endlein & Federle [46] used custom-built force platforms with μN-resolution to measure the reaction force of a single leg of an ant during climbing. While 3D force platforms are among the few methods that allow simultaneous measurement of frictional and adhesive forces, it is typically impossible to measure the contact area during attachment due to constraints of the setup design space. Increasing the spatial resolution of the force platform to enable contact stress measurements would require multiple individual sensors for each adhesive pad, which quickly becomes impractical due to growing costs and time needed for calibration and data analysis.

Optic tactile sensors (Figure 2B) have been developed to measure contact areas and forces of adhesive pads during locomotion. By enabling the visualization of contact area, this method addresses one of the major limitations of force platforms. Such sensors exploit optical phenomena, like light refraction, to highlight areas where an adhesive organ comes into contact with a substrate, and measure substrate deformations to quantify contact forces. Earlier optic contact area sensors used in insect studies worked with photo-elastic gelatin [49], making use of polarizing filters to measure substrate deformation and, as a result, ground reaction forces. This method, however, is limited in substrate selection.

Later optic sensors exploit frustrated total internal reflection (FTIR), a technique first developed by Betts et al. in 1980 [83]. FTIR works by trapping a beam of light inside of a transparent substrate of high refractive index compared to air, e.g., glass, which has a refractive index of 1.5 compared to 1.0 for air. By shining light into the substrate at a shallow angle, the light will reflect internally and, when an object comes into contact with the substrate, the relative reflective index will be lowered locally, allowing light to escape and highlight the contact area. FTIR is limited by camera resolution. Stride frequencies of up to 100 strides per second require adhesion to be established and reversed within milliseconds [8]. Therefore, capturing the dynamics of such events requires a high temporal resolution, which in cameras typically conflicts with the high spatial resolution that is needed to record adhesion events in small animals like insects (e.g., the leg of a mosquito has a diameter of 50 μm [36]). Having both high temporal and spatial resolutions requires efficient data processing procedures and cameras with high quantum efficiency sensors. This makes tactile optic sensors and FTIR good alternatives for slow and large animals.

Eason et al. [52] developed an advanced FTIR-based sensor to measure the adhesive stress distribution of a gecko foot during climbing. This sensor makes use of a polymeric sensing membrane covered in flexible pyramidal bumps, named taxels, placed atop an acrylic waveguide. When force is applied to the membrane, the taxels buckle and the contact area between the sensing membrane and the waveguide increases, causing more light to scatter. This way measurable light intensity is related to the applied pressure, allowing the mapping of stress distributions during contact at high spatial and temporal resolutions (about 60 taxels per mm^2 at 60 Hz). FTIR has mostly been used for tree frogs in a completely free animal experiment [51] and in combination with rotation platforms [10,15]. Federle & Endlein [50] have also successfully used FTIR to image contact area in ants, measuring areas of several hundreds of μm^2 at frame rates of up to 250 Hz.

Rotation platforms (Figure 2C) provide a way to vary the orientation of an animal relative to the gravitational field. After the animal is placed on a horizontal platform, the platform is rotated around a horizontal axis until the animal is pulled off by gravity. The angle of the platform at which the animal drops off can be used to quantify adhesive force, with a completely inverted platform coinciding with an adhesive force equal to (or greater than) the animal's weight. For this reason, rotation platforms are limited to animals whose safety factor (SF: the ratio of attachment force to body weight) is one or lower. Rotation platforms have been used to study tree frogs [9,10,15,18,55], salamanders [84], beetles [53], mirid bugs [54], and ticks [37]. This technique can be relatively easily combined with FTIR to measure contact areas and determine average stresses. Moreover, the rotation platform is minimally invasive (i.e., animals are unconstrained) and the substrate can be easily exchanged or modified (e.g., covered with a liquid film; [9]). Because of the typically high SF of insects and arachnids, rotation platforms are not well-suited for adhesion force measurements in these animals, but can instead be used to compare the probability of attachment to different substrates [37]. While rotation platforms are ideally suited for studies with animals with SF of one or lower, there are no explicit upper or lower limits for the magnitudes of forces they can measure. In Table 1, the lower bound for the measurable force range coincides with previous measurements on mirid bugs [54].

Adhesion force centrifuges (Figure 2D.i) are the most frequently used alternative to rotation platforms for insects. When used to measure adhesive forces, the studied animal is placed on the side of a drum or vertical platform attached to a horizontal arm. The drum or arm is then rotated around a vertical axis at increasing angular velocities (typically up to 3000 revolutions per minute) until the centrifugal force exceeds the adhesive force and the studied animal detaches. This method is effective for animals with high SFs. Force centrifuges are able to record forces in a range between around {500 µN, 500 mN} (lower bound: motor precision, upper bound: maximum motor rotation speed). These ranges could be expanded by optimizing the centrifuge motor. Centrifuging techniques used to measure adhesion forces were first introduced by Dixon et al. [85] and later used to study ants [40,56–58], moths [59], and stick insects [41].

The effectiveness of force centrifuges is limited by subject mass. Since centrifugal forces are directly proportional to subject mass, special care needs to be placed on the structural robustness of the setup when scaling up. Moreover, the high centrifugal forces required to overcome high adhesion result in greater impact forces after release, which increases the risk of injury for test subjects. This makes using force centrifuges for heavier animals ethically challenging. Centrifuges are best suited for insect studies, or for animals with masses in the range of $\{1\ \text{mg}, 1\ \text{g}\}$. Gorb et al. [61] concluded that the influence of aerodynamic drag on force measurements using centrifuges is negligible in insects; however, aerodynamic forces may become significant for larger or non-streamlined specimens.

Friction force centrifuges (Figure 2D.ii) use the principle of controlling centrifugal forces for measuring static and dynamic friction forces, similar to adhesion force centrifuges. In friction force centrifuge measurements, subjects are placed on top of a horizontal disk or drum that rotates around a vertical axis. A laser or camera is used to monitor the subjects' distance from the center of the disk. Measuring the tangential acceleration and the centrifugal force component, the friction force can be calculated. Keeping the rotational velocity constant after static friction is overcome by the centrifugal force, dynamic friction can be calculated by tracking the sliding displacement and deriving acceleration. Like adhesion force centrifuges, friction force centrifuges are most effective for insects with high SFs and low body mass. Most friction force centrifuge experiments are based on a setup developed by Gorb et al. [61]. This or a similar setup has been used to study ants [86], beetles [62,63], coddling moths [64], sawfly larvae [60], and syrphid flies [61].

2.2. *Local Forcing*

Whole Animal Measurements

Adhesion force tethers (Figure 3A.i) provide a simple way to quantify adhesive forces. They are cheap and easy to set up, but invasive, as they require a strain gauge or scale to be attached to the animal, which may trigger unnatural postures or unwanted reactions due to induced stress. For example, a tethered study on the Tokay gecko *Gekko gecko* was carried out by Pugno et al. [71]. While the study clearly showed a decreasing trend in adhesive force over multiple trials due to foot damage, it underestimated the adhesive capacity of the gecko by more than a factor of 30. The authors suggested this to be the result of 'imperfections' on the toes; however, another likely explanation lies in the forced posture. The subject's limbs were pulled to unnatural angles wherein it was unable to fully engage its adhesive structures. While it is challenging to prevent such effects, synchronised video recordings of the animal may help monitor for induced changes in posture and for effects of tether location on measured attachment forces.

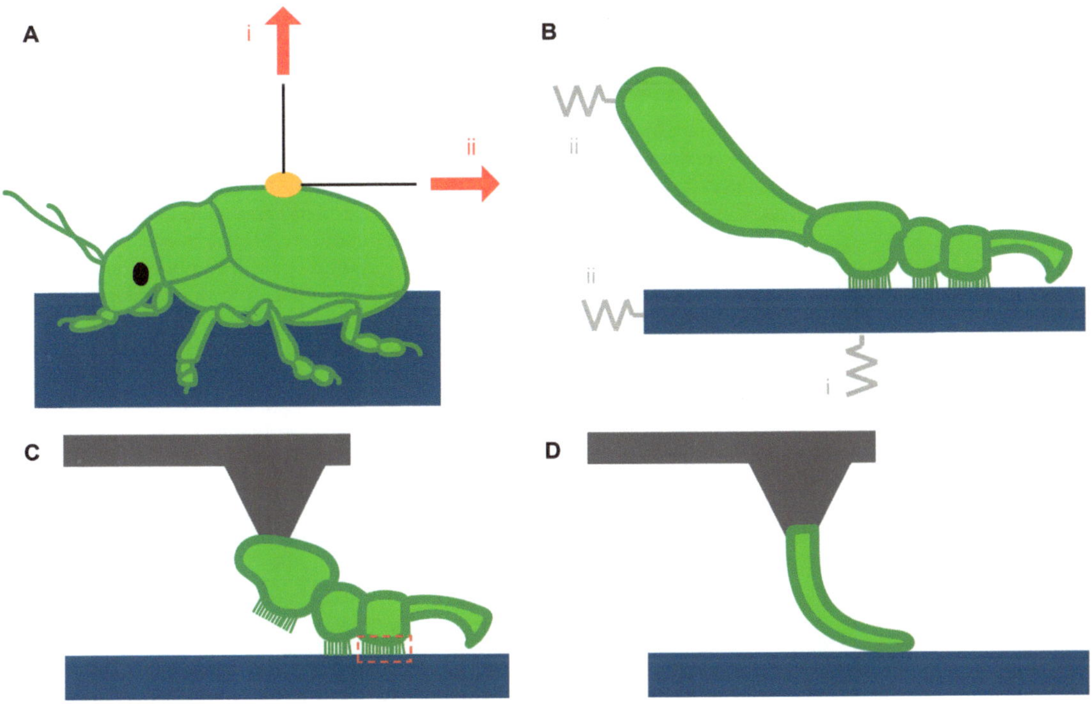

Figure 3. Measuring attachment forces with local forcing. (**A**) Tethered experiments where a wire is attached to an animal to measure (i) adhesion or (ii) friction forces. (**B**) Measurements on a limb using force transducers (FTs) to measure (i) adhesion and (ii) friction (or shear) forces. Typically, the shear force is controlled and adhesion measured [8,65]. (**C**,**D**) Atomic force microscopy (AFM) used to measure adhesion of a (**C**) limb and (**D**) its sub-structure, e.g., a seta. Typically, the limb or sub-structure (green) is attached to the AFM probe (grey) and then brought into contact with a substrate (blue) [36].

Friction force tethers (Figure 3A.ii) are the most commonly used method to measure friction forces. Like in adhesion force tether experiments, the studied animal is attached with a wire to a strain or force gauge. Alternatively, it is possible to pull on an animal positioned on a force sensor. The method was first used by Walker et al. [66] in a study on blowflies, in which the substrate was pulled while the animal remained stationary. This way, dynamic friction was measured for various pulling directions. Later studies measured static friction by making the animal walk over the substrate, pulling on the force transducer. Reviewed studies suggest a measurable force range of {200 μN, 10 mN} due to sensor limitations. All studies reviewed made use of the same force transducer: 10 g capacity, Biopac Systems Ltd., Santa Barbara, CA, USA. Bounds could be expanded by using force transducers of a higher capacity or sensitivity.

The method has been used as means of validation for force centrifuge tests [57], as well as to study the attachment of insects to various substrates. Examples include studying the effects of free surface energy [67], substrate roughness [68], or substrate chemistry [69] on insect friction. Additionally, tethered animal trials have been used to study the attachment capacity of insects to various plants [30,31,70] and flower petals [32].

2.3. Limbs and Below

In fundamental studies into the physicochemical basis of biological attachment, the behavior of the animal should be controlled for, such as when investigating how sub-structures enable adhesion and friction, or how adhesion and friction together contribute to attachment. In these cases, it makes sense to isolate the body part or sub-structure of interest. Doing so increases the controllability of the experiment and enables the measurement of forces in greater detail than in whole animal studies. Moreover, when we exclude confounding factors due to animal behavior from the experiment, we can more accurately estimate the maximum capacity of an adhesive system.

Force transducers (FTs; Figure 3B) are widely used to measure adhesion and friction of the toes, pads, and sub-structures (like fibres, setae, or spatulae) of geckos, frogs, and insects. Different configurations have been developed in various studies, but most of them are either uniaxial or biaxial FTs. Uniaxial FTs in limb studies are mostly used to measure adhesive forces. Biaxial FTs, mounted to a translation stage in combination with a closed loop controller, can be used to keep adhesive forces constant to isolate frictional forces or measure adhesive forces while applying shear loads.

Several types of sensors have been used. Uniaxial FTs typically rely on fibre optic springs [65,72] or piezoelectric sensors [19]. Spinner et al. [73] used an uniaxial FT to measure friction forces, by sliding the feet of a chameleon over a rod attached to the FT. Biaxial FTs mostly rely on strain gauges placed in perpendicular directions [2,3,18,74–76]. Force transducers are able to record forces in a range between {80 μN,100 mN} (lower bound: sensor precision [41], upper bound: sensor limitation of the 10g force transducers used). One study by Autumn et al. [77] used a 3-axis force sensor to measure the friction force of an array of setae from a gecko for various loading directions. One study by Gillies et al. [78], also on a gecko, used a 6-axis force sensor, though this was presumably due to availability. Keeping the amount of measuring axes to a minimum is beneficial since it reduces the amount of calibration needed, controller complexity, financial costs, and data processing.

Atomic force microscopy (AFM; Figure 3C,D) is an indispensable method in bioadhesion research, either to measure adhesive or frictional forces directly or functioning in a supportive role. AFM relies on the optical or piezoresistive sensing of the deflection of a cantilever, which is brought into contact with a substrate. AFM can measure adhesion forces with a resolution of 70 pN [79–81]. This makes AFM suitable to measure adhesive forces in a range of around {200 pN,1 μN} (lower bound: roughly three times the precision of 70 pN [79], upper bound: maximum force in flexible probe range [87]). AFM is not limited to a specific animal group or animal weight because it measures at a very small spatial range, e.g., at the (sub-)setal range. AFM has, for example, been used to measure the adhesive capacity of gecko setae [3,79], capillary forces on the terminal plates of fly setae [82], the friction profile across individual substrate features on the toe pads of tree frogs [88], and the adhesion of a mosquito limb on rough substrates [36].

3. Discussion

In the previous section, we presented a broad overview of existing methods to study the attachment of terrestrial animals (see Table 1 for a summary). When deciding on a method for a new study, one should consider a few questions. What parameters need to be measured (e.g., force, contact area, stress)? What are the magnitudes of the parameters to be measured? Is the method suitable for the animal of interest? Does the method provide the freedom to choose and/or vary experimental conditions (e.g., substrate characteristics)? Does the method limit the behavior of the animal? Are there alternative methods available for the study?

In this section we present relevant considerations when selecting a method. First we consider some of the limitations of the most prevalent methods with respect to scale and subject, e.g., species and body part. Then, the trade-offs in selecting a method for a study are discussed. Lastly, we present outlooks for future development and the general implications of animal adhesion studies in science and society.

3.1. Limitations

When deciding on a method, it is critical to consider the size of the animal and magnitudes of the attachment forces it can generate. Figure 4 shows a regime map of the most common adhesion and friction force measurement methods. Only AFM, 2D (biaxial) FTs, tethers, rotation platforms, and force centrifuges are included. Force platform studies are excluded because they include both whole animal and limb measurements, as well as 1D, 2D, and 3D force measurements, and so are difficult to compare. To our knowledge, there are insufficient previous studies ($n < 3$) available in the literature to make meaningful estimates of the regimes of photo-elastic gelatin and optic tactile sensors. However, their limitations were discussed in the previous section.

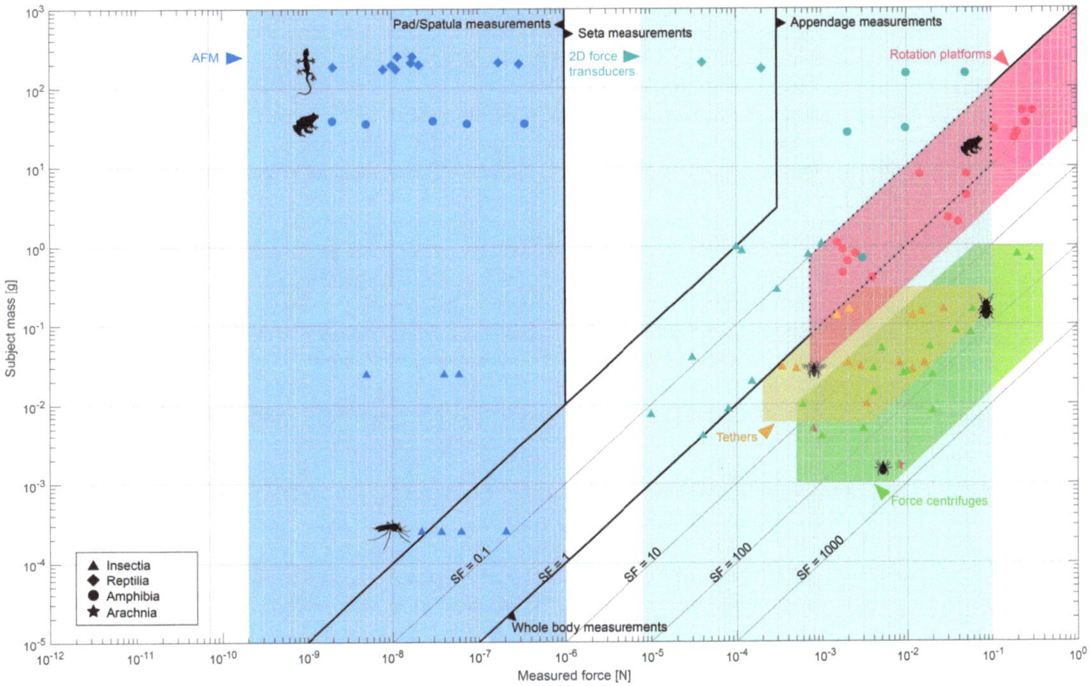

Figure 4. Ranges of common adhesion and friction force measurement techniques: AFM (blue), 2D (biaxial) force transducers (turquoise), force centrifuges (green), rotation platforms (pink) and tethers (orange). Data points indicate animal mass and measured force per study, with the symbols denoting taxonomic class. Diagonal lines indicate constant safety factor (SF) lines. Thick black lines denote boundaries between measurements on sub-structures (pad/spatula and setae, respectively), limbs, and whole animals. The area in between the dotted lines shows an overlap of the ranges of limb and whole body measurements. Reviewed studies investigated animals that range across six orders of magnitude in mass, and reported forces that range across nine orders of magnitude. Two studies within the 'force centrifuges' region are shown with two colors, indicating the study made use of two methods, namely rotation platforms and force centrifuges.

In Figure 4, measured force is plotted against subject mass as reported in the reviewed studies. The data shows two distinct trends: (1) whole animal studies follow constant safety factor (SF) lines, and (2) body part measurements are limited by sensor precision. The measurable ranges are also outlined in Table 1 and in more detail in the foregoing section.

As noted before, rotation platform limits are explained well by the animal's SF, which should be considered during the design of an experiment. For the other whole animal force measurement methods, tethers and force centrifuge measurements, SF bounds are suggested as well by the reviewed studies. Tethered studies are not effective when SF < 1, since animals that can not sustain their own weight through friction will likely start slipping when pulling their own body weight. There is a considerable overlap between tethers and force centrifuge studies, suggesting both are capable of studying the same animal species and expected SF or animal weight does not need to be considered when choosing between the two. However, Federle et al. [57] report higher adhesion forces for ants when measured using a centrifuge compared to a tether. They speculate that tethers (i.e., local forcing) may affect an animal's posture and natural response more than a centrifuge (i.e., global forcing).

Considering body part measurements, there is a clear gap between AFM and FTs. The bounds of these methods are set by sensor limitations. Reviewed studies and sensor limitations suggest a gap in the $\{1\ \mu N, 10\ \mu N\}$ range, above the maximum flexible probe range of AFM and below the sensor precision of FTs. When forces in this range are expected extra consideration should be taken in designing the measurement setup. Notably, both methods are suitable for any type of animal and are not limited by animal weight because these methods are used to investigate limbs or their sub-structures.

3.2. Trade-Offs in Study Design

In addition to animal size and expected force magnitudes, there are other factors to consider when deciding on a method to measure bioadhesion. First, one needs to determine if measurements should be carried out on whole animals or their limbs or sub-structures. Measurements with whole animals are influenced by behavior (e.g., motivation) and body kinematics (e.g., posture). However, investigating behavior may shed light on the postures and kinematics that animals use to promote attachment. For example, observations on tree frogs found that when attaching to overhanging substrates they spread their limbs away from their body to presumably minimize the angle between their limbs and substrate to prevent peeling [15].

While some behaviors promote attachment, there are others that may hinder it. Bioadhesion measurements only work when animals attach to substrates and do not jump or fly away. Insects capable of flight may need to be incapacitated by gluing or trimming their wings to prevent escape. In their study with moths, Al Bitar et al. [59] had to cut the insects' wings to prevent them from fleeing during measurements using force centrifuges. Such modifications allow measuring of attachment forces, but may affect the animal's behavior and response to external stimuli.

For fundamental studies into the physicochemical basis of attachment, bioadhesion measurements are best carried out with individual limbs or their sub-structures, where animal behavior can be controlled for. These measurements enable control over kinematics and mechanics, and thus may provide a deeper insight into the mechanisms underlying the generation of adhesion and friction. For example, previous work using individual limbs has found that the adhesive pads of geckos, tree frogs, and insects are shear-sensitive and generate increased adhesion under enhanced shear loading [8]. The linear relationship between shear force and adhesive force would be impossible to observe with whole animals. By working with individual limbs and biaxial FTs, the shear forces were controlled while adhesive forces measured.

In another example, the adhesive forces generated by a single gecko seta were carefully measured using AFM [2]. Then, the measured forces were compared with predictions from an analytical model of van der Waals forces (i.e., the interaction forces between the molecules on the seta and the substrate) to test if such intermolecular forces underpin gecko adhesion [3]. This finding motivated the development of gecko-inspired, micro-structured adhesives that stick without glue by also exploiting van der Waals forces [89]. Therefore, bioadhesion studies using limbs or their sub-structures have the potential to generate fundamental knowledge of great importance for the design of biomimetic adhesives.

As stated in the introduction, in order to measure attachment performance, an animal needs to experience an external force that works against the adhesion and friction it can generate. This external force can be applied globally, as a field, or locally, and the way it is applied can significantly influence the study. Global forcing is typically done using gravitational or centrifugal forces. These force fields act on the whole animal uniformly and simulate the forcing that an animal may experience when attaching to vertical or overhanging substrates. Local forcing acts on individual body parts. While such forcing is not typically experienced by animals in day-to-day life, it enables the isolation of individual limbs (and their sub-structures) and provides minimalist ways to measure maximum attachment performance, e.g., tethered studies require only a thin wire and force sensor.

Finally, the parameters that need to be measured and controlled, i.e., the dependent and independent variables, respectively, should be identified. Table 1 outlines the dependent and independent variables that were measured and controlled in previous studies. Based on this, tethers, force transducers, and AFM are the most versatile methods. They enable variation and control of independent variables, especially substrate properties and interaction kinematics as well as mechanics. Force platforms and optical methods are the most limited with respect to independent variables. This is primarily because the substrates cannot be controlled or varied due to requirements dictated by the methods, e.g., force platforms have sensors embedded and optical methods require substrate transparency.

3.3. Beyond Adhesion and Friction Measurements

While this review focuses primarily on techniques used for measuring forces, there are other parameters that need to be measured to fully grasp the attachment of a given animal. Theoretical models of contact mechanics and attachment can help identify underlying physicochemical mechanisms, but require validation through comparisons with experimental observations. Typically, the models predict adhesion and friction forces that can be compared to measured values; however, the models also depend on additional parameters as inputs.

One particularly important parameter needed in theoretical models of contact mechanics is the distance between the adhesive pad (or its sub-structures) and substrate. The magnitude of this distance could help determine which types of interactions are dominant or negligible. For example, for 10-μm spherical particles under dry conditions, electrostatic forces from the net charge on the particles dominate for distances greater than 100 nm, electrostatic forces from local charge patches dominate for distances between 10 and 100 nm, and van der Waals forces dominate for distances less than 10 nm [90]. Furthermore, if there is fluid present, measuring fluid film thickness can help determine if the fluid acts like a lubricant or enhances friction.

These distances can be measured through interference reflection microscopy (IRM). This technique was first developed to measure how close cells are to substrates [91], but was later used with tree frog toe pads [76,92]. In tree frogs, it was found that while mucus is present on the toe pads, parts of the surface features on frog toes are in quasi-direct contact with the substrate, with separation distances between 0 and 35 nm [76], indicating a potential contribution of van der Waals forces or other 'dry' interactions in tree frog attachment. Additionally, it was found that there is an intermediate fluid film thickness ($\sim$200 nm) that enhances friction compared to a fully wet (lubricating) or fully dry state [92].

Fluids covering the contact surface are an inherent part of many bioadhesive systems. For example, tree frog toes—as the whole amphibian body—are covered with a watery mucus [93], and insects secrete a viscous emulsion onto their adhesive pads. While these fluids help to prevent skin and cuticle from drying out and may have anti-bacterial and anti-fungal properties [94,95], their implications in bioadhesion are still being investigated. The physical and chemical properties of these fluids have been measured using various techniques. To measure the fluid's viscosity, methods were adopted from the field of rheology. For tree frog mucus, laser optical tweezers were used to measure the viscous force exerted on a trapped particle by the mucus [76]. The viscosity of insect pad fluid was measured by placing small tracer particles in a drop of the fluid and recording the dampening of the particle's Brownian motion (or thermal fluctuations) through the fluid's viscosity [96].

For chemical characterisation of the fluid, several techniques have been used. In tree frogs, cryo-histochemistry, attenuated total reflectance-infrared spectroscopy, and sum frequency generation spectroscopy have been used. From the measurements, it was found that the mucus on the toe pads is chemically similar to the mucus secreted by other body parts, including the belly [93]. In insects, gas chromatography and mass spectrometry have been used to characterize the chemical composition of their secreted fluids [97]. From this characterization, it was found that, like in tree frogs, the fluid secretions on the adhesive pads are chemically similar to those secreted throughout the rest of the body [98].

Surface tension is another important physical property of a bioadhesive fluid, as the capillary forces associated with it can be dominant at small spatial scales. However, to our knowledge, this property so far has been measured only indirectly through contact angle measurements [58,99,100]. Contact angle, or the angle between the substrate and fluid meniscus, quantifies the 'wettability' of a fluid on a substrate. For insects and tree frogs, this contact angle has been found to be quite small (~10°) on a wide variety of substrates, so the adhesive fluid appears to be highly wetting regardless of substrate chemistry [58,99,100]. Recent studies of insects have made assumptions of the surface tension of the fluid given that it is comprised of hydrocarbons [101,102]. This assumed approximate value sufficed for these studies since the models provided leading order analyses of the capillary interactions. For more detailed and accurate models, direct characterization will be required.

The material properties of the pad tissues, setae, or spatulae are also important for understanding bioadhesion. Animals stick to a wide variety of substrates, including smooth and rough ones. For rough substrates, the adhesive pads should conform to asperities in order to form a large area of close contact. A pad's ability to conform to rough substrates is dictated by its physical properties, especially its stiffness or Young's modulus. This property can be measured using micro- or nano-indentation, where the adhesive pad is compressed by a small probe and its stress response is measured, or using optical techniques, like confocal laser scanning microscopy [103]. Using such techniques, it has been found that setae on the adhesive pads of beetles are stiffer at the base and softer at the tip [103]. Similarly, the smooth adhesive pads of insects exhibit softer tissues in the outer layers and stiffer tissues underneath [104]. On the other hand, for tree frogs, it was found that the outer layers of the toe pads are stiffer than internal tissues [105,106].

Pad stiffness not only influences conformability, but may also affect the strength of adhesion. Classical experiments measuring the adhesion between a spherical indenter and flat substrate found that adhesion increases with material stiffness [107]. Similarly, the attachment force of fiber-reinforced adhesives such as gecko toes is proportional to the tensile stiffness of the fiber-reinforcement [108]. Therefore, there seems to be a trade-off between having soft pad tissues to conform to rough substrates and having stiff tissues to generate strong adhesion. In geckos and tree frogs, blood sacks have been observed immediately underneath the adhesive skin surface. Blood pressure may be controlled in these sacks to help tune pad stiffness [106,109]. Having such control could enable geckos and tree frogs to easily conform to rough or non-flat substrates using soft tissues and then stiffen the tissues to promote strong adhesion. A similar mechanism has been exploited by synthetic adhesives that use phase changing liquid metals [110].

AFM is a very versatile method that allows more than just contact force measurements. Many studies that investigate the effects of substrate properties on attachment use AFM to measure roughness, or to image surface sub-structures. Alternatives for measuring surface roughness of biological samples, such as scanning electron microscopy (SEM), are prone to artefacts from the preparation steps, such as shrinkage or drying, and are not suitable for living animals [105]. AFM can also be used for indentation experiments. Micro-indentation using FTs with a motorized stage is sufficient for larger structures, such as whole tree frog toes [106]. However, for smaller structures, AFM is required, for example to measure the stiffness of epithelial cells and local friction profiles over single pillars on tree frog toes [88], or the stiffness of the adhesive tarsal setae of ladybird beetles [103].

While the physical and chemical properties of adhesive pads and their fluid secretions are important for developing physicochemical models of adhesion and friction, the ways in which contact is established and released, i.e., pad and limb kinematics, can significantly influence attachment and detachment. Previous work has found that animals may be able to control adhesion by varying shear forces [8,65]. In addition to controlling adhesion via shear, tree frogs have been observed to spread out their limbs away from their body in response to increased loads [15]. By spreading their limbs, they not only promote shearing but also decrease the angle between their limbs and substrate. Just like in sticky tapes, minimizing this angle may prevent peeling. For insects, it has been found that attachment and detachment occur at different time scales [111,112]. Specifically, adhesive pads move quicker during detachment, which is believed to help conserve the secreted fluid. A faster separation velocity ensures that less fluid is deposited on the substrate. Additionally, a slower approach during attachment may help generate intimate contact and reduce the gap between pad and substrate to increase adhesion and friction forces [112].

3.4. Perspectives

Based on the reviewed data, we could map established force measurement methods to show their effectiveness and limitations, as summarized in Table 1 and Figure 4. From this analysis, we find that studying attachment for the large and slow no longer poses a problem. The frontier lies at the small and fast. Measuring small and fast processes still poses a considerable challenge given the trade-offs in spatial and temporal resolutions for cameras and sensors. There is renewed interest in optical methods during the past decade [9,10,47,51]. With visual data processing technologies, data storage and transfer capacities, and optic systems ever improving, optics-based methods seem promising, like the optic-tactile sensor developed by Eason et al. [52] to directly measure adhesive stress.

Quantifying adhesive and frictional stresses can help reveal the true performance of biological adhesives, since it provides a scale-independent measure of adhesion and friction and captures the exact contact stress distribution. Typically, adhesive pads are asymmetric and limbs are rarely oriented completely parallel or perpendicular to a substrate; therefore, forces are applied with offsets that induce moments and cause imbalances in contact stress distribution. Direct measurements of contact stress distribution can pinpoint where stress concentrations occur to reveal how the adhesive may fail and how limb kinematics influence adhesion and friction. However, to our knowledge, optic-tactile sensors are the only ones capable of contact stress measurements at the moment. Measuring adhesive and frictional stresses across various animals could contribute significantly to our understanding of the scaling of adhesive performance in biological systems [13].

In this review, we have largely skipped over micro-electromechanical sensors (MEMS). Interest in MEMS for measuring attachment seemingly faded in the past decade, but MEMS might be key in exploring the realm of fast and small. A MEMS force plate for studying insect locomotion developed by Bartsch et al. [113,114] has barely been cited in actual animal studies. The same holds for a biaxial MEMS cantilever design by Lin & Tramer [115]. This raises the question: is MEMS irrelevant to bioadhesion research, or have developments in MEMS design gone unnoticed in bioadhesion research?

Bioadhesion has always been a fascinating subject to study for biologists and engineers alike. Their work over the last decades resulted in various insights into these remarkable mechanics, attracting an ever-increasing interest from various other disciplines. Electrical engineers, (soft) roboticists, medical engineers, material scientists, and ecologists all benefit from discoveries in bioadhesion and work to tackle multidisciplinary problems, such as protecting honey bees, preventing animal pests, or developing new soft grippers for various applications.

Author Contributions: Conceptualization, L.M.v.d.B., J.K.A.L. and G.J.A.; writing—original draft preparation, L.M.v.d.B.; writing—review and editing, J.K.A.L. and G.J.A. All authors have read and agreed to the published version of the manuscript.

Funding: This research was funded by the Soft Robotics Consortium financed by the 4TU.Federation Project 4TU-UIT-335.

Data Availability Statement: Not applicable. All data is provided in the text, figures, and table.

Acknowledgments: We thank Dimitra Dodou for the helpful suggestions and discussions.

Conflicts of Interest: The authors declare no conflict of interest.

Glossary

The following terms are used in this manuscript:

Adhesion	[Newtons; N] the attractive contact force acting perpendicular to the substrate.
Friction	[Newtons; N] the contact force resisting motion parallel to the substrate.
Static friction	[Newtons; N] the friction force acting on a stationary object.
Dynamic friction	[Newtons; N] the friction force acting on a sliding object.
Contact area	[square meters; m^2] the area of an adhesive in direct contact with a substrate.
Adhesive stress (Tenacity)	[Newtons per square meter; N/m^2] the adhesion force per unit contact area. It provides a scale-independent representation of adhesive capacity.
Shear stress	[Newtons per square meter; N/m^2] the friction force per unit contact area.
Fluid viscosity	[Newton seconds per square meter; $N\text{-}s/m^2$] the resistance of a fluid to shearing. For example, honey is 10,000 times more viscous than water.
Substrate roughness	[nanometer; nm] the average height of the bumps, features, and asperities on a substrate.
Substrate energy	[milli-Newton per meter; mN/m] the excess energy that a surface of a material has compared to its bulk. If a substrate has high energy, then, generally, liquids and solids interact strongly with it.
Surface tension	[milli-Newton per meter; mN/m] the force (per unit length) acting tangential to a liquid-air interface. It is what enables insects to stand on the water surface and drives water drops to become spherical.
Young's modulus (Stiffness)	[Pascals; Pa] the physical property that represents how easily a material can stretch or deform.

References

1. Langowski, J.K.A.; Dodou, D.; Kamperman, M.; van Leeuwen, J.L. Tree frog attachment: Mechanisms, challenges, and perspectives. *Front. Zool.* **2018**, *15*, 1–21. [CrossRef] [PubMed]
2. Autumn, K.; Liang, Y.A.; Hsieh, S.T.; Zesch, W.; Chan, W.P.; Kenny, T.W.; Fearing, R.; Full, R.J. Adhesive force of a single gecko foot-hair. *Nature* **2000**, *405*, 681–685. [CrossRef] [PubMed]
3. Autumn, K.; Sitti, M.; Liang, Y.A.; Peattie, A.M.; Hansen, W.R.; Sponberg, S.; Kenny, T.W.; Fearing, R.; Israelachvili, J.N.; Full, R.J. Evidence for van der Waals adhesion in gecko setae. *Proc. Natl. Acad. Sci. USA* **2002**, *99*, 12252–12256. [CrossRef]
4. Gao, H.; Wang, X.; Yao, H.; Gorb, S.; Arzt, E. Mechanics of hierarchical adhesion structures of geckos. *Mech. Mater.* **2005**, *37*, 275–285. [CrossRef]
5. Gorb, S.N. The design of the fly adhesive pad: Distal tenent setae are adapted to the delivery of an adhesive secretion. *Proc. R. Soc. Lond. Series B Biol. Sci.* **1998**, *265*, 747–752. [CrossRef]
6. Gorb, S.; Beutel, R. Evolution of locomotory attachment pads of hexapods. *Naturwissenschaften* **2001**, *88*, 530–534. [CrossRef]
7. Gorb, S.N. Uncovering insect stickiness: Structure and properties of hairy attachment devices. *Am. Entomol.* **2005**, *51*, 31–35. [CrossRef]
8. Federle, W.; Labonte, D. Dynamic biological adhesion: Mechanisms for controlling attachment during locomotion. *Philos. Trans. R. Soc. B* **2019**, *374*, 20190199. [CrossRef]

9. Endlein, T.; Barnes, W.J.P.; Samuel, D.S.; Crawford, N.A.; Biaw, A.B.; Grafe, U. Sticking under wet conditions: The remarkable attachment abilities of the torrent frog, *Staurois guttatus*. *PLoS ONE* **2013**, *8*, e73810. [CrossRef]
10. Langowski, J.K.; Rummenie, A.; Pieters, R.P.; Kovalev, A.; Gorb, S.N.; van Leeuwen, J.L. Estimating the maximum attachment performance of tree frogs on rough substrates. *Bioinspir. Biomim.* **2019**, *14*, 025001. [CrossRef]
11. Gorb, S. *Attachment Devices of Insect Cuticle*; Kluwer Academic: Dordrecht, The Netherlands, 2001.
12. Song, Y.; Dai, Z.; Wang, Z.; Ji, A.; Gorb, S.N. The synergy between the insect-inspired claws and adhesive pads increases the attachment ability on various rough surfaces. *Sci. Rep.* **2016**, *6*, 1–9. [CrossRef]
13. Labonte, D.; Clemente, C.J.; Dittrich, A.; Kuo, C.Y.; Crosby, A.J.; Irschick, D.J.; Federle, W. Extreme positive allometry of animal adhesive pads and the size limits of adhesion-based climbing. *Proc. Natl. Acad. Sci. USA* **2016**, *113*, 1297–1302. [CrossRef] [PubMed]
14. Persson, B.; Gorb, S. The effect of surface roughness on the adhesion of elastic plates with application to biological systems. *J. Chem. Phys.* **2003**, *119*, 11437–11444. [CrossRef]
15. Endlein, T.; Ji, A.; Samuel, D.; Yao, N.; Wang, Z.; Barnes, W.J.P.; Federle, W.; Kappl, M.; Dai, Z. Sticking like sticky tape: Tree frogs use friction forces to enhance attachment on overhanging surfaces. *J. R. Soc. Interface* **2013**, *10*, 20120838. [CrossRef] [PubMed]
16. Amador, G.J.; Endlein, T.; Sitti, M. Soiled adhesive pads shear clean by slipping: A robust self-cleaning mechanism in climbing beetles. *J. R. Soc. Interface* **2017**, *14*, 20170134. [CrossRef]
17. Amador, G.J.; Hu, D.L. Cleanliness is next to godliness: Mechanisms for staying clean. *J. Exp. Biol.* **2015**, *218*, 3164–3174. [CrossRef] [PubMed]
18. Crawford, N.; Endlein, T.; Barnes, W.J.P. Self-cleaning in tree frog toe pads; a mechanism for recovering from contamination without the need for grooming. *J. Exp. Biol.* **2012**, *215*, 3965–3972. [CrossRef]
19. Hansen, W.R.; Autumn, K. Evidence for self-cleaning in gecko setae. *Proc. Natl. Acad. Sci. USA* **2005**, *102*, 385–389. [CrossRef]
20. Hawkes, E.W.; Christensen, D.L.; Han, A.K.; Jiang, H.; Cutkosky, M.R. Grasping without squeezing: Shear adhesion gripper with fibrillar thin film. In Proceedings of the 2015 IEEE International Conference on Robotics and Automation (ICRA), Seattle, WA, USA, 26–30 May 2015; pp. 2305–2312.
21. Song, S.; Majidi, C.; Sitti, M. Geckogripper: A soft, inflatable robotic gripper using gecko-inspired elastomer micro-fiber adhesives. In Proceedings of the 2014 IEEE/RSJ International Conference on Intelligent Robots and Systems, Chicago, IL, USA, 14–18 September 2014; pp. 4624–4629.
22. Zhou, M.; Tian, Y.; Sameoto, D.; Zhang, X.; Meng, Y.; Wen, S. Controllable interfacial adhesion applied to transfer light and fragile objects by using gecko inspired mushroom-shaped pillar surface. *ACS Appl. Mater. Interfaces* **2013**, *5*, 10137–10144. [CrossRef]
23. Henrey, M.; Ahmed, A.; Boscariol, P.; Shannon, L.; Menon, C. Abigaille-III: A versatile, bioinspired hexapod for scaling smooth vertical surfaces. *J. Bionic Eng.* **2014**, *11*, 1–17. [CrossRef]
24. Kim, S.; Spenko, M.; Trujillo, S.; Heyneman, B.; Santos, D.; Cutkosky, M.R. Smooth vertical surface climbing with directional adhesion. *IEEE Trans. Robot.* **2008**, *24*, 65–74.
25. Murphy, M.P.; Kute, C.; Mengüç, Y.; Sitti, M. Waalbot II: Adhesion recovery and improved performance of a climbing robot using fibrillar adhesives. *Int. J. Robot. Res.* **2011**, *30*, 118–133. [CrossRef]
26. Bergeles, C.; Yang, G.Z. From passive tool holders to microsurgeons: Safer, smaller, smarter surgical robots. *IEEE Trans. Biomed. Eng.* **2013**, *61*, 1565–1576. [CrossRef] [PubMed]
27. Glass, P.; Cheung, E.; Sitti, M. A legged anchoring mechanism for capsule endoscopes using micropatterned adhesives. *IEEE Trans. Biomed. Eng.* **2008**, *55*, 2759–2767. [CrossRef] [PubMed]
28. Kwon, J.; Cheung, E.; Park, S.; Sitti, M. Friction enhancement via micro-patterned wet elastomer adhesives on small intestinal surfaces. *Biomed. Mater.* **2006**, *1*, 216. [CrossRef] [PubMed]
29. Langowski, J.K.A.; Sharma, P.; Shoushtari, A.L. In the soft grip of nature. *Sci. Robot.* **2020**, *5*, eabd9120. [CrossRef] [PubMed]
30. Salerno, G.; Rebora, M.; Gorb, E.; Gorb, S. Attachment ability of the polyphagous bug *Nezara viridula* (Heteroptera: Pentatomidae) to different host plant surfaces. *Sci. Rep.* **2018**, *8*, 1–14. [CrossRef]
31. Salerno, G.; Rebora, M.; Kovalev, A.; Gorb, E.; Gorb, S. Contribution of different tarsal attachment devices to the overall attachment ability of the stink bug *Nezara viridula*. *J. Comp. Physiol. A* **2018**, *204*, 627–638. [CrossRef] [PubMed]
32. Bräuer, P.; Neinhuis, C.; Voigt, D. Attachment of honeybees and greenbottle flies to petal surfaces. *Arthropod-Plant Interact.* **2017**, *11*, 171–192. [CrossRef]
33. Amador, G.J.; Hu, D.L. Sticky solution provides grip for the first robotic pollinator. *Chem* **2017**, *2*, 162–164. [CrossRef]
34. Féat, A.; Federle, W.; Kamperman, M.; Murray, M.; van der Gucht, J.; Taylor, P. Slippery paints: Eco-friendly coatings that cause ants to slip. *Prog. Org. Coatings* **2019**, *135*, 331–344. [CrossRef]
35. Féat, A.; Federle, W.; Kamperman, M.; van der Gucht, J. Coatings preventing insect adhesion: An overview. *Prog. Org. Coatings* **2019**, *134*, 349–359. [CrossRef]
36. Pashazanusi, L.; Lwoya, B.; Oak, S.; Khosla, T.; Albert, J.N.; Tian, Y.; Bansal, G.; Kumar, N.; Pesika, N.S. Enhanced adhesion of mosquitoes to rough surfaces. *ACS Appl. Mater. Interfaces* **2017**, *9*, 24373–24380. [CrossRef] [PubMed]
37. Voigt, D.; Gorb, S. Functional morphology of tarsal adhesive pads and attachment ability in ticks *Ixodes ricinus* (Arachnida, Acari, Ixodidae). *J. Exp. Biol.* **2017**, *220*, 1984–1996. [CrossRef] [PubMed]
38. Irschick, D.J.; Austin, C.C.; Petren, K.; Fisher, R.N.; Losos, J.B.; Ellers, O. A comparative analysis of clinging ability among pad-bearing lizards. *Biol. J. Linn. Soc.* **1996**, *59*, 21–35. [CrossRef]

39. Smith, J.M.; Barnes, W.J.; Downie, J.R.; Ruxton, G.D.; et al. Structural correlates of increased adhesive efficiency with adult size in the toe pads of hylid tree frogs. *J. Comp. Physiol. A* **2006**, *192*, 1193–1204. [CrossRef]
40. Labonte, D.; Federle, W. Scaling and biomechanics of surface attachment in climbing animals. *Philos. Trans. R. Soc. B Biol. Sci.* **2015**, *370*, 20140027. [CrossRef]
41. Labonte, D.; Struecker, M.Y.; Birn-Jeffery, A.V.; Federle, W. Shear-sensitive adhesion enables size-independent adhesive performance in stick insects. *Proc. R. Soc. B* **2019**, *286*, 20191327. [CrossRef] [PubMed]
42. Autumn, K.; Hsieh, S.T.; Dudek, D.M.; Chen, J.; Chitaphan, C.; Full, R.J. Dynamics of geckos running vertically. *J. Exp. Biol.* **2006**, *209*, 260–272. [CrossRef]
43. Dai, Z.; Wang, Z.; Ji, A. Dynamics of gecko locomotion: A force-measuring array to measure 3D reaction forces. *J. Exp. Biol.* **2011**, *214*, 703–708. [CrossRef]
44. Hanna, G.; Jon, W.; Barnes, W.J. Adhesion and detachment of the toe pads of tree frogs. *J. Exp. Biol.* **1991**, *155*, 103–125. [CrossRef]
45. Reinhardt, L.; Weihmann, T.; Blickhan, R. Dynamics and kinematics of ant locomotion: Do wood ants climb on level surfaces? *J. Exp. Biol.* **2009**, *212*, 2426–2435. [CrossRef]
46. Endlein, T.; Federle, W. On heels and toes: How ants climb with adhesive pads and tarsal friction hair arrays. *PLoS ONE* **2015**, *10*, e0141269. [CrossRef] [PubMed]
47. Endlein, T.; Ji, A.; Yuan, S.; Hill, I.; Wang, H.; Barnes, W.J.P.; Dai, Z.; Sitti, M. The use of clamping grips and friction pads by tree frogs for climbing curved surfaces. *Proc. R. Soc. B Biol. Sci.* **2017**, *284*, 20162867. [CrossRef] [PubMed]
48. Ji, A.; Yuan, S.; Endlein, T.; Hill, I.D.; Wang, W.; Wang, H.; Jiang, N.; Zhao, Z.; Barnes, W.J.P.; Dai, Z. A force-measuring and behaviour-recording system consisting of 24 individual 3D force plates for the study of single limb forces in climbing animals on a quasi-cylindrical tower. *Bioinspir. Biomim.* **2019**, *14*, 046004. [CrossRef]
49. Full, R.; Yamauchi, A.; Jindrich, D. Maximum single leg force production: Cockroaches righting on photoelastic gelatin. *J. Exp. Biol.* **1995**, *198*, 2441–2452. [CrossRef] [PubMed]
50. Federle, W.; Endlein, T. Locomotion and adhesion: Dynamic control of adhesive surface contact in ants. *Arthropod Struct. Dev.* **2004**, *33*, 67–75. [CrossRef] [PubMed]
51. Hill, I.D.; Dong, B.; Barnes, W.J.P.; Ji, A.; Endlein, T. The biomechanics of tree frogs climbing curved surfaces: A gripping problem. *J. Exp. Biol.* **2018**, *221*, jeb168179. [CrossRef]
52. Eason, E.V.; Hawkes, E.W.; Windheim, M.; Christensen, D.L.; Libby, T.; Cutkosky, M.R. Stress distribution and contact area measurements of a gecko toe using a high-resolution tactile sensor. *Bioinspir. Biomim.* **2015**, *10*, 016013. [CrossRef]
53. Gorb, E.; Gorb, S. Attachment ability of the beetle *Chrysolina fastuosa* on various plant surfaces. *Entomol. Exp. Appl.* **2002**, *105*, 13–28. [CrossRef]
54. Voigt, D.; Gorb, E.; Gorb, S. Plant surface–bug interactions: *Dicyphus errans* stalking along trichomes. *Arthropod-Plant Interact.* **2007**, *1*, 221–243. [CrossRef]
55. Barnes, W.J.; Oines, C.; Smith, J.M. Whole animal measurements of shear and adhesive forces in adult tree frogs: Insights into underlying mechanisms of adhesion obtained from studying the effects of size and scale. *J. Comp. Physiol. A* **2006**, *192*, 1179–1191. [CrossRef] [PubMed]
56. Brainerd, E. Adhesion force of ants on smooth surfaces. *Am. Zool.* **1994**, *34*, 128.
57. Federle, W.; Rohrseitz, K.; Holldobler, B. Attachment forces of ants measured with a centrifuge: Better 'wax-runners' have a poorer attachment to a smooth surface. *J. Exp. Biol.* **2000**, *203*, 505–512. [CrossRef] [PubMed]
58. Federle, W.; Riehle, M.; Curtis, A.S.; Full, R.J. An integrative study of insect adhesion: Mechanics and wet adhesion of pretarsal pads in ants. *Integr. Comp. Biol.* **2002**, *42*, 1100–1106. [CrossRef] [PubMed]
59. Al Bitar, L.; Voigt, D.; Zebitz, C.P.; Gorb, S.N. Tarsal morphology and attachment ability of the codling moth *Cydia pomonella* L.(Lepidoptera, Tortricidae) to smooth surfaces. *J. Insect Physiol.* **2009**, *55*, 1029–1038. [CrossRef]
60. Voigt, D.; Gorb, S.N. Attachment ability of sawfly larvae to smooth surfaces. *Arthropod Struct. Dev.* **2012**, *41*, 145–153. [CrossRef] [PubMed]
61. Gorb, S.; Gorb, E.; Kastner, V. Scale effects on the attachment pads and friction forces in syrphid flies (Diptera, Syrphidae). *J. Exp. Biol.* **2001**, *204*, 1421–1431. [CrossRef]
62. Lüken, D.; Voigt, D.; Gorb, S.N.; Zebitz, C. Die Tarsenmorphologie und die Haftfähigkeit des Schwarzen Batatenkäfers *Cylas puncticollis* (Boheman) auf glatten Oberflächen mit unterschiedlichen physiko-chemischen Eigenschaften. *Mitteilungen Der Dtsch. Ges. Fur Allg. Und Angew. Entomol.* **2009**, *17*, 109–113.
63. Grohmann, C.; Blankenstein, A.; Koops, S.; Gorb, S.N. Attachment of *Galerucella nymphaeae* (Coleoptera, Chrysomelidae) to surfaces with different surface energy. *J. Exp. Biol.* **2014**, *217*, 4213–4220.
64. Al Bitar, L.; Voigt, D.; Zebitz, C.P.; Gorb, S.N. Attachment ability of the codling moth *Cydia pomonella* L. to rough substrates. *J. Insect Physiol.* **2010**, *56*, 1966–1972. [CrossRef] [PubMed]
65. Labonte, D.; Federle, W. Rate-dependence of 'wet' biological adhesives and the function of the pad secretion in insects. *Soft Matter* **2015**, *11*, 8661–8673. [CrossRef] [PubMed]
66. Walker, G.; Yulf, A.; Ratcliffe, J. The adhesive organ of the blowfly, *Calliphora vomitoria*: A functional approach (Diptera: Calliphoridae). *J. Zool.* **1985**, *205*, 297–307. [CrossRef]

67. Gorb, E.; Voigt, D.; Eigenbrode, S.D.; Gorb, S. Attachment force of the beetle *Cryptolaemus montrouzieri* (Coleoptera, Coccinellidae) on leaflet surfaces of mutants of the pea *Pisum sativum* (Fabaceae) with regular and reduced wax coverage. *Arthropod-Plant Interact.* **2008**, *2*, 247–259. [CrossRef]
68. Gorb, E.; Hosoda, N.; Miksch, C.; Gorb, S. Slippery pores: Anti-adhesive effect of nanoporous substrates on the beetle attachment system. *J. R. Soc. Interface* **2010**, *7*, 1571–1579. [CrossRef] [PubMed]
69. England, M.W.; Sato, T.; Yagihashi, M.; Hozumi, A.; Gorb, S.N.; Gorb, E.V. Surface roughness rather than surface chemistry essentially affects insect adhesion. *Beilstein J. Nanotechnol.* **2016**, *7*, 1471–1479. [CrossRef]
70. Salerno, G.; Rebora, M.; Piersanti, S.; Gorb, E.; Gorb, S. Mechanical ecology of fruit-insect interaction in the adult Mediterranean fruit fly *Ceratitis capitata* (Diptera: Tephritidae). *Zoology* **2020**, *139*, 125748. [CrossRef]
71. Pugno, N.; Lepore, E.; Toscano, S.; Pugno, F. Normal adhesive force-displacement curves of living geckos. *J. Adhes.* **2011**, *87*, 1059–1072. [CrossRef]
72. Jiao, Y.; Gorb, S.; Scherge, M. Adhesion measured on the attachment pads of *Tettigonia viridissima* (Orthoptera, Insecta). *J. Exp. Biol.* **2000**, *203*, 1887–1895. [CrossRef]
73. Spinner, M.; Westhoff, G.; Gorb, S.N. Subdigital setae of chameleon feet: Friction-enhancing microstructures for a wide range of substrate roughness. *Sci. Rep.* **2014**, *4*, 1–9. [CrossRef]
74. Drechsler, P.; Federle, W. Biomechanics of smooth adhesive pads in insects: Influence of tarsal secretion on attachment performance. *J. Comp. Physiol. A* **2006**, *192*, 1213–1222. [CrossRef] [PubMed]
75. Zhou, Y.; Robinson, A.; Steiner, U.; Federle, W. Insect adhesion on rough surfaces: Analysis of adhesive contact of smooth and hairy pads on transparent microstructured substrates. *J. R. Soc. Interface* **2014**, *11*, 20140499. [CrossRef] [PubMed]
76. Federle, W.; Barnes, W.; Baumgartner, W.; Drechsler, P.; Smith, J. Wet but not slippery: Boundary friction in tree frog adhesive toe pads. *J. R. Soc. Interface* **2006**, *3*, 689–697. [CrossRef] [PubMed]
77. Autumn, K.; Dittmore, A.; Santos, D.; Spenko, M.; Cutkosky, M. Frictional adhesion: A new angle on gecko attachment. *J. Exp. Biol.* **2006**, *209*, 3569–3579. [CrossRef]
78. Gillies, A.G.; Henry, A.; Lin, H.; Ren, A.; Shiuan, K.; Fearing, R.S.; Full, R.J. Gecko toe and lamellar shear adhesion on macroscopic, engineered rough surfaces. *J. Exp. Biol.* **2014**, *217*, 283–289. [CrossRef] [PubMed]
79. Huber, G.; Gorb, S.N.; Spolenak, R.; Arzt, E. Resolving the nanoscale adhesion of individual gecko spatulae by atomic force microscopy. *Biol. Lett.* **2005**, *1*, 2–4. [CrossRef]
80. Huber, G.; Mantz, H.; Spolenak, R.; Mecke, K.; Jacobs, K.; Gorb, S.N.; Arzt, E. Evidence for capillarity contributions to gecko adhesion from single spatula nanomechanical measurements. *Proc. Natl. Acad. Sci. USA* **2005**, *102*, 16293–16296. [CrossRef]
81. Huber, G.; Gorb, S.N.; Hosoda, N.; Spolenak, R.; Arzt, E. Influence of surface roughness on gecko adhesion. *Acta Biomater.* **2007**, *3*, 607–610. [CrossRef] [PubMed]
82. Langer, M.G.; Ruppersberg, J.P.; Gorb, S. Adhesion forces measured at the level of a terminal plate of the fly's seta. *Proc. R. Soc. Lond. Ser. B Biol. Sci.* **2004**, *271*, 2209–2215. [CrossRef]
83. Betts, R.; Duckworth, T.; Austin, I.; Crocker, S.; Moore, S. Critical light reflection at a plastic/glass interface and its application to foot pressure measurements. *J. Med. Eng. Technol.* **1980**, *4*, 136–142. [CrossRef] [PubMed]
84. O'Donnell, M.K.; Deban, S.M. The effects of roughness and wetness on salamander cling performance. *Integr. Comp. Biol.* **2020**, *60*, 840–851. [CrossRef] [PubMed]
85. Dixon, A.; Croghan, P.; Gowing, R. The mechanism by which aphids adhere to smooth surfaces. *J. Exp. Biol.* **1990**, *152*, 243–253. [CrossRef]
86. Federle, W.; Baumgartner, W.; Holldobler, B. Biomechanics of ant adhesive pads: Frictional forces are rate-and temperature-dependent. *J. Exp. Biol.* **2004**, *207*, 67–74. [CrossRef] [PubMed]
87. Park, S.J.; Goodman, M.B.; Pruitt, B.L. Analysis of nematode mechanics by piezoresistive displacement clamp. *Proc. Natl. Acad. Sci. USA* **2007**, *104*, 17376–17381. [CrossRef] [PubMed]
88. Kappl, M.; Kaveh, F.; Barnes, W.J.P. Nanoscale friction and adhesion of tree frog toe pads. *Bioinspir. Biomim.* **2016**, *11*, 035003. [CrossRef]
89. Geim, A.K.; Dubonos, S.; Grigorieva, I.; Novoselov, K.; Zhukov, A.; Shapoval, S.Y. Microfabricated adhesive mimicking gecko foot-hair. *Nat. Mater.* **2003**, *2*, 461–463. [CrossRef] [PubMed]
90. Walton, O.R. Review of adhesion fundamentals for micron-scale particles. *KONA Powder Part. J.* **2008**, *26*, 129–141. [CrossRef]
91. Gingell, D.; Todd, I. Interference reflection microscopy. A quantitative theory for image interpretation and its application to cell-substratum separation measurement. *Biophys. J.* **1979**, *26*, 507–526. [CrossRef]
92. Zhang, L.; Chen, H.; Guo, Y.; Wang, Y.; Jiang, Y.; Zhang, D.; Ma, L.; Luo, J.; Jiang, L. Micro–nano hierarchical structure enhanced strong wet friction surface inspired by tree frogs. *Adv. Sci.* **2020**, *7*, 2001125. [CrossRef]
93. Langowski, J.K.; Singla, S.; Nyarko, A.; Schipper, H.; van den Berg, F.T.; Kaur, S.; Astley, H.C.; Gussekloo, S.W.; Dhinojwala, A.; van Leeuwen, J.L. Comparative and functional analysis of the digital mucus glands and secretions of tree frogs. *Front. Zool.* **2019**, *16*, 1–17. [CrossRef]
94. Beattie, A.J.; Turnbull, C.; Hough, T.; Jobson, S.; Knox, R.B. The vulnerability of pollen and fungal spores to ant secretions: evidence and some evolutionary implications. *Am. J. Bot.* **1985**, *72*, 606–614. [CrossRef]
95. Clarke, B.T. The natural history of amphibian skin secretions, their normal functioning and potential medical applications. *Biol. Rev.* **1997**, *72*, 365–379. [CrossRef] [PubMed]

96. Peisker, H.; Heepe, L.; Kovalev, A.E.; Gorb, S.N. Comparative study of the fluid viscosity in tarsal hairy attachment systems of flies and beetles. *J. R. Soc. Interface* **2014**, *11*, 20140752. [CrossRef] [PubMed]
97. Vötsch, W.; Nicholson, G.; Müller, R.; Stierhof, Y.D.; Gorb, S.; Schwarz, U. Chemical composition of the attachment pad secretion of the locust *Locusta migratoria*. *Insect Biochem. Mol. Biol.* **2002**, *32*, 1605–1613. [CrossRef]
98. Geiselhardt, S.F.; Geiselhardt, S.; Peschke, K. Comparison of tarsal and cuticular chemistry in the leaf beetle *Gastrophysa viridula* (Coleoptera: Chrysomelidae) and an evaluation of solid-phase microextraction and solvent extraction techniques. *Chemoecology* **2009**, *19*, 185–193. [CrossRef]
99. Drotlef, D.M.; Stepien, L.; Kappl, M.; Barnes, W.J.P.; Butt, H.J.; del Campo, A. Insights into the adhesive mechanisms of tree frogs using artificial mimics. *Adv. Funct. Mater.* **2013**, *23*, 1137–1146. [CrossRef]
100. Kaimaki, D.M.; Andrew, C.N.; Attipoe, A.E.; Labonte, D. The physical properties of the stick insect pad secretion are independent of body size. *J. R. Soc. Interface* **2022**, *19*, 20220212. [CrossRef]
101. Gernay, S.; Federle, W.; Lambert, P.; Gilet, T. Elasto-capillarity in insect fibrillar adhesion. *J. R. Soc. Interface* **2016**, *13*, 20160371 [CrossRef]
102. Gilet, T.; Heepe, L.; Lambert, P.; Compère, P.; Gorb, S.N. Liquid secretion and setal compliance: The beetle's winning combination for a robust and reversible adhesion. *Curr. Opin. Insect Sci.* **2018**, *30*, 19–25. [CrossRef] [PubMed]
103. Peisker, H.; Michels, J.; Gorb, S.N. Evidence for a material gradient in the adhesive tarsal setae of the ladybird beetle *Coccinella septempunctata*. *Nat. Commun.* **2013**, *4*, 1–7. [CrossRef]
104. Scholz, I.; Baumgartner, W.; Federle, W. Micromechanics of smooth adhesive organs in stick insects: Pads are mechanically anisotropic and softer towards the adhesive surface. *J. Comp. Physiol. A* **2008**, *194*, 373–384. [CrossRef] [PubMed]
105. Scholz, I.; Barnes, W.J.P.; Smith, J.M.; Baumgartner, W. Ultrastructure and physical properties of an adhesive surface, the toe pad epithelium of the tree frog, *Litoria caerulea* White. *J. Exp. Biol.* **2009**, *212*, 155–162. [CrossRef] [PubMed]
106. Barnes, W.J.; Goodwyn, P.J.P.; Nokhbatolfoghahai, M.; Gorb, S.N.; et al. Elastic modulus of tree frog adhesive toe pads. *J. Comp. Physiol. A* **2011**, *197*, 969–978. [CrossRef] [PubMed]
107. Johnson, K.L.; Kendall, K.; Roberts, a. Surface energy and the contact of elastic solids. *Proc. R. Soc. Lond. A Math. Phys. Sci.* **1971**, *324*, 301–313.
108. Gilman, C.A.; Imburgia, M.J.; Bartlett, M.D.; King, D.R.; Crosby, A.J.; Irschick, D.J. Geckos as springs: Mechanics explain across-species scaling of adhesion. *PLoS ONE* **2015**, *10*, e0134604. [CrossRef]
109. Russell, A.P. A contribution to the functional analysis of the foot of the Tokay, *Gekko gecko* (Reptilia: Gekkonidae). *J. Zool.* **1975** *176*, 437–476. [CrossRef]
110. Ye, Z.; Lum, G.Z.; Song, S.; Rich, S.; Sitti, M. Phase change of gallium enables highly reversible and switchable adhesion. *Adv. Mater.* **2016**, *28*, 5088–5092. [CrossRef]
111. Niederegger, S.; Gorb, S. Tarsal movements in flies during leg attachment and detachment on a smooth substrate. *J. Insect Physiol.* **2003**, *49*, 611–620. [CrossRef]
112. Gernay, S.M.; Labousse, S.; Lambert, P.; Compère, P.; Gilet, T. Multi-scale tarsal adhesion kinematics of freely-walking dock beetles. *J. R. Soc. Interface* **2017**, *14*, 20170493. [CrossRef] [PubMed]
113. Bartsch, M.; Federle, W.; Full, R.; Kenny, T. Small insect measurements using a custom MEMS force sensor. In Proceedings of the TRANSDUCERS'03. 12th International Conference on Solid-State Sensors, Actuators and Microsystems. Digest of Technical Papers (Cat. No.03TH8664), Boston, MA, USA, 8–12 June 2003; Volume 2, pp. 1039–1042.
114. Bartsch, M.S.; Federle, W.; Full, R.J.; Kenny, T.W. A multiaxis force sensor for the study of insect biomechanics. *J. Microelectromech Syst.* **2007**, *16*, 709–718. [CrossRef]
115. Lin, H.T.; Trimmer, B.A. A new bi-axial cantilever beam design for biomechanics force measurements. *J. Biomech.* **2012**, *45*, 2310–2314. [CrossRef] [PubMed]

biomimetics

Article

Underwater Attachment of the Water-Lily Leaf Beetle *Galerucella nymphaeae* (Coleoptera, Chrysomelidae)

Constanze Grohmann, Anna-Lisa Cohrs and Stanislav N. Gorb *

Department of Functional Morphology and Biomechanics, Zoological Institute, Kiel University, Am Botanischen Garten 9, 24118 Kiel, Germany; cgrohmann@zoologie.uni-kiel.de (C.G.); anna.cohrs92@hotmail.com (A.-L.C.)
* Correspondence: sgorb@zoologie.uni-kiel.de

Abstract: While the reversible attachment of artificial structures underwater has moved into the focus of many recent publications, the ability of organisms to walk on and attach to surfaces underwater remains almost unstudied. Here, we describe the behaviour of the water-lily leaf beetle *Galerucella nymphaeae* when it adheres to surfaces underwater and compare its attachment properties on hydrophilic and hydrophobic surfaces underwater and in the air. The beetles remained attached to horizontal leaves underwater for a few minutes and then detached. When the leaf was inclined, the beetles started to move upward immediately. There was no difference in the size of the tarsal air bubble visible beneath the beetles' tarsi underwater, between a hydrophilic (54° contact angle of water) and a hydrophobic (99°) surface. The beetles gained the highest traction forces on a hydrophilic surface in the air, the lowest on a hydrophobic surface in air, and intermediate traction on both surfaces underwater. The forces measured on both surfaces underwater did not differ significantly. We discuss factors responsible for the observed effects and conclude that capillary forces on the tarsal air bubble might play a major role in the adhesion to the studied surfaces.

Keywords: adhesion; underwater; contact angle; insect; biomechanics; locomotion

Citation: Grohmann, C.; Cohrs, A.-L.; Gorb, S.N. Underwater Attachment of the Water-Lily Leaf Beetle *Galerucella nymphaeae* (Coleoptera, Chrysomelidae). *Biomimetics* **2022**, *7*, 26. https://doi.org/10.3390/biomimetics7010026

Academic Editor: Ille C. Gebeshuber

Received: 7 January 2022
Accepted: 11 February 2022
Published: 14 February 2022

1. Introduction

In adhesion science, the topic of underwater attachment has moved into the focus of many publications in recent years. Permanent underwater adhesion using glues is well known for numerous aquatic animals [1,2] and this knowledge was even transferred to the synthesis of biomimetic glues [3–6]. Also, reversible artificial adhesive systems based on surface nano- and microstructures for underwater application were recently produced [7–9]. However, reversible adhesion by animals underwater was the concern of only a few studies. Hosoda and Gorb (2012) studied beetles inhabiting a terrestrial environment [10]. Other researchers focused on the underwater adhesive performance of the hair-like structures of, e.g., diving beetles or mussels, without considering the attachment of the organism as a whole [11,12]. In summary, many questions on the attachment of organisms remain unanswered. For example, the adhesion to wet substrates as well as the roles of capillary adhesion and nanobubbles still need to be clarified [13].

Smooth and hairy attachment devices occur in different insect taxa, but nubby structures can also be found [14,15]. Adult beetles usually possess a hairy attachment system [16] and the attachment is promoted by the secretion of a mixture of hydrocarbons, fatty acids, and alcohols onto the contact area [17,18]. Different physico-chemical principles rule underwater adhesion compared with attachment in the air. Hosoda and Gorb (2012) discovered that, due to the hairy microstructure of the beetles' tarsi, an air bubble is stably kept underwater beneath the feet of the green dock beetle *Gastrophysa viridula* [10]. They postulated that these bubbles de-wet substrates and allow direct contact between the pad fluid and the dried solid substrate surface. Additionally, the air bubble itself produces capillary adhesion at its perimeter. However, *G. viridula* is a terrestrial beetle that lives and feeds on the common sorrel *Rumex acetosa* and can only submerge its feet underwater during heavy rain In the present study, we aimed at studying the reversible underwater attachment system in a species that depends more strongly on a good attachment performance underwater We have therefore chosen the limnic water-lily leaf beetle *Galerucella nymphaeae* (Linnaeus 1758), which lives and forages on floating leaves of the waterlilies *Nuphar* sp. and *Nymphaea* sp. [19]. This beetle species is regularly exposed to wet and submerged surfaces, i.e., the leaves of its host plants are submerged or flooded, for example, by water birds trampling them down. Especially after rain, patches filled with water often remain on the leaves surfaces for quite a long time.

In the present paper, we first asked how *G. nymphaeae* behaves when it is submerged while attaching to a surface. Second, we tested whether the beetles' tarsal air bubbles differ in size on hydrophilic (54° contact angle) and hydrophobic (99° contact angle) surfaces Third, we assessed the traction forces of *G. nymphaeae* on these two types of surfaces. Finally we measured the buoyancy forces of *G. nymphaeae* and *G. viridula* and included these data in our discussion of the factors that might be responsible for the observed underwater adhesion mechanism.

2. Materials and Methods

2.1. Tarsal Morphology

Galerucella nymphaeae beetles were collected from floating leaves of *Nuphar lutea* at a small artificial pond located in the botanical garden of Kiel University, Germany in July and August 2016. *Gastrophysa viridula* beetles were collected from *Rumex obtusifolius* plants in the surrounding areas of Kiel during the same months. For scanning electron microscopy (SEM), tarsi of *G. nymphaeae* were air-dried, attached to the SEM stubs, and sputtered with a ~20 nm thick layer of gold-palladium. Images were taken with a SEM Hitachi S4800 (Hitachi High-Technologies, Tokyo, Japan) at an accelerating voltage of 3–5 kV.

2.2. Behaviour Underwater

Adult individuals of *G. nymphaeae* were observed on floating leaves of *N. lutea* in the botanical garden of Kiel University. We documented some general behavioural patterns of beetles in their natural environment.

2.2.1. Horizontal Leaf

To assess the behaviour of *G. nymphaeae* when it is pressed down underwater, floating leaves of *N. lutea* with a beetle attaching to its upper side were pressed 10 cm deep under water. The leaves were held horizontally the entire time. The behaviour of the beetles was observed and the period of time until they detached from the leaves was measured. We took the data of seven beetles that were already running when the lily pad was pushed down underwater. A further seven individuals were sitting still when pushed underwater Each beetle was measured three to four times; the recovery time between two runs was one minute. In total, 25 measurements were recorded with running beetles and the same amount with beetles standing still.

2.2.2. Sloped Leaf

A similar kind of experiment was performed with leaves that were pressed down at an angle of 60° to the water's surface. We observed and recorded the behaviour of nine beetles that were running and a further nine individuals that were standing still. Each beetle was tested once.

For the following experiments in the lab, adult individuals of *G. nymphaeae* were collected in the botanical garden of Kiel University. We kept them underneath a piece of *N. lutea* leaf in a Petri dish containing a moist tissue. The tissue and the leaf were renewed after a couple of days. When no experiments were being performed, the Petri dish was stored in a fridge at approximately 10 °C.

2.3. Experiment 1: The Formation of the Subtarsal Air Bubbles on Different Surfaces

To determine the size of the air bubble trapped beneath the tarsi of *G. nymphaeae* underwater, we allowed the beetles to attach to a glass slide that was then pushed into a Petri dish filled with tap water. We immediately took images from underneath the beetles' feet with a stereo microscope (Leica MZ 205A, Leica GmbH, Wetzlar, Germany) that was mounted upside down. The Petri dish was fixed above the objective lens of the inverted microscope. Using coaxial illumination, the contact area between the tarsi or the air bubble and the glass slide could be seen and photographed through an ocular camera (BMS Eyepiece & C-mount camera, 5 Megapixel, Breukhoven, The Netherlands) (Figure 3). We compared the area of the air bubbles in beetles standing on hydrophilic surfaces with those on hydrophobic surfaces. To make glass slides hydrophilic or hydrophobic, we silanised them with (3-Aminopropyl)triethoxysilane (Carl Roth, Karlsruhe, Germany) and Dichlorodimethylsilane (Merck Schuchardt OHG, Hohenbrunn, Germany) and gained contact angles of water on these surfaces of 54.22° $\pm$ 2.30° and 99.38° $\pm$ 1.87°, respectively (n = 20 each; measured with a contact angle measurement device OCA 200, Dataphysics, Filderstadt, Germany). To measure the area of the bubble, we used Photoshop CS 5 (Adobe Systems GmbH, Munich, Germany).

In some cases, the bubbles trapped under the beetles' feet were connected to a bubble that surrounded the entire body. This occurred in equal numbers (n = 8) on both surfaces and often when the beetle's tarsus was close to its body. We excluded these cases from our dataset.

We took images of up to three different, randomly chosen legs of 21 *G. nymphaeae* individuals on both surfaces, summing up to a sample size of 34 for the 54° surface and 28 for the 99° surface. We then measured the time until the beetles detached from the different glass surfaces. Finally, we calculated how the size of the air bubble changed from the beginning of the experiment compared with the moment directly before beetle detachment. This last experiment was performed for 19 and 17 specimens on the CA = 54° and CA = 99° surfaces, respectively.

2.4. Experiment 2: Traction Force Measurements of the Beetles Walking on Different Surfaces

To test the beetles' attachment abilities on different surfaces underwater and on land, we used a load cell force transducer (BIOPAC systems, Goleta, CA, USA) with a 25 g sensor. The sensor was clamped vertically above the test surface. One end of a human hair was fixed to the sensors' end and the other end was glued to the elytra of *G. nymphaeae* with the aid of a droplet of molten beeswax. When a beetle walked, it pulled the sensor via the hair. The resulting force was recorded during a 60 s long pulling period and the peak force was determined and used for later analyses.

We tested pulling forces that the beetles generated on hydrophilic and hydrophobic glass slides in a random order. Contact angles of water on these surfaces averaged 54.22° $\pm$ 2.30° (mean $\pm$ sd, n = 21) and 99.38° $\pm$ 1.87°, respectively (measured with the contact angle measurement device OCA 200, Dataphysics, Filderstadt, Germany). All but two beetles (n = 33) were tested twice on each surface, in air (n = 33) and with the surface and the beetles submerged in the tap water (n = 31).

2.5. Experiment 3: Measurement of the Buoyancy Force

In our discussion, we compare the traction forces of *G. nymphaeae* with previous findings for *G. viridula* [10]. In order to assess whether differences in the traction forces between these two species are caused by their different buoyancy forces, we measured the forces that are needed to push individual beetles of *G. nymphaeae* and *G. viridula* underneath the water's surface. We therefore used a wire loop and measured the forces with a load cell force transducer with a 10 g sensor (BIOPAC systems, Goleta, CA, USA). We used distilled water that we degassed beforehand in an ultrasonic bath (Bandelin Sonorex RK 52, BANDELIN electronic, Berlin, Germany) at a frequency of 35 kHz. Due to the agility of the beetles, we froze them at −80 °C for at least 10 min and thawed them at room temperature immediately before the measurements. The specimens were carefully laid on the water surface and pushed down in a controlled manner using a motorised micromanipulator that moved the sensor and the wire loop at the speed of 600 μm per second. We measured the maximal forces that occurred (i) while the beetles were pushed down and deformed the water surface but still had contact to the air and (ii) when the beetles' bodies were completely underwater. The same individuals were pushed down in a 0.1% solution of the surfactant Triton X (Sigma-Aldrich Chemie GmbH, Steinheim, Germany) as a control. The Triton X solution was also degassed in the ultrasonic bath with the same frequency prior to the experiment. We measured forces of 20 individual beetles of each species in distilled water and directly afterwards in the aqueous solution of Triton X.

3. Results

3.1. Tarsal Morphology

The ventral sides of the first and second tarsomeres of *G. nymphaeae* are covered with long pointed setae, while the third tarsomere bears spatula-shaped setae (Figure 1). We did not detect any differences between the attachment structures of female and male beetles and therefore do not differentiate between sexes in this study (Figure S5, Supplementary Materials). The elytra of the water-lily leaf beetle are covered with long hairs (Figure 1d,e)

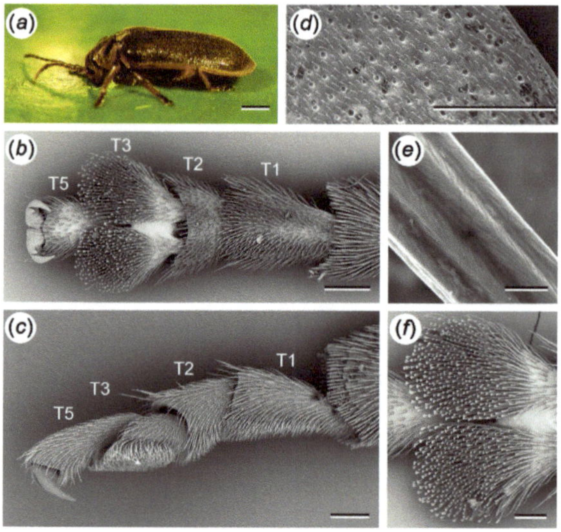

Figure 1. Appearance and tarsal attachment organs of adult *G. nymphaeae*. (**a**) *G. nymphaeae* on the water-lily surface. (**b**) Tarsus, ventral. (**c**) Tarsus, lateral. (**d**) Elytra. (**e**) Hair coverage on the elytra. (**f**) Third tarsal segment, ventral. (**d**–**f**) SEM images. T1–T5: Tarsal segments; T4 is reduced. Scale bars: (**a**) = 1 mm; (**b**,**c**) = 100 μm; (**d**) = 500 μm; (**e**) = 1 μm, (**f**) = 50 μm. (**a**) Courtesy of Andreas Blankenstein. (**b**,**c**,**f**) Reprinted with permission from ref. [15]. Copyright The Authors of the ref. [15] under the license of Company of Biologists Publication Agreement.

3.2. Behaviour Underwater

We observed that *G. nymphaeae* avoids walking through wet patches on the leaves. When a beetle falls off into the water, it swims actively to the nearest leaf or to the rim of the pond. When water was sprayed over the beetles to imitate rain, they stopped moving and withstood the "shower". This applies to formerly running and still-standing beetles.

3.2.1. Horizontal Leaf

When pressing *G. nymphaeae* underneath the water's surface, a visible bubble encased the elytra of nearly each inactive beetle; active ones had this bubble in roughly 60% of the observations. Underwater, all individuals of G. *nymphaeae* opened their elytra after a while, and a bubble under the abdomen became visible. Independently of the time spent underwater, some of the beetles waggled their legs before detaching from the leaf.

Those beetles that were already running on the leaf in air continued to run when it was pressed into the water, and they detached after an average of 2:57 (1:18–3:58) min. (median value, interquartile range; Figure 2). Only a single beetle detached after more than 5 min underwater, and it did so in three out of four repetitions.

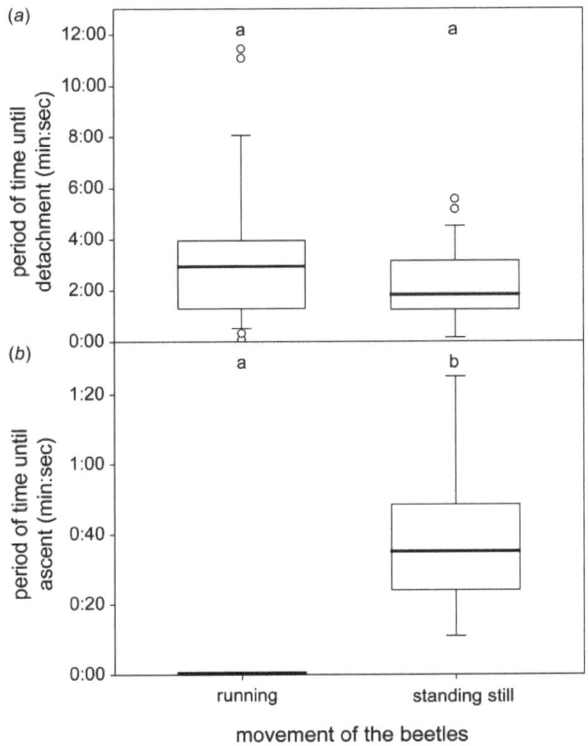

Figure 2. Period of time until G. *nymphaeae* detaches from the horizontal leaf (upper plot) or ascends to the top of the inclined leaf (lower plot) underwater. The plots show the medians (lines within the boxes), 25th and 75th percentiles (ends of boxes), 10th and 90th percentiles (error bars), and outlying values (circles). Different letters above two boxes indicate significant differences between two groups. Upper plot: seven running and a further seven beetles standing still were tested 3–4 times (n = 25 for each test). Lower plot: nine running and a further nine beetles standing still were tested once (n = 9 for each test). Different letters (**a**,**b**) within each subplot indicate presence of statistically significant difference between samples.

Beetles that stood still in the air remained roughly half as long underwater until they detached (1:49, 1:15–3:10 min), but we recognized a great variation in the time periods even within a single individual. The differences between running beetles and those standing still were not significant (p = 0.28, Mann–Whitney Rank Sum Test).

3.2.2. Sloped Leaf

All formerly running beetles started to move upward within 5 s on a sloped water-lily pad underwater (Figure 2). Those that were standing still before started to ascend after 0:35 (0:24–0:49) min.

3.3. The Formation of the Subtarsal Air Bubbles on Different Surfaces

In general, we discovered a great variety of different shapes and sizes of tarsal air bubbles, both between individual beetles and within the same individual (Figure 3). The size of the bubble did not differ significantly on either of the two glass plates with 54° and 99° contact angles, respectively (Mann–Whitney Rank Sum Test, $p = 0.101$; Figure 4). Similarly, the lapse of time the beetles spent underwater until they boosted themselves to the water surfaces and the difference in the bubble size at the beginning and immediately before the beetles detach did not depend on the surface (Mann–Whitney Rank Sum Test, $p = 0.969$ and $p = 0.547$, respectively; Figure 4).

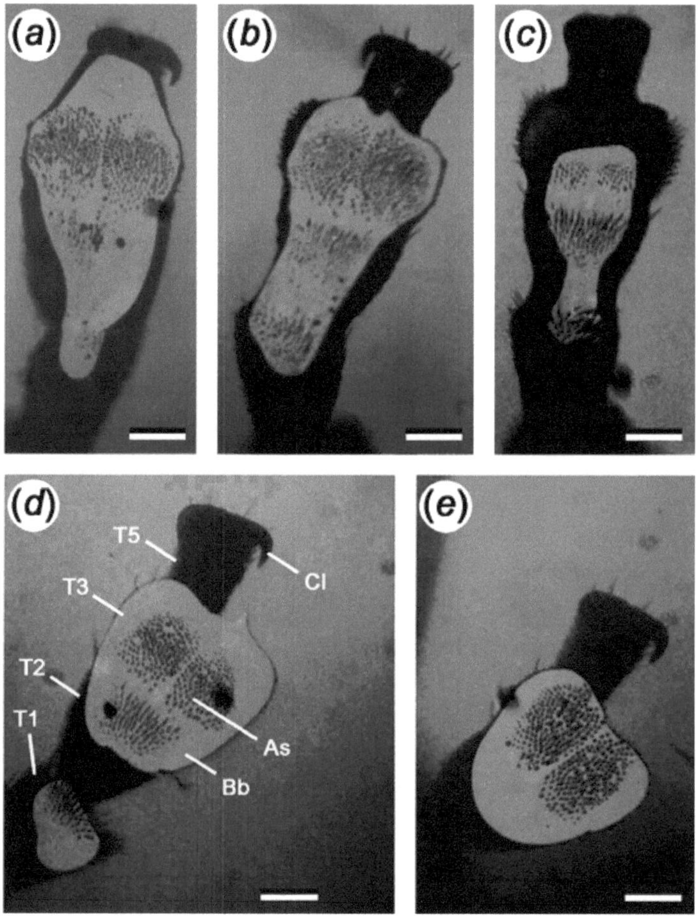

Figure 3. Images of the tarsal air bubble of *G. nymphaeae* beetles standing on glass slides underwater. Due to coaxial illumination, the air appears white and the contact area of attachment hairs on the glass slide appears black. As: Attachment setae. Bb: Subtarsal air bubble. Cl: Claw. T1–T5: Tarsal segments. Scale bars: (**a–e**): 0.1 mm.

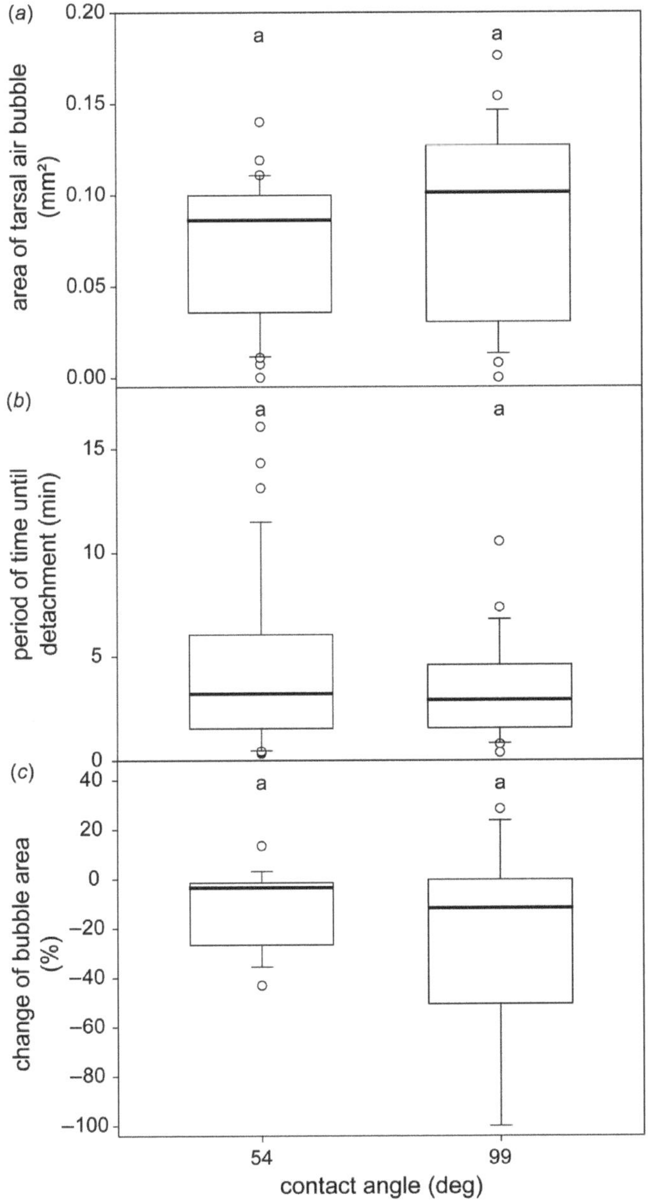

Figure 4. The subtarsal air bubble on different surfaces. Area of the tarsal air bubble taken immediately after *G. nymphaeae* was submerged on hydrophilic and hydrophobic glass slides (**a**), time until beetle detachment (**b**), and difference of the bubble area between the beginning of the experiment and just before the beetle detached (**c**). The plots show the medians (lines within the boxes), 25th and 75th percentiles (ends of boxes), 10th and 90th percentiles (error bars) and outlying values (circles). The same letter above the two boxes in each experiment indicates no significant differences between the two groups. Sample size: 1–3 different randomly chosen legs of 21 beetles; upper and middle plot: $n = 34$ and $n = 28$, lower plot: $n = 19$ and $n = 17$ for the 54° and 99° surfaces, respectively. Same letters (a, a) within each subplot indicate absence of statistically significant difference between samples.

3.4. Traction Force Measurements of the Beetles Walking on Different Surfaces

In air, the maximal traction force the beetles gained was almost four times higher on glass slides with a 54° contact angle (3.69 ± 1.92 mN, mean ± sd) compared to slides with a 99° contact angle (1.00 ± 0.81 mN) (Figure 5; One Way Repeated Measures ANOVA with Holm–Sidak pairwise comparisons, $p < 0.001$). In contrast, no significant differences in traction forces were found between the two surfaces when the beetles were underwater (54° surface: 1.79 ± 1.27 mN vs. 99° surface: 1.80 ± 0.78 mN; $p = 0.975$).

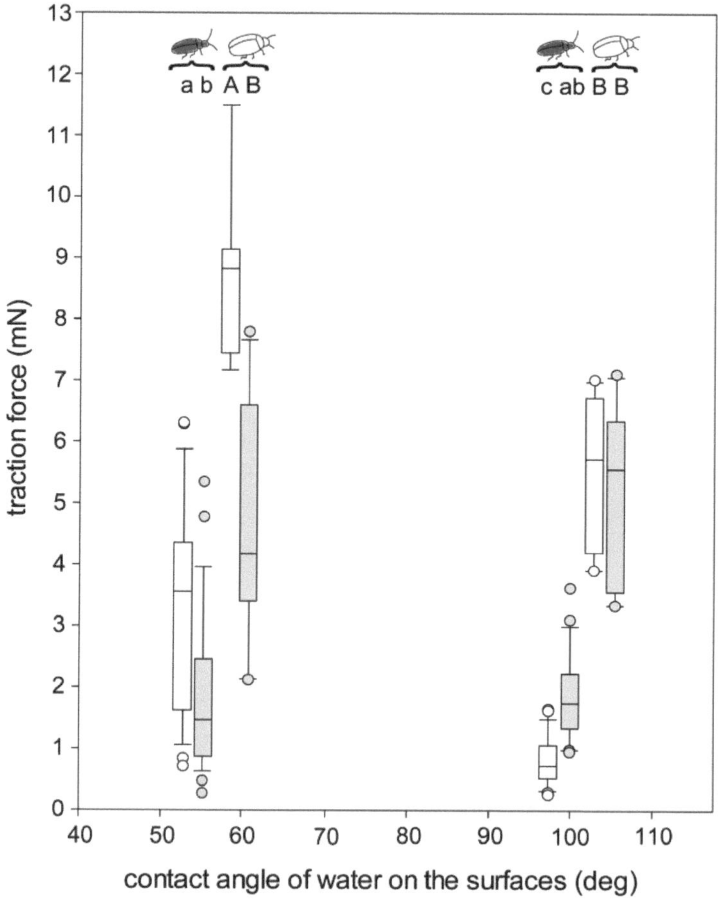

Figure 5. Traction forces of *G. nymphaeae* on different surfaces in air and underwater. For comparison data for *G. viridula* are added from [10]. The plots show the medians (lines within the boxes), 25th and 75th percentiles (ends of boxes), 10th and 90th percentiles (error bars), and outlying values. White boxes and circles: in air; grey boxes and circles: underwater. Different letters above two boxes (separately within small letters and within large letters) indicate significant differences between two groups (multiple pairwise comparisons). Pairs of boxes at 54° and 99° (black beetle icon): *Galerucella nymphaeae*. Pairs of boxes at 59° and 104° (white beetle icon): *Gastrophysa viridula*. Each beetle (n = 33 for *G. nymphaeae*, n = 29 for *G. viridula*) was tested once in air and once underwater; two of these *G. nymphaeae* beetles were solely tested in air. Different letters (a, b or A, B) indicate statistically significant difference between samples.

3.5. Measurement of the Buoyancy Force

When the beetle was pushed down slightly, the tension film of the water surface was deformed and the beetle and/or the air bubble that encased it still had contact with the air. In this situation, the force to push *G. nymphaeae* down was 1.6 times greater than the force needed for *G. viridula* (0.78 ± 0.11 mN and 0.49 ± 0.10 mN, respectively; $p < 0.001$, *t*-test). When there was no longer contact with the air and the beetle was surrounded by water, the force needed to push the beetle down was less than a tenth of the previous value for both species (0.07 ± 0.01 and 0.03 ± 0.01 mN, respectively; $p \leq 0.001$, Mann–Whitney Rank Sum Test).

When being pushed in the aqueous solution of Triton X, the beetles, especially *G. viridula*, sank almost immediately to the ground of the plastic jar. Within each species, the force needed to press the beetle down did not depend on the beetle's weight (*G. nymphaeae*: bubble still having contact to the air: $R^2 = 0.05$, linear regression, beetle fully encased by water: $R^2 < 0.01$; *G. viridula*: $R^2 = 0.02$ and $R^2 = 0.06$, respectively).

4. Discussion

4.1. Tarsal Morphology

We found similarly shaped attachment hairs on the tarsi of both sexes of *G. nymphaeae*. While the males of many chrysomelid species have discoidal tips of attachment setae, to securely attach to the females' smooth elytra during copulation [20], we assume the lack of such specialised hairs is due to the presence of closely spaced hairs on the female and male elytra of *G. nymphaeae* beetles [15].

4.2. Behaviour Underwater

We observed no difference whether the beetles were running or standing still; from both initial situations, they continued to run or stand on the leaf for some minutes after it was submerged horizontally into water. This proves that, although the beetles in general avoid flooded patches on leaves, they can attach well and even run underwater. The individuals that were moving tended to walk to the rim of the leaf. This behaviour enables the beetles to grip and climb upward at the leaf's edge by using the action of their contralateral legs. If a leaf was pushed at an incline underwater, beetles that were running beforehand immediately started to walk and ascend the leaf. Those that were standing started to walk and ascend the leaf in most cases within 1 min. This behaviour is a strategy for escaping from the submerged situation. It remains, however, unclear whether this behaviour is common for chrysomelid beetles or whether it has evolved in *G. nymphaeae*, owing to its limnic habit.

4.3. Subtarsal Air Bubbles on Different Surfaces

In contrast to our initial hypothesis, the area of the tarsal air bubble did not differ between hydrophilic and hydrophobic surfaces. Possible differences might be masked by the fact that on each surface, the size of the bubble varied greatly (Figures 3 and 4), and using our technique, we were able to detect and analyse the bubble only under a single tarsus at a time. According to our hypothesis, a larger tarsal air bubble on a hydrophobic surface should prolong the period of time until the beetle detaches from the surface. Similarly, the bubble area should decrease to a larger extent on the hydrophobic surfaces. However, in both cases, we did not detect any differences between the two surfaces (Figure 4). There could be two reasons for this. On the one hand, as already mentioned, possible differences might be covered by the large variability of the bubble shape and volume on a single tarsus. In turn, the bubble size of a single tarsus might be balanced by the bubble sizes below the other five tarsi of the beetle. On the other hand, the period of time until the beetle detaches might be ruled by other factors, such as the amount of oxygen stored in the air bubble encasing the beetle elytra.

4.4. Traction Forces on Different Surfaces and Buoyancy

4.4.1. Traction Forces on Hydrophilic and Hydrophobic Surfaces in Air

In air, *G. nymphaeae* revealed stronger traction forces on the hydrophilic compared with the hydrophobic surfaces (Figure 5). Such differences might be attributed to the chemical composition of the attachment fluid that seems to form a stronger contact between an attachment hair and the surface on hydrophilic compared with hydrophobic surfaces [15]. It was previously shown that a film of water is present on hydrophilic surfaces in air. Capillary forces between this water layer and a probe tip lead to enhanced friction forces, depending on the film's thickness, if compared with a solid–solid contact [21,22]. Additionally, to the fluid's chemical and physical properties, interactions between the film of water on the surface and the attachment devices could lead to the observed stronger traction forces on the hydrophilic surface compared with the hydrophobic surface in air. We do not expect such a film of water on the hydrophobic surfaces in air, as only at a humidity of >90% is a monolayer of water formed on such surfaces [23].

4.4.2. Traction Forces on the Hydrophilic Surface in Air and Underwater

In contrast, on the hydrophilic surface, *G. nymphaeae* revealed weaker traction forces underwater when compared with those measured in air. As long as hairs attach to a dry de-wetted surface underwater, the role of the fluid should be exactly the same as in air. The fluid resembles the composition of the beetles' cuticular lipids [17,18] and it is unlikely that the fluid can be adapted to the substrate by the beetle in such a short period of time. Hence, the relatively low traction forces underwater compared with those in air can either be explained by (i) a reduced number of attachment hairs that form contact to the surface due to the fact that only hairs within the tarsal air bubble can form a contact, (ii) buoyancy forces, reducing beetle load force to the substrate due to air reservoirs in and around the beetles' bodies, especially on the hairy elytra of *G. nymphaeae*, or (iii) capillary forces at the tarsal air bubble that were found to be negative on hydrophilic surfaces underwater increasing with an increasing hydrophobicity of the surface [24].

4.4.3. Traction Forces on the Hydrophobic Surface in Air and Underwater

On the hydrophobic surface, however, a contrary pattern occurs for *G. nymphaeae* with stronger traction forces underwater when compared with those in air. The underlying factors for this effect might be as follows. (i) Capillary forces between the tarsi and the water film on the surface should not play a role here, as we expect only a monolayer of water on the hydrophobic surface in air [23]. (ii) A contact reduction between setae and the surface should lead to a decrease in the traction force. (iii) The same result is to be expected due to the presence of buoyancy forces. (iv) The remaining factor that might be responsible for the higher traction forces underwater on the hydrophobic surface, in comparison to the experiment in air, is increased positive capillary forces at the tarsal air bubble.

We performed our experiments with functionalised glass surfaces with water contact angles of 54° and 99°. We can assume that on surfaces with other contact angles, different combinations of factors might play a role in beetle adhesion. For example, on surfaces with extremely low contact angles, the presence of a thick film of water might impede the contact formation between the tarsi and the surface [25]. It was previously shown that traction forces of the chrysomelid beetle *G. viridula* were strongly reduced underwater on a surface with a contact angle of 43° compared with 59°, and structured polymers did not adhere at all underwater to surfaces with a roughly 20° contact angle [10]. On surfaces without such extremely low contact angles, the surface might be totally de-wetted below the air bubbles [10], enabling contact formation.

There are at least two further studies in which the attachment to surfaces with contact angles similar to ours were assessed. For example, our findings for *G. nymphaeae* are almost similar to those previously obtained for *G. viridula* (Figure 5; [10]). However, while we measured higher traction forces underwater, compared with those in air on the hydrophobic surface, no differences were detected for *G. viridula* on this surface. The traction forces of G.

viridula were in general higher than those of *G. nymphaeae*, probably due to the higher weight of *G. viridula* in air and its lower buoyancy forces underwater. Shear forces of the tokay gecko (*Gecko gecko*) showed a surprising similarity to our results on a 50° glass surface and a 97° polytetrafluoroethylene surface, although, on a 94° surface (octadecyltrichlorosilane self-assembled monolayer formed on the surface of glass), totally different shear forces of geckos were measured [26].

5. Conclusions

From our data, we conclude that for the analysed contact angles of 54° and 99°, capillary forces at the tarsal air bubble seem to play a role in the unexpected higher traction forces of *G. nymphaeae* on the 99° surface underwater, when compared to those in air. We must keep in mind that we did not consider many further factors that might play an additional role. These are, for example, the normal force [27], the presence of surface asperities [26,28], the presence and thickness of a water film on the surface [21], and the degrees of hydrophilicity and hydrophobicity. Strong attachment and friction performance underwater, similar to those found in animal adhesive pads, is of importance for biologically inspired solutions in biomedical engineering and wearable flexible electronics, etc. [29–35].

Supplementary Materials: The following are available online at https://www.mdpi.com/article/10.3390/biomimetics7010026/s1, Table S1: Period of time until *G. nymphaeae* detaches from the horizontal leaf underwater or ascends to the top of the inclined leaf underwater, Table S2: The subtarsal air bubble on different surfaces, Table S3: Traction force of *G. nymphaeae* in air and underwater, Table S4: Buoyancy forces of *G. nymphaeae* and *G. viridula*, Figure S1: Female (mid leg) and male (hind leg) setae at high magnification, SEM images.

Author Contributions: Conceptualization, S.N.G.; methodology, C.G., A.-L.C. and S.N.G.; validation, C.G. and A.-L.C.; investigation, A.-L.C. and C.G.; resources, S.N.G.; data curation, C.G., A.-L.C. and S.N.G.; writing—original draft preparation, C.G. and A.-L.C.; writing—review and editing, S.N.G.; visualization, C.G. and A.-L.C.; supervision, S.N.G.; project administration, S.N.G.; funding acquisition, S.N.G. All authors have read and agreed to the published version of the manuscript.

Funding: Funding was provided by the Deutsche Forschungsgemeinschaft (grant GO 995/34-1) to S.N.G. within the the framework of the DFG special priority program "Soft Material Robotic Systems".

Institutional Review Board Statement: Not applicable.

Informed Consent Statement: Not applicable.

Data Availability Statement: Datasets generated during the study are available in the Supplementary Materials.

Acknowledgments: We would like to thank the staff of the Botanical Garden Kiel for permission to collect beetles and leaves. E. Appel, S. Eisenkölbl, E.V. Gorb, L. Heepe, N. Hosoda, A. Kovalev, L. Oesert and J. Wolff helped with measurement techniques and fruitful discussions.

Conflicts of Interest: The authors declare no conflict of interest.

References

1. Danner, E.W.; Kan, Y.; Hammer, M.U.; Israelachvili, J.N.; Waite, J.H. Adhesion of mussel foot protein *Mefp-5* to mica: An underwater superglue. *Biochemistry* **2012**, *51*, 6511–6518. [CrossRef] [PubMed]
2. Nakano, M.; Kamino, K. Amyloid-like conformation and interaction for the self-assembly in barnacle underwater cement. *Biochemistry* **2015**, *54*, 826–835. [CrossRef] [PubMed]
3. Stewart, R.J. Protein-based underwater adhesives and the prospects for their biotechnological production. *Appl. Microbiol. Biotechnol.* **2011**, *98*, 27–33. [CrossRef] [PubMed]
4. White, J.D.; Wilker, J.J. Underwater bonding with charged polymer mimics of marine mussel adhesive proteins. *Macromolecules* **2011**, *44*, 5085–5088. [CrossRef]
5. Zhong, C.; Gurry, T.; Cheng, A.A.; Downey, J.; Deng, Z.; Stultz, C.M.; Lu, T.K. Strong underwater adhesives made by self-assembling multi-protein nanofibres. *Nat. Nanotech.* **2014**, *9*, 858–866. [CrossRef]
6. Nishida, J.; Higaki, Y.; Takahara, A. Synthesis and characterization of barnacle adhesive mimetic towards the underwater adhesion. *Chem. Lett.* **2015**, *44*, 1047–1049. [CrossRef]

7. Heepe, L.; Kovalev, A.E.; Gorb, S.N. Direct observation of microcavitation in underwater adhesion of mushroom-shaped adhesive microstructure. *Beilstein J. Nanotechnol.* **2014**, *5*, 903–909. [CrossRef]
8. Stark, A.Y.; McClung, B.; Niewiarowski, P.H.; Dhinojwala, A. Reduction of water surface tension significantly impacts gecko adhesion underwater. *Integr. Comp. Biol.* **2014**, *54*, 1026–1033. [CrossRef]
9. Tian, D.; Guo, Z.; Wang, Y.; Li, W.; Zhang, X.; Zhai, J.; Jiang, L. Phototunable underwater oil adhesion of micro/nanoscale hierarchical-structured ZnO mesh films with switchable contact mode. *Adv. Funct. Mater.* **2014**, *24*, 536–542. [CrossRef]
10. Hosoda, N.; Gorb, S.N. Underwater locomotion in a terrestrial beetle: Combination of surface de-wetting and capillary forces. *Proc. R. Soc. Lond. B* **2012**, *279*, 4236–4242. [CrossRef]
11. Chen, Y.; Shih, M.-C.; Wu, M.-H.; Yang, E.-C.; Chi, K.-J. Underwater attachment using hairs: The functioning of spatula and sucker setae from male diving beetles. *J. R. Soc. Interface* **2014**, *11*, 20140273. [CrossRef] [PubMed]
12. Andrade, G.R.; de Araújo, J.L.F.; Nakamura Filho, A.; Guañabens, A.C.P.; de Carvalho, M.D.; Cardoso, A.V. Functional surface of the golden musse's foot: Morphology, structures and the role of cilia on underwater adhesion. *Mater. Sci. Eng. C* **2015**, *54*, 32–42. [CrossRef] [PubMed]
13. Stark, A.Y.; Klittich, M.R.; Sitti, M.; Niewiarowski, P.H.; Dhinojwala, A. The effect of temperature and humidity on adhesion of a gecko-inspired adhesive: Implications for the natural system. *Sci. Rep.* **2016**, *6*, 30936. [CrossRef] [PubMed]
14. Beutel, R.G.; Gorb, S.N. Ultrastructure of attachment specializations of hexapods (Arthropoda): Evolutionary patterns inferred from a revised ordinal phylogeny. *J. Zool. Syst. Evol. Res.* **2001**, *39*, 177–207. [CrossRef]
15. Grohmann, C.; Blankenstein, A.; Koops, S.; Gorb, S.N. Attachment of *Galerucella nymphaeae* (Coleoptera, Chrysomelidae) to surfaces with different surface energy. *J. Exp. Biol.* **2014**, *217*, 4213–4220. [CrossRef]
16. Stork, N.E. A scanning electron microscope study of tarsal adhesive setae in the Coleoptera. *Zool. J. Linn. Soc.* **1980**, *68*, 173–306. [CrossRef]
17. Ishii, S. Adhesion of a leaf feeding ladybird *Epilachna vigintioctomaculata* (Coleoptera: Coccinellidae) on a vertically smooth surface. *Appl. Entomol. Zool.* **1987**, *22*, 222–228. [CrossRef]
18. Geiselhardt, S.F.; Federle, W.; Prüm, B.; Geiselhardt, S.; Lamm, S.; Peschke, K. Impact of chemical manipulation of tarsal liquids on attachment in the Colorado potato beetle, *Leptinotarsa decemlineata*. *J. Insect Physiol.* **2010**, *56*, 398–404. [CrossRef]
19. Smirnov, N.N. Nutrition of *Galerucella nymphaeae* L. (Chrysomelidae), mass consumer of water-lily. *Hydrobiologia* **1960**, *15*, 208–224. [CrossRef]
20. Stork, N.E. Experimental analysis of adhesion of *Chrysolina polita* (Chrysomelidae, Coleoptera) on a variety of surfaces. *J. Exp. Biol.* **1980**, *88*, 91–107. [CrossRef]
21. Scherge, M.; Li, X.; Schaefer, J.A. The effect of water on friction of MEMS. *Tribol. Lett.* **1999**, *6*, 215–220. [CrossRef]
22. Opitz, A.; Scherge, M.; Ahmed, S.I.-U.; Schaefer, J.A. A comparative investigation of thickness measurements of ultra-thin water films by scanning probe techniques. *J. Appl. Phys.* **2007**, *101*, 064310. [CrossRef]
23. Huber, G.; Mantz, H.; Spolenak, R.; Mecke, K.; Jacobs, K.; Gorb, S.N.; Arzt, E. Evidence for capillarity contributions to gecko adhesion from single spatula nanomechanical measurements. *Proc. Natl. Acad. Sci. USA* **2005**, *102*, 16293–16296. [CrossRef] [PubMed]
24. Peng, Z.L.; Wang, C.; Chen, S.H. Effects of surface wettability on gecko adhesion under water. *Colloids Surf. B* **2014**, *122*, 662–668. [CrossRef]
25. Bauer, U.; Scharmann, M.; Skepper, J.; Federle, W. 'Insect aquaplaning'on a superhydrophilic hairy surface: How *Heliamphora nutans* Benth. pitcher plants capture prey. *Proc. R. Soc. B* **2013**, *280*, 20122569. [CrossRef]
26. Stark, A.Y.; Badge, I.; Wucinich, N.A.; Sullivan, T.W.; Niewiarowski, P.H.; Dhinojwala, A. Surface wettability plays a significant role in gecko adhesion underwater. *Proc. Natl. Acad. Sci. USA* **2013**, *110*, 6340–6345. [CrossRef]
27. Pushkarova, R.A.; Horn, R.G. Surface forces measured between an air bubble and a solid surface in water. *Colloids Surf. A* **2005**, *261*, 147–152. [CrossRef]
28. England, M.W.; Sato, T.; Yagihashi, M.; Hozumi, A.; Gorb, S.N.; Gorb, E.V. Surface roughness rather than surface chemistry essentially affects insect adhesion. *Beilstein J. Nanotechnol.* **2016**, *7*, 1471–1479. [CrossRef]
29. Wang, Y.; Zhang, L.; Guo, Y.; Gan, Y.; Liu, G.; Zhang, D.; Chen, H. Air bubble bridge-based bioinspired underwater adhesion. *Small* **2021**, *17*, 2103423. [CrossRef]
30. Zhang, L.; Chen, H.; Guo, Y.; Wang, Y.; Jiang, Y.; Zhang, D.; Ma, M.; Luo, J.; Jiang, L. Wet surface attachment: Micro-nano hierarchical structure enhanced strong wet friction surface inspired by tree frogs. *Adv. Sci.* **2020**, *7*, 2070116. [CrossRef]
31. Ma, Y.; Ma, S.; Wu, Y.; Pei, X.; Gorb, S.N.; Wang, Z.; Liu, W.; Zhou, F. Remote control over underwater dynamic attachment/detachment and locomotion. *Adv. Mater.* **2018**, *30*, e1801595. [CrossRef] [PubMed]
32. Voigt, D.; Konrad, W.; Gorb, S.N. A universal glue: Underwater adhesion of the secretion of the carnivorous flypaper plant Roridula gorgonias. *Interface Focus* **2015**, *5*, 20140053. [CrossRef] [PubMed]
33. Kizilkan, E.; Heepe, L.; Gorb, S.N. Underwater adhesion of mushroom-shaped adhesive microstructure: An air-entrapment-effect. In *Biological and Biomimetic Adhesives: Challenges and Opportunities*; Santos, R., Aldred, A., Gorb, S.N., Flammang, P., Eds.; RSC Publishing: Cambridge, UK, 2013; pp. 65–71.
34. Ditsche-Kuru, P.; Koop, J.H.E.; Gorb, S.N. Underwater attachment in current: The role of setose attachment structures on the gills of the mayfly larvae *Epeorus assimillis* (Ephemeroptera, Heptageniidae). *J. Exp. Biol.* **2010**, *213*, 1950–1959. [CrossRef] [PubMed]
35. Varenberg, M.; Gorb, S.N. A beetle-inspired solution for underwater adhesion. *J. R. Soc. Interface* **2008**, *5*, 383–385. [CrossRef]

MDPI AG
Grosspeteranlage 5
4052 Basel
Switzerland
Tel.: +41 61 683 77 34

Biomimetics Editorial Office
E-mail: biomimetics@mdpi.com
www.mdpi.com/journal/biomimetics

www.ingramcontent.com/pod-product-compliance
Lightning Source LLC
Chambersburg PA
CBHW041401210926
51850CB00013B/3147